GROUNDED THEORY

GROUNDED THEORY

THE PHILOSOPHY, METHOD, AND WORK OF BARNEY GLASER

Edited by
VIVIAN B. MARTIN AND ASTRID GYNNILD

BrownWalker Press
Boca Raton

Grounded Theory: The Philosophy, Method, and Work of Barney Glaser

BrownWalker Press
Boca Raton, Florida • USA
2012

ISBN-10: 1-61233-515-2 *(paper)*
ISBN-13: 978-1-61233-515-5 *(paper)*

ISBN-10: 1-61233-516-0 *(ebook)*
ISBN-13: 978-1-61233-516-2 *(ebook)*

www.brownwalker.com

Chapter 16 was originally published as a chapter, Generating Formal Theory, in *Doing Formal Theory: A Proposal* (2006), by Barney Glaser. Mill Valley, CA: Sociology Press.

Cover image "Abstract Tree" © Gavin Dunt | Dreamstime.com
Cover design by Shereen Siddiqui

Library of Congress Cataloging-in-Publication Data

Grounded theory : the philosophy, method, and work of Barney Glaser / edited by Vivian B. Martin and Astrid Gynnild.
p. cm.
Includes bibliographical references.
ISBN-13: 978-1-61233-515-5 (pbk. : alk. paper)
ISBN-10: 1-61233-515-2 (pbk. : alk. paper)
1. Grounded theory. 2. Glaser, Barney G. I. Martin, Vivian B., 1956- II. Gynnild, Astrid, 1959- III. Title.

H61.24.G76 2011
001.4'2092--dc23

2011028615

Acknowledgments

A lot of enthusiasm and support helped us bring this project to life despite delays. A special thanks goes to all our contributors for sticking by us even when the plans were vague and publication an uncertainty. Dr. Helen Scott helped pull together some manuscripts and deal with the challenges of working with writers living on four continents. In Jeff Young we found an open-minded, supportive publisher. Christie Mayer was a patient, eagle-eyed production editor. Sonya Martin is a whiz on Microsoft Word and saved us many hours of frustration. Maxine Martin's editorial assistance was crucial as we pushed toward the deadline. Reviewers in an early phase of this project helped us take another look at the bigger picture. We thank everyone who helped push us toward a quality publication. Of course, this project would not be possible without the trust of Dr. Barney G. Glaser, whose spirit of discovery inspires us to move ever forward.

Table of Contents

INTRODUCTION: MENTORING A METHOD

Astrid Gynnild
University of Bergen

Vivian B. Martin
Central Connecticut State University

This book is about a mentor, his method, and the application of its principles. Grounded theory is reportedly the most invoked method in qualitative research worldwide, and the book is an edited collection with contributions from researchers from nine countries and four continents. The book is an attempt to clarify, elaborate on, and extend aspects of what Barney G. Glaser, the co-discoverer of the method, views as *classic grounded theory*. Though it was Glaser who drafted most of the book that codified the method, *The Discovery of Grounded Theory*, his original conception of the method is not always understood; the method has undergone several splits since its discovery in the mid-1960s. In the last two decades, Glaser has written and published nearly a dozen books about the method in an effort to correct the record and explicate grounded theory.

This collection of original research articles and essays by grounded theorists who have studied with Glaser is for researchers, novice and experienced, interested in understanding more about classic grounded theory, how it can be learned and taught, be it through face-to-face, long-distance or written mentoring. In particular, the project is aimed at what grounded theory researcher and scholar Phyllis Stern has dubbed minus mentors, academics who want to learn to do grounded theory but who do not have immediate access to face-to-face mentors. Anecdotal evidence suggests that most PhD students who choose grounded theory as the methodological approach for their dissertation start out as minus mentees in the sense that their professors do not know the method or ways of teaching it. As a consequence, both students and professors need to search for information elsewhere, not the least in readily accessible edited collections online.

Glaser's influence as a mentor ultimately ties together the chapters in this book, something we came to realize on a chilly but pleasant January day as we sat in a London Starbucks, crunched between two small round tables on which we were sorting strips of yellow paper. We had scratched the titles of the chapters in this book on the paper and, while sorting, it became more and more evident that *mentorship* tied together most of the work in this project. We also saw that mentorship in grounded theory involves a wide variety of professional developmental relationships. To introduce this book we want to say something about the multifaceted process of sowing a research idea at one point in history, and growing it with the assistance of research networkers who in turn mentor using methods initiated close to half a century ago.

Mentoring grounded theory

Mentoring grounded theory cuts across disciplines, countries, and generations of grounded theory researchers. It demonstrates the interwoven complexity of developmental processes of research professionals in the digital age. Both of the discoverers of grounded theory brought to the method strong influences from some of sociology's "great men." For Strauss, it was Everett Hughes, whose lineage traced back to founders of the Chicago School; for Glaser it was Robert Merton and Paul Lazarsfeld at Columbia. These were traditional academic relationships, but the spread of grounded theory has been shaped by different models that increasingly span the globe and take on properties that can tell us much about teaching, learning, and developing new forms of academic mentorship.

Grounded theory has been both spread and stunted by mentoring relationships. As other authors have noted, the method differs from many others in that it is very much tied to its co-discoverers, who have been active in it for long periods. In Strauss's case, mentoring continued up until his death in 1996. Barney Glaser, on the other hand, has written most of his books on grounded theory in the 21st century, and is still a mentoring role model for the rapidly growing, worldwide network of grounded theory. The founders' active engagement in mentoring a method that developed in two different directions no doubt contributes to some of the battles of legitimacy. But it also meant, at least in the early years, that people not associated with the discoverers created extensions and workarounds that have sometimes muddied the execution of the method.

Despite these confusions, the spirit of grounded theory and its explicit aim to give researchers autonomy from the stronghold of extant theory has continued to attract many. How these newcomers find the method is itself part of the mentorship process. As we discovered while putting together this project, coming to grounded theory is an interesting outcome of guidance from trusted persons. Both of us had been introduced to grounded theory by relatives, both academics, who had come across it through their own personal academic relationships. We began thinking about the role of door-openers for the method and have since observed that they work on both a local and global scale. For example, one contributor to this book, Evert Gummesson, a professor emeritus at the University of Stockholm, discovered Glaser and his work and struck up a relationship that brought Glaser to Europe for the first time in 25 years. It eventually lead to Gummesson recommending Glaser for an Honorary Doctorate at Stockholm. Importantly, this recognition helped the method spread in Sweden and the rest of Europe from the 1990's.

As a young PhD student, Barney Glaser studied with sociologists such as Merton and Lazarsfeld in typical student-teacher relationships. Later, after engaging in a peer-to-peer mentoring relationship with Anselm Strauss, Glaser separated in order to redefine the direction of grounded theory and become autonomous in his professional role. Throughout this process he not only developed his own ways of mentoring PhD students from outside of academia, but he also acts as an experienced role model who provides encouragement, advice, coaching and moral support to learners who want to move on with the classic grounded theory protocol.

We suggest that grounded theory is mentored through cultivating competence of grounded theory networks over extended periods of time. The cultivating processes take place in writing, and in virtual and face-to-face academic encounters. The ultimate goal of Barney Glaser's mentoring is first, to help PhD candidates get their degrees, and second, to support the manifestation of peer mentoring in multiple networks. Peer mentoring multiplies rapidly when a sufficient number of grounded theorists having reached a level of method proficiency at which they are able to inspire, support and critique each other in ways that are profitable for both parties. Peer mentoring is closely related to peer reviewing, and both aspects of being among equals have manifested in the work with this book.

The proposition of cultivating competence in multiple networks is based on an analysis of the chapters in this anthology combined with data collection elsewhere in grounded theory environments. The key dimensions of cultivating GT competencies in multiple networks concern the teaching of troubleshooting skills; providing psychosocial support and building bonds of friendship; facilitating informal network meeting points such as grounded theory seminars and online discussion groups; and establishing professional authority through running a peer-reviewed journal and ensuring the accessibility of Glaser's many books on the method. In reality, the books serve as asynchronous, written mentoring of grounded theory. These are key issues in Glaser's strategy for mentoring the method.

By investing personally and professionally in the build-up of a competent peer-to-peer network, the number of competent performers of the method is likely to increase, and the results of Glaser's long-term focus on grounded theory seed planting in Europe, North America, Asia, and Australia are evident. As will be demonstrated throughout the book, many experienced grounded theorists around the world have become engaged in different ways of using and mentoring the method. Some are engaged in formal teaching and mentoring of the method in academic institutions; some have built a whole career on mentoring and further development of grounded theory approaches; some spend their leisure time teaching the method in grounded theory seminars; some are mentoring others through the participation in discussion forums online; some take on more formal one-to-one obligations of mentoring via the Internet; some contribute by peer-to-peer sharing of new ways of doing data collection; and some contribute by analyzing philosophical frameworks and revisiting early history of the method. All together, these different approaches help build up a multifaceted knowledge base for mentoring grounded theory. Degrees of involvement in mentoring and time expenditure on grounded theory activities vary, but mentors in evolving peer networks do have at least two things in common: They are intrinsically motivated to help others learn the method and persistent in developing their own maturation with the method.

All of the contributors to this book studied with Glaser. A few were taught grounded theory in an early phase of Glaser's career. Most have taken up study with him in the last decade as part of his troubleshooting seminars around the world to help novice and experienced researchers better understand the method. A core of the people contributing to the book first met at such a seminar in Paris in May 2002. They were from the US, Ireland, Sweden, and elsewhere. The network they have been part of is an example of the type of peer mentoring enabled and sustained internationally. Since they met with Glaser at different stages of his career, the mentoring processes they went through also varied from formal mentoring initiated through course participation at a university to very informal mentoring either face-by-face at some of Glaser's seminars or through long distance calls or emails.

Nevertheless, since the discovery of grounded theory, it appears that many researchers, from different disciplines and different continents, have generated their own grounded theories without ever attending a grounded theory seminar or experiencing a learning milieu where principles of grounded theory were taught or discussed. The term "minus-mentoring," introduced earlier, points to the fact that even though all PhD candidates have one or more official supervisors at their own institution, many of these supervisors do not have any competency in grounded theory. In practical work, therefore, the candidate might be considered a minus-mentee.

In their chapter on the role minus mentoring has played in remodeling grounded theory ("Learning Methodology Minus Mentorship," chapter 4), Antoinette McCallin, Alvita K. Nathaniel, and Tom Andrews write that "although mentoring is seldom discussed as an essential component of grounded theory research training, mentorship advances understanding." With references to literature in the field, the authors argue that mentorship is invaluable to the learner to engage in excellence and scholarship. Mentoring not only concerns methodological support; it also socializes and informs new researchers into a community of scholars. When the novice scholars and their supervisors are uninformed, they rewrite the method with their own misunderstandings. According to McCallin, Nathaniel, and Andrews, part of the problem is that:

> novice researchers seldom appreciate the fine distinctions of the method until they are engaged in the research process. If a researcher is under supervision from a qualitative researcher who is not familiar with grounded

> theory at all, the learner is exposed to qualitative generalizations that are at odds with what is a specialist methodology.

The authors write that another problem is caused by the fact that even when novice grounded theory researchers search for grounded theory in journals and in books, they easily get into trouble. Much of what is written is methodologically inaccurate: "All too often grounded theory research published in well-respected international journals bears little resemblance to the original methodology."

This dilemma highlights a problem with literature as a mentoring tool in academia. Is the information accurate? Can the article or book be trusted? In contrast to many method books, Glaser's books guide researchers throughout the whole process of doing a grounded theory study. When reading his books for the first time, researchers across disciplines are often astonished by the wealth of details in grounded theory trajectories that are presented; they experience that the books speak to their own dilemmas and challenges while generating theory. Glaser has met the need for books by establishing the Sociology Press and the Grounded Theory Institute, which ensures that the books would always be accessible.

As Glaser's theorizing is based on systematic analysis of large amounts of data, the relevancy of the written information is evident. Yet after he withdrew from academia in the late 1980s, and thereby also withdrew from formal mentoring in established university settings, there was a void that needed to be filled. The number of emails that Glaser received from researchers on several continents indicated that the books of grounded theory were important. But even though the method was thoroughly discussed in writing, this form of "long distance" mentoring was not enough. The minus-mentees needed something more. As a consequence, Glaser developed the grounded theory seminars, which provide a new learning cycle for PhD candidates who grappled with the method. These seminars offered an opportunity for learners of the method to experience face-to-face mentoring in grounded theory.

The mentoring provided at these seminars will usually be complementary to the more discipline-specific mentoring provided at home institutions. However, as grounded theory is mainly inductive and experiential, it is necessary that future mentors have personal experience in doing the method. As a role model for those who aspire to become grounded theory mentors, Barney Glaser embodies what many PhD candidates need during their doctoral process: a mentor who is clear and upfront on the roles of both the mentee and the mentor; mentors who can supervise them through the process, not in detail, but who demonstrate an empathetic understanding of what they are struggling with. In other words, psychosocial aspects of mentoring tend to be just as important as the strictly skill-dependent aspects of it. These aspects of informal and formal mentoring are supported by existing research (Henderson Daniel, 2006, de Janasz and Sullivan 2001).

Mentoring grounded theory is carried out through a multiplicity of competent support networks. From networking that has come out of the many grounded theory seminars, grounded theory mentoring has been spread to many countries in the world. With the introduction of the Internet and virtual communication, the pace of dissemination has increased. Within these large networks, the first issue that takes considerable amounts of time is building competency in theorizing based on grounded theory principles. The contributors of this book have all specialized in different substantive areas with grounded theory. They come from many disciplines and many countries. As a basic principle, this multiplicity takes care of several issues. For PhD candidates it simplifies the process of establishing contact with several mentors. As long as the mentoring is informal and a supplement to the formal supervising that a candidate receives at the home institution, the multiplicity of mentors that might be contacted takes away some of the time pressure that would otherwise fall on one person. Moreover, with several potential mentors at hand, it is more likely that the individual researcher connects with people who match his or her specific needs.

Mentoring principles in grounded theory are rooted in data about the needs of mentees, and influenced by Barney Glaser's vision of wanting to give people their sense of being themselves. As Glaser points out in the conversation with Astrid Gynnild, grounded theory might be applied not only on academic research projects; learning the method gives researchers "their creativity, their independence, their autonomy, their contribution, their self-satisfaction and their motivational joy." In the research literature, mentoring is repeatedly referred to as having two primary functions for mentees: the career-related function and the psychosocial function. The career-related function establishes the mentor "as a coach who provides advice to enhance the mentee's professional performance and development," whereas the psychosocial function establishes the mentor as a "role model and support system for the mentee (Daniel et al, 2006, p 6). Moreover, mentoring is typically theorized as a process. Daniel et al (2006) and Russell (2004) conceptualize mentoring as a four-stage process in which an initiation stage, a cultivation stage, a separation stage, and a redefinition stage are identified. According to the general mentoring glossary, the cultivation stage is a phase in which "career and psychosocial functions are enhanced during the 2 to 5 years after the mentoring is initiated" (Russell, 2004).

From the perspective of mentoring a method presented in this chapter, there are no such specific limitations of the cultivating stage. It is definitely true that all researchers, no matter what method, mentor, or project they are engaged with at any time of their career, will go through stages of initiation, cultivation, separation, and redefinition in mentoring relationships. This goes for mentees as well as for mentors. However, Barney Glaser's life work indicates that to successfully mentor a method, cultivation approaches should preferably last for several decades and there should be an increase in the multiplicity of approaches.

A person who takes on the task to mentor a method, of course, does so through mentoring other persons. In practice, this fact means that even though the mentor is steadily cultivating his project throughout the years, both he and potential mentees are likely to experience many moments and situations of no return where relationships are bound to break in some way or other. Glaser broke out from his early career relationships with people like Merton and Lazarsfeld at Columbia and later broke away from the method's co-founder Strauss, although they managed to keep their personal relationship. More precisely, Glaser claims that Strauss was the one who broke out, as he at one stage in the method's history deviated from its established definitions.

Many students of Glaser have also separated from the original version of the method throughout the years. Initiation, cultivation, separation, and redefinition are stages that all humans go through not only in relationships in general; it is a necessary trajectory on the way from being a child to becoming an adult. In this perspective, the most interesting question to be posed is not whether or when anybody is going to break out from a network or an academic field of interest. The most interesting issue to analyze is what happens during and after redefining a relationship, be it to a mentor or a mentee. According to the mentoring glossary, redefinition is "the phase after the separation phase when the mentoring relationship may end or change qualities" (Russell 2004, p. 609).

Some contributors to this anthology were mentored by Glaser but later separated from his approach in order to establish their own grounded theory branch. Others again have redefined and levelled up to peer-to peer relationships in their grounded theory networks. In this way, open networks manage to become self-supportive mentoring environments. Mentoring can be task-focused, relationship-based, and both. Mentoring grounded theory as a method exemplifies a binary approach to enhancing professional performance and development and ensuring a gradual expansion of the approach. The wish to mentor others often surfaces when people are midway in their careers, when they are at a peer-to-peer stage in their own professional development (Allen, 2002). Allen also points out that mentors who are intrinsically motivated for a task are likely to be more focused on psychosocial dimensions of mentoring than others. People who are focused on mentoring others usually exhibit an other-oriented empathy that relates to psychosocial mentoring. In the case of mentoring the grounded theory method, the other-oriented empathy conveyed is also grounded in the data.

The necessity of multiple mentorships, moment mentoring, and complementary mentoring is likely to increase in an academic world where researchers are working more and more on collaborative projects and loose networks around the globe. When mentoring a method, individuals who choose to work within the definitions of the method go through a process of being cultivated to cultivating. This joint project is one such outcome. The book is divided into four parts focused on the most important dimensions of mentoring a method: teaching, doing, explicating and documenting its roots and epistemology, and advancing it.

How this book is organized

While editing this project we had some debate over whether to use the "classic" grounded theory modifier throughout to draw the line dividing the version of grounded theory advanced here and the work of others. In some ways, using the modifier to describe Glaser's version of grounded theory--he does not like Glaserian, as it is sometimes called– cedes ground and allows remodeled work to carry the banner of grounded theory without challenge. As a result, we chose to go forth with simply Grounded Theory in the title. At various points in the book, however, it is necessary for authors to use "classic" to clarify distinctions that set Glaser's vision apart from those of others. The need to make classic grounded theory's strictures especially explicit is critical in chapter 1, Odis Simmons's "Why Classic Grounded Theory," which opens Part I Teaching Grounded Theory. Simmons, who studied with both Glaser and Strauss, explains how he came to understand that grounded theory as advanced by Glaser differed from Strauss's version and why he came to prefer the conceptualization power of Glaser's classic grounded theory. Simmons also cuts through discussions that seek to categorize classic grounded theory as objectivist rather than constructivist and shows how these labels put grounded theory in unnecessary boxes.

The Grounded Theory Institute's Troubleshooting Seminar was the vehicle through which most contributors to the book, including the co-editors, came to know Glaser and one another. In chapter 2, "Atmosphering for Conceptual Discovery," Astrid Gynnild examines the process of participating in the seminar and presents a theory of learning that, she hypothesizes, can be extended beyond the grounded theory seminar to better understand empowering learning situations. Through interviews, observations, and other data, she theorizes the conscious strategies Glaser has used to help PhD candidates from around the world complete grounded theory dissertations.

Despite the success of grounded theory dissertations in many types of institutions, there are still special challenges PhD candidates doing classic grounded theory dissertations face at their home institutions. Andy Lowe, who has helped a number of candidates navigate common pitfalls, and Wendy Guthrie, who completed a dissertation with Lowe's guidance, provide a blueprint in chapter 3, "Getting Through the PhD Process using Classic GT: A supervisor/researcher perspective," which anticipates issues ranging from the original proposal and IRB issues to the final write-up. The fourth chapter rounding out Part I is "Learning Methodology Minus Mentorship," which was cited earlier in this introduction. McCallin, Nathaniel, and Andrews not only address the scholarly concerns when a method is misshaped by those who are not adequately informed, but also highlight the moral and ethical issues raised by such representations.

Part II Doing Grounded Theory provides a range of instruction in ways to expand data collection, think about doing grounded theory, and research extending some of the original work on awareness of dying. In chapter 5, "Conducting Grounded Theory online," Helen Scott leads readers through the process of conducting grounded theory interviews online, a form of data collection becoming more popular as more researchers become aware of the rich possibilities online. In chapter 6, "Using Video Methods in Grounded Theory," Lisbeth Nilsson takes grounded theory's "all is data" dictum further with a discussion of collecting data with video, a technique she refined while conducting research among disable people who could not speak and participate in the interview as it is traditionally known. Video of work with patients in her work as an occupational therapist became the source for codes and building of a theory that has won respect internationally. In chapter 7, "Developing Grounded Theory Using Focus Group Interviews," Cheri Ann Hernandez clears up some of the misapprehensions grounded theorists might have about using focus groups, which she argues can aid grounded theorists so long as they are mindful of some of the issues she outlines.

The use of qualitative software to aid the completion of grounded theory projects is a sticky topic, one that is too often dismissed without adequate discussion of how the process really works. In chapter 8, "The Utility and Efficacy of Qualitative Research Software in Grounded Theory Research," Michael K. Thomas goes beyond the flat-out rejections or the uncritical embrace of such tools with an overview that lays out the options and limitations from the point of view of a classic grounded theorist who has used some of the leading software programs and can critique in context. Chapter 9, "De-Tabooing Dying in Western Society: From Awareness to Control in the Dying Situation," by Hans Thulesius, extends the *Awareness of Dying* work that played a critical role in launching grounded theory. The data in the emerging theory points to processes in which the main concern of the patient, now aware of his terminality, wants control of dying, whether it be through the right to die or active participation in decisions for palliative care.

Grounded theory's status as a method embraced worldwide means that scholars for whom English is not the first language discover grounded theories that they sometimes write up for native and English audiences. Conceptualizing in one language and writing in another has certain advantages and challenges. Moreover, translating grounded theory books, particularly the seminal texts, can leave translators searching hard for the right ride. These are among the ideas Massimiliano Tarozzi explores in chapter 10, "On Translating Grounded Theory: When Translating is Doing."

Part III Historical and Philosophical Grounding is an opportunity to better understand the historical roots of grounded theory, as well as some of its philosophical underpinnings. Chapter 11, "Lessons for a Lifetime: Learning Grounded Theory from Barney Glaser," by Kathy Charmaz, who studied with both Strauss and Glaser, interviewed classmates from the period to provide a picture of what it was like to have Glaser as a professor. Most associated with developing a model that departs from classic grounded theory, Charmaz, nonetheless, in describing Glaser's assistance at her dissertation defense, illustrates the important role Glaser played as a mentor for that time in her life.

Those who study with Glaser know he eschews the "rhetorical wrestle" and implores students to get on with the work of doing grounded theory. Nevertheless, the nature of scholarship is such that epistemological discussions cannot be ignored, and that is what Alvita Nathaniel provides in chapter 12, "An Integrated Philosophical Framework that fits Grounded Theory." Her goal here is not to retrofit a philosophical framework on top of grounded theory; rather, her aim is to show grounded theory's affinity with the pragmatism of Charles Peirce. Despite the lack of an explicit connection between Glaser's grounded theory and pragmatism, Nathaniel argues that grounded theory's strictures can be understood philosophically when situated alongside pragmatism.

In actuality, several influences shaped Barney Glaser's method. In chapter 13, "Creative Autonomy: Early Influences in the Emergence of Classic Grounded Theory Methodology," Judith A. Holton discusses and documents the autobiographical and intellectual influences that Glaser brought together to wrest his autonomy from academic structures and create what became grounded theory. The chapter fits into fruitful discussion with chapter 14, "Grounded Glaser," by Evert Gummesson, who seeks to capture some of the personal qualities of Glaser that are instructive to social science in general. Chapter 15, "Living the Ideas: A Biographical Interview with Barney Glaser," by Astrid Gynnild, is a breathing illustration of some of the previous chapters as readers are given a question and answer format to follow along as Glaser talks about life, love, and grounded theory.

Part IV Advancing the Method is forward-looking even as it necessarily reaches back. Chapter 16, "Generating Formal Grounded Theory," is a reprint of chapter 5 from Glaser's book *Doing Formal Theory* (2007). Glaser wrote the book to answer the many questions grounded theory researchers have about formal theory, which, given the paucity of such theories, seems to intimidate many. Picking up on the challenge of formal theory, Tom Andrews's "Reflection on Generating a Formal Grounded Theory," chapter 17, shares some of the hard-won insights he has gained while developing a formal theory in visualizing worsening progression.

Critical to the development of grounded theories, substantive and formal, will be greater utilization of quantitative data. Grounded theory is not a qualitative method despite the fact that it is mainly used by qualitative researchers. Glaser always envisioned the use of both kinds of data, but many grounded theories do not have training in quantitative methods. Chapter 18, "From Theoretical Generation to Verification using Structural Equation Modeling," by Mark S. Rosenbaum discusses how structural equation model can be used to test grounded theories. Rosenbaum acknowledges that Glaser does not think grounded theories need to be taken through traditional verification protocols; however, Rosenbaum points to the preference for verification in some disciplines, arguing often requisite for publication. He further argues that structural equation modeling comes out of the same qualitative math analysis that Paul Lazarsfeld was doing when Glaser was studying with him. Part IV ends with chapter 19, "The Power of an Enduring Concept," by Vivian B. Martin, who uses her developing theory on discounting awareness to demonstrate the continuing power of the original theory of awareness contexts and to tease out the unfinished theory development suggested in *Awareness of Dying*, as well as some of Glaser's earlier pre-grounded theory work. She argues that early Glaser work and suggestions for the secondary analysis of qualitative research are necessary for grounded theorists to embrace if they are to build more formal theories.

It is our hope that grounded theorists at all stages of competence will find something useful to incorporate into their grounded theory practice. Much is said here about the desire to get good information into the hands of minus mentors, but this book is also for the many skilled GT researchers around the globe who are searching for more insights, inspiration, and ideas to move on with their own GT projects.

References

Allen, T. D. (February 2003). Mentoring others: A dispositional and motivational approach *Journal of Vocational Behavior 62*(1), pp 134-154

Daniel, J. H. et al (2006). *Introduction to Mentoring. A Guide for Mentee and Mentors.* Centering on Mentoring, Presidential Task Force, American Psychology Association

De Janasz, S. C. and Sullivan S. D. (April 2001). Multiple mentoring in academe. Developing the professorial network. *Journal of Vocational Behavior, 64*(2), pp 263-283.

Russell, M.S. (2004). Mentoring. In Goethals et al, *Encyclopedia of Leadership*, Sage Publications, Thousand Oaks, CA. Vol. 3 pp.992-5.

PART I

TEACHING GROUNDED THEORY

I

Why Classic Grounded Theory

Odis E. Simmons
Fielding Graduate University

Since Glaser and Strauss first introduced grounded theory in their 1967 book, *The Discovery of Grounded Theory* (hereafter referred to simply as *Discovery*), it has been both praised and criticized. Some critics disfavor the whole idea of grounded theory. These critics tend to be strongly invested in the tenets of positivistic approaches to social/behavioral research and have an unfavorable view of approaches other than their own. Criticisms coming from those of a positivist bent are usually along the lines of grounded theory not meeting the requirements of the standard model of science; in other words, it is not sufficiently positivistic.

These criticisms are also likely to come from social/behavioral scientists who have a superficial or distorted view of grounded theory that is skewed by their inability to fathom that a method that doesn't meet their preferred canons of science might be scientifically legitimate. Given that to me a proper science must above all be true to its subject matter and there are doubts that positivistic social/behavioral sciences are able to cover all facets of their subject matter, particularly the role of "meaning making," it is clear that other approaches to the study of human behavior, including grounded theory, have much to offer. Criticisms of grounded theory have also been lodged by non-positivist social scientists as well. Ironically, these criticisms often claim that grounded theory is *too* positivistic. I will leave it to others to engage in these particular "rhetorical wrestles" (Glaser, 1998. pp. 35-46).

The criticisms of concern to me here are criticisms lodged by those who are favorable towards grounded theory, including many who consider themselves to be grounded theorists, but see ontological and epistemological problems with Glaser and Strauss' original conceptualization, now commonly referred to as "classic," or sometimes, "Glaserian" grounded theory. These critics usually assert that Glaser and Strauss' original conception of grounded theory has positivist, objectivist underpinnings (Bryant, 2002a, 2002b; Mills, Bonner & Francis, 2006; Urquhart, 2002), although some critics approach their criticism from other angles (Haig, 1995; Kools, McCarthy, Durham & Robrecht, 1996). Some of these critics offer ways in which they think the method could be improved. Martin (2007) provided a pointed, critical response to those who have made this claim.

Charmaz (2000, 2006) has aptly characterized these approaches as "constructivist grounded theory." Mills et al. (2006) have referred to them as "evolved" grounded theory. The term "evolved" implies that grounded theory as originally conceived by Glaser and Strauss needed to be refined or improved. In my view these "remodeled" (Glaser, 2003) versions of grounded theory thoroughly miss the primary purpose of the method, which is to ground theory in *data*, not the speculations and imaginations of researchers, theorists, and ideologists.

The positivist, objectivist criticism is lodged more frequently at Glaser than at Strauss. I think in large part this is inferred from Glaser having studied and received his Ph.D. in sociology at Columbia University, which had a strong objectivist and positivist bent, whereas Strauss studied and received his Ph.D. in sociology at the University of Chicago, which had a strong qualitative, participant observation, symbolic interactionist bent. However Glaser was a strongly independent student. He was inspired by and fed off the ideas of his professors, but did not adopt them as his own. To the contrary, he was at times rather critical of them. It was they who inspired his term "theoretical capitalist," a not particularly reverent expression. Yet, he transformed many of their ideas into classic grounded theory.[1]

Strauss remained closer to the ideas of his graduate school professors. This played a role in the divergent directions Glaser and Strauss took following the publication of *Discovery*. Having been a student of both Glaser and Strauss in the early 1970s I witnessed this first hand. To be sure, I was not the only Glaser and Strauss student who saw the divergences in their methodological approaches. Charmaz, for one, said that she saw their differences early on.

Because Strauss was not in residence until several months after I began my studies in the Graduate Program in Sociology at the University of California, San Francisco, I had taken two terms of Glaser's Analysis seminar, in which he taught grounded theory. By then, I had a solid familiarity with Glaser's rendering of grounded theory, including where he had taken it beyond what was written in *Discovery*. In his seminars and conversations with his students (because he and I commuted together, I had the pleasure of many such conversations), Glaser continued to develop his ideas about grounded theory, which eventually appeared in *Theoretical Sensitivity*.

When Strauss returned from his time away, I enrolled in his methods seminar. During the first session of this seminar I immediately recognized that there were major differences between the way Strauss and Glaser portrayed grounded theory, particularly regarding how to analyze data. Glaser emphasized the importance at the outset of minimizing preconceptions, including preconceived questions and categories. Strauss began providing us with both. Strauss' perspective and much of his language was drawn from the interests, questions, and categories of the symbolic interactionist perspective, in which he had been so thoroughly immersed during his years at the University of Chicago. Having studied under some notable sociologists with a qualitative, symbolic interactionist bent, such as Stanford Lyman, John Lofland,

Fred Davis, and Leonard Schatzman, I was well familiar with that approach to sociological research.

Although Strauss used some of the terminology (constant comparison, theoretical sampling, etc.) from *Discovery*, the content of what he was saying did not always match the understanding of grounded theory that I had gained from reading *Discovery* and participating in classes and individual conversations with Glaser. Because I assumed that the ideas of Glaser and Strauss would be consistent, I was a bit taken aback. I also noticed a distinct difference from Glaser in the language of research Strauss used as well as even his view on the scope of grounded theory. Strauss spoke of grounded theory as merely a style of qualitative analysis. He held this view at least until 1987 when in the Preface to his book, *Qualitative Analysis for Social Scientists* he referred to grounded theory as, "a particular style of qualitative analysis of data (*grounded theory*)" (italics in original) (Strauss, 1987 p. xi). In the Introduction to this book he also wrote of grounded theory:

> So, it is not really a specific method or technique. Rather it is a style of doing qualitative analysis that includes a number of distinct features, such as theoretical sampling, and certain methodological guidelines, such as the making of constant comparisons and the use of a coding paradigm, to ensure conceptual development and density. (p. 5)

From the beginning, Glaser spoke of it as a full, systematic research method. Today it is commonly considered to be a full research method. To be sure, I learned a great deal from Strauss. He was a brilliant sociologist and a kind, supportive, and thoughtful person, which is one of the main reasons why I selected him as my dissertation committee chair. Glaser also served on my dissertation committee. Fortunately, their different understandings about grounded theory never got in the way of my research. They were both very supportive of my methodological approach, which was guided by grounded theory as laid out in *Discovery*.

Since my graduate school days I have used and taught grounded theory as it was articulated in *Discovery* and the subsequent writings of Glaser because it produces theory that is more completely grounded in data, which makes it more suitable for action. In classic grounded theory, throughout the process, everything must "earn" its way into a theory through constant comparison of data rather than being imported from other sources.

Following *Discovery*, Glaser wrote numerous books expanding and articulating classic grounded theory in greater detail (Glaser, 1978, 1992, 1996, 1998, 2001, 2003, 2005, 2006, 2008, 2009) as well as editing a series of grounded theory anthologies (Glaser, 1993, 1994, 1995, 1996), as well as one co-edited with Holton (2007). Strauss wrote one grounded theory related book as sole author, although grounded theory was not in the title (Strauss, 1987).[2] With Corbin, he coauthored the First and Second Editions (Strauss died before the Second Edition was completed) of a grounded theory book, entitled *Basics of Qualitative Research* (1990, 1998). Because Strauss and Corbin's take on grounded theory diverged considerably from *Discovery* and Glaser's expanded articulation of the methodology introduced in that book, Glaser wrote an incisive, thorough retort to Strauss and Corbin in his 1992 *Basics of Grounded Theory Analysis*. In his retort he wrote:

> As the reader has seen in this book, Anselm and I have profoundly different views of Grounded Theory. What started out as a book of corrections ended up showing that Strauss indeed has used a different methodology all along, probably from the start in 1967 and it was not obvious until our more recent articulations and formulations. (Glaser, 1992, p. 122)

The divergences between Glaser and Strauss and the proliferation of "constructivist," "remodeled" versions of grounded theory get right to the heart of the title of this piece, "Why Classic Grounded Theory." When I personally discovered grounded theory in 1967 as a sociology undergraduate student, I found it to be very appealing firstly because, as Burawoy (1988, p.275) later put it, "Grounded theory is a populist sociology, a way in which all of us can turn our data into the very best scientific theory," and secondly because of the promise of grounded theories for "...whoever wishes to apply them, to bring about change, a controllable theoretical foothold..." (Glaser & Strauss, 1965, p. 268). Pursuing this promise, I have developed methods that extend the logic and process of classic grounded theory into the action arena (Simmons, 1994, Simmons & Gregory, 2003).[3] In fact, using (classic) grounded theory in real world applications, and teaching others how to do it, has been the primary interest and ambition of my professional life.

It stands to reason that the more grounded a theory is the more of a controllable theoretical foothold it will provide and therefore the more valuable it will be as a basis for action. In short, because as a method classic grounded theory produces a denser, richer, grounded theory more completely grounded in data without extant influences than the various forms of "remodeled," "constructivist," or "evolved" versions of grounded theory, including Strauss' version, it is more suitable for designing action for change.

As I often tell my students, "You are unlikely to attain your 'what ought to be' unless you have a clear, accurate understanding of 'what is'." Classic grounded theory is better equipped than other forms of grounded theory to produce a clear, accurate understanding of what is. Even if it unknowingly incorporates what staunch constructivists would view as unavoidable constructivist elements, because of the emphasis on minimizing preconceptions, "what is actually happening in the data" (Glaser, 1978, p. 57) and "what is really going on" (Glaser, 1998, p.12) are more likely to prevail. Although for the particular purposes of social scientists constructivist elements may be suitable, successful action requires that a theory be as true to the data as possible.

Constructivism Versus Objectivism

Most if not all of the variations of grounded theory that Glaser refers to as "remodeled" and Mills et al. refer to as "evolved" can be subsumed under Charmaz's "constructivist grounded theory" category. Charmaz (2000, 2006) juxtaposes constructivist grounded theory with what she terms "objectivist grounded theory." In the glossary of Charmaz (2006), she defines constructivism as:

> ...a social scientific perspective that addresses how realities are made. This perspective assumes that people, including researchers, construct the realities in which they participate...Constructivists acknowledge that their interpretation of the studied phenomenon is itself a construction.
>
> She defines objectivist grounded theory as:
> ...a grounded theory approach in which the researcher takes the role of a dispassionate, neutral observer who remains separate from the research participants, analyzes their world as an outside expert, and treats research relationships and representation of participants as unproblematic. Objectivist grounded theory is a form of positivist qualitative research and thus subscribes to many of the assumptions and logic of the positivist tradition.

She sees these categories as ideal types and evidently does not propose them as entirely discrete.

> I juxtapose constructivist and objectivist forms of ground theory here for clarity; however, whether you judge a specific study to be constructivist or objectivist depends on the extent to which its key characteristics conform to one tradition or the other. (Charmaz, 2006, p. 130)

She appears to place both Strauss and Glaser in the objectivist category:

> In this approach, the conceptual sense the grounded theorist makes of data derives from them; meaning inheres in the data and the grounded theorist discovers it (see, for example Corbin & Strauss, 1990; Glaser, 1978; Glaser & Strauss, 1967). (Charmaz, 2006, p. 131)

Mills et al. (2006) categorized different approaches to grounded theory as either "traditional" or "evolved." Unlike Charmaz, they place Glaser and Strauss in different categories. They place Glaser in the traditional category, which seems to parallel Charmaz's objectivist category. They place Strauss and Corbin in the evolved category: "Here, we discuss the development of constructivist grounded theory from its beginnings in the work of Strauss and Corbin through to the work of sociologist Kathy Charmaz" (p. 2).

This type of categorization may lend general understanding to the differences between classic and other forms of grounded theory, but they may also lead to confusion, particularly when the placement of Glaser and Strauss and Corbin is at odds between schemes. They are also reflective of the age-old dualistic rhetorical wrestle in the social/behavioral sciences between positivistic and interpretive approaches to scientific understanding. Like this debate, they run the risk of oversimplification, particularly for less discerning observers. The multiplicity of contradictory interpretations of grounded theory and particularly classic grounded theory highlight the extent to which each critic's chosen ontological/epistemological/methodological paradigm serves as constructivist foundation for their standpoint.

If pressed to categorize, I would place Strauss and Corbin in the constructivist/evolved category and Glaser in neither. I think Strauss and Corbin along with other remodeled forms of grounded theory fit the constructivist category because they incorporate unnecessary, preconceived elements into data collection and analysis. As Glaser (1978, 1992, 1998, 2001, 2003) points out, this easily results in multiple forms of forcing during the research process. Glaser insisted that from the beginning everything in the process must "earn" its way. Although he makes this point throughout his many grounded theory publications, see particularly Glaser (2003, Chapter 4).

The grounded theory method was originally designed to minimize preconceptions, so as to ensure that theories systematically emerge directly from data. Researcher driven constructivist elements were designed out of the process to ensure that theories remain as fully grounded in data as possible, throughout. For this reason, I view the term "constructivist grounded theory" as an oxymoron. In his response to Charmaz, Glaser (2002) referred to it as "a misnomer."

A more appropriate term might be "quasi-grounded theory," which would be similar to the distinction made between the experimental method and modified versions of that method. Like quasi-experimental methods in relation to the experimental method, remodeled/constructivist/evolved approaches to grounded theory use the jargon (Glaser, 2009) and some of the standards, and rules of classic grounded theory, but contain significant variations that can influence, weaken, and unground the product.

However, quasi-experimental methods were designed to approximate experimental research in situations where the pure experimental method would produce problems of a serious ethical, logistical, and/or practical nature. This is not the case with constructivist (remodeled, evolved) forms of grounded theory. The choice to use them does not stem from requirements of the research situation, but rather from ontological assumptions. And, just as quasi-experimental methods produce less convincing findings than the pure method, remodeled grounded theory weakens the purpose of the classic method, which is to provide a research process that will enable researchers to generate a theory that is systematically grounded in data at all stages, to the extent humanly possible.

A key phrase here is, "to the extent humanly possible." A staunch constructivist position would hold that all human knowledge, no matter how derived, is a human construct, necessarily involving interpretation. On the most fundamental ontological, epistemological level, this may be the case. Although it may be the case that it is impossible to research and understand anything without incorporating something of oneself into it, that is not legitimate justification for choosing to purposefully incorporate one's theoretical paradigm, political or professional ideology, theoretical speculations, and/or personal predilections into the research design and process—unless of course one hopes to lend them unearned credence. As Glaser (2002, p. 6) stated, "Constructionism is used to legitimate forcing. It is like saying that if the researcher is going to be part of constructing the data, then he/she may as well construct it his way." He asserts that this "makes the researcher's interactive impact on the data more important than the participants."

The staunch constructivist position holds that those who think they can do research without preconceptions or incorporating something of themselves into the research are merely fooling themselves. This essentially tautological argument places all human meaning, including science, into one category from which no one can escape. Although at the most fundamental level it may be true, a category that subsumes or explains everything really explains nothing because it doesn't provide for differentiation or nuance. A more moderate position on this would prove to be less of an analytical trap.

Unless you assume a staunch, obdurate, purist position, I don't think classic grounded theory fits suitably into an either/or constructivist-objectivist dichotomy. Charmaz seems to allow for this possibility when she states (also quoted above), "whether you judge a specific study to be constructivist or objectivist depends on the extent to which its key characteristics conform to one tradition or the other" (Charmaz, 2006, p. 130). But, in contrast to my view here, she places classic grounded theory fairly solidly in the objectivist category.

However, classic grounded theory effectively incorporates *reasonable* elements of both without being either. Although its research process is as minimally constructivist as possible its view and treatment of multiple perspectives, meaning making, and such tend towards constructivism, as does Glaser's (2002) recognition that the grounded theory researcher's perspective is but one perspective amongst many. Unlike Glaser, true objectivists see common sense meanings and interpretations as competitive with but inferior to those of positivistic science.

Its acceptance of common general sociological concepts such as structure and process and its charge to identify and name "underlying or latent" patterns (Glaser, 2002, p. 5) does not require a belief in objectivist "assumptions about truth" or "an external reality" (Charmaz in Puddephatt, 2006, p. 5). Charmaz appears to recognize this, as she states,

> I think any good sociologist is always looking for hidden assumptions. So, the notion of finding what is under the surface, what is tacit, what is liminal, and what certain positions tend to assume, is, I think, very congenial to a constructivist approach. I don't think that it is necessarily objectivist. To presuppose that there are certain structures would be objectivist. However, to take a look, and a rather deep look, and try to find things that are latent, would be quite consistent with social constructionism and grounded theory. (p. 10)

To "presuppose that there are certain structures" would constitute an enormous preconception, which would be directly inconsistent with classic grounded theory and Glaser's position. Given this, I'm not clear how Charmaz's statement differs from Glaser's position. One can conceptually name latent patterns structure or process without seeing them as "prior" or endowing them with "thing-like" qualities (Durkheim 1938). With his general theoretical code "basic social process" (Glaser, 1978) of which there are two types, "basic social psychological" and "basic social structural" (Glaser & Holton, 2005), Glaser allows that there are *emergent* liminal patterns and structure that are abstract of time, place and people. However, he does not preconceive *certain* structures or processes (p.5, p. 11).

If the faith classic grounded theory places in the ability of skilled, well-trained researchers to minimize preconceptions, remain honest to the data, name rather than interpret, let it emerge, resist forcing, maintain a firm commitment to the ideal of objectivity (by this I do not mean objectivism), and its aim of "generalizing, simplifying, parsimonious statements, and universalizing in abstract terms so that it cuts across fields" (Charmaz in Puddephatt, 2006, p. 5) is objectivist, it is certainly a decidedly kinder, gentler form of objectivism than that of Durkheim and other positivists who took their cues from him.

I think because of its highly rigorous, systematic empirical nature, and its "qualitative math" logic to which Glaser (1998, 2003) acknowledges much is owed quantitative methods, classic grounded theory has some of the flavor and tenor of positivistic social science. But, other than these properties it shares little with that paradigm, certainly not its underlying ontological/epistemological assumptions. At times, Glaser has been highly critical of positivistic aspects of quantitative methods. However, he used what he saw as their more favorable aspects to inform his development of grounded theory.

Although, Charmaz (2006) pointed out, "It has gained acceptance from quantitative researchers who sometimes adopt it in their projects that use mixed methods..." (p. 9). She further stated, "Grounded theory not only became known for its rigor and usefulness, but also for *its* (italics in the original) positivistic assumptions." In saying this, she seems to have suggested that positivists recognize at least some kinship with classic grounded theory because of commonly held underlying assumptions.

It appears to me that social scientists of many stripes seem to be reading assumptions into Glaser and Strauss' and Glaser's words, as if they failed to say what they actually meant. I think it is inappropriate to use criteria and interpret classic grounded theory (or any other method) from one's own preferred perspective. It can and should be understood on its own terms. As Glaser (1992, 1998, 2003) maintained, it is not qualitative (which tends toward constructivism) or quantitative (which tends towards objectivism); it is a unique, general, inductive, concept/theory generating method that may borrow from both but is neither, unless you force fit it into a dualistic conception.

Constructivism

Opportunities for introducing constructivist elements from the researcher are present during both data collection and analysis. However classic grounded theory is designed to prevent this, to the extent possible.

Data collection

Most grounded theorists collect their own data, for the most part using non-structured, open-ended interviewing because it has high conceptual yield. If done skillfully, this type of interviewing can be conducted in such a way as to greatly minimize opportunities for the interviewer to inject constructivist elements into the interview. As Glaser (2002) pointed out, "If the data is garnered through an interview guide that forces and feeds interviewee responses then it is constructed to a degree by interviewer imposed interactive bias. But...with the passive, non-structured interviewing or listening of the GT interview-observation method, constructivism is held to a minimum" (p. 10).

Typically, the researcher begins the interview, without an interview guide, by posing a "grand tour" question or inquiry. A grand tour question is designed to convey to the respondent that they are being invited to discuss *what is relevant to them* (not the researcher) about the general topic area, on their terms. A good grand tour question will be phrased in the most open-ended manner possible. Through theoretical sampling,[4] grand tour questions become increasingly more specific as the theory emerges. But, even as data collection needs become more and more selective as a theory emerges, questions are still asked in the least leading manner possible. Subsequent questions are formulated in relation to the respondent's reply to the previous question. These measures keep intentional researcher constructivist elements and relevancies from creeping into the interview. As Glaser (2002) said,

> Much GT interviewing is a very passive listening and then later during theoretical sampling focused questions to other participants during site spreading and based on emergent categories. It is hard for mutual constructed interpretations to characterize this data even though the data may be interpretive...(p. 5)

As the research process continues, theoretical sampling guides data collection by informing the researcher about "what data to collect next and where to find them" (Glaser & Strauss, 1967, p. 45). In this manner, interview topics are based on what has already been discovered from previous data and the emerging theory. This offers another barrier to the introduction of researchers' preconceived or received constructivist elements.

If a researcher enters the field with preconceived questions or categories, as non-classic forms of grounded theory allow, emergence and grounding are derailed. It has always intrigued me that some people who see themselves as grounded theorists would make this choice, particularly given that it is not necessary. As I said above, even if at a deep ontological, epistemological level some degree of constructivism is impossible to avoid, that does not justify inviting it! A critical tenet of grounded theory is to minimize, not encourage, preconceptions.

Analysis

Regarding analysis, at its most essential level classic grounded theory simply requires that researchers look for patterns in data, any kind of data,[5] name (conceptualize) those patterns, identify relationships between the conceptualized patterns, and write them into a theory, maintaining grounding in the data throughout. Considering we have only words with which to work (some might think that numbers enable us to circumvent words, but numbers are empty abstractions until they are bestowed with words), these are the least interpretive/constructivist procedures possible. Special attention is paid to achieving the best possible fit between the pattern being named and the word or phrase selected to represent it. As I tell my students, "The reader of your theory will not have access to your data, so to them, the words or phrases you select will *be* the data."

In general, classic grounded theory requires only what is feasible—that researchers remain as open as possible and make a serious attempt to suspend and minimize preconceptions, and at all times remain honest to the data. Classic grounded theory can give the researcher confidence in addressing the epistemological question, "how do you know this?" because at any point the process can be deconstructed back to the data. And, unlike constructivist forms of grounded theory the intentional use of unearned extant questions, concepts, ideas, theories, ideologies, frameworks, and such is proscribed.

Objectivism

Regarding objectivism and classic grounded theory, like Charmaz, many critics claim that Glaser and Strauss in *Discovery* and Glaser since (i.e. classic grounded theory) take an objectivist (realist, positivist) position because they assume an underlying reality that can be discovered, using (classic) grounded theory. I think this is ironically a constructivist interpretation in which meanings are projected into *Discovery* and Glaser's subsequent writings that simply aren't there. It is telling that these authors make general assertions about objectivism in classic grounded theory with a paucity of clear, specific examples from the texts, although it isn't difficult to find statements that indicate the contrary. Glaser (2002) stated clearly that:

> GT is a perspective based methodology and people's perspectives vary. And as we showed in "Awareness of Dying" (Glaser & Strauss, 1965) participants have multiple perspectives that are varyingly fateful to their action. Multiple perspectives among participants is often the case and then the GT researcher comes along and raises these perspectives to the abstract level of conceptualization hoping to see the underlying or latent pattern, another perspective. (p. 5)

Latent or underlying patterns and related multiple perspectives do not an underlying obdurate, objective reality make! The terms "latent pattern" or "underlying pattern" are not synonymous with "objective reality."[6] If Glaser and Strauss in *Discovery* and Glaser in his subsequent articulations of classic grounded theory intended for the method to be about discovering objective reality, they would have stated that overtly. They certainly were not unaware of this age old debate in the social sciences. Authors who discern objective reality in the words of Glaser and Strauss evidently see evidence of a latent pattern of which Glaser and Strauss were not aware. This implies that Glaser and Strauss didn't even understand their own position.

Furthermore, in the foregoing quote, Glaser makes it clear that he sees the abstractions of grounded theorists simply to be another perspective. He does not say or imply that he sees the abstractions of grounded theorists to be "*the*" perspective. Those who take the position that an objective reality exists see scientific renderings as epistemologically sound reflections of that reality and common sense, subjective interpretations as epistemologically flawed. Unlike objectivism, classic grounded theory is not about discovering an obdurate, objective reality independent of subjective realities; it is about discovering, conceptualizing, and explaining patterned subjective realities, with full recognition that meanings are continuous, emergent social constructions. This is contrary to Durkheim's (1938) objectivist assertion that "society is prior."

Many of these patterns are fundamental to human experience and durable, although subject to change and variation over time and setting—all of which classic grounded theory considers. Humans engage in "cultivating relationships" (Simmons, 1993), "doing time" (Lee, 1993), "shoring up" (Patnode, 2004), "weathering change" (Raffanti, 2005), "leveraging" Spieker Slaughter (2006), "navigating a new experience "(Vander Linden, 2006 & 2008), "working the system" (Stillman, 2007), "maximizing potential" (Maddy, 2008), and myriad other basic, persistent patterns of behavior. These are what Glaser was referring to as "underlying or latent patterns." Of course, people whose behavior contributes to a latent pattern can easily become or be made to become aware of them. For example in my study of the relationship between milkmen and their customers (Simmons, 1993), I discovered the core category, "cultivating relationships." Prior to my sharing the concept with the milkmen from whom I had collected my data, they were unaware that they were cultivating relationships and that this was an essential part of their jobs. When I later pointed it out, they immediately understood. And, with the informal theoretical foothold that I had provided them, they devised enhanced cultivating strategies and became even better at it. What had been latent easily became apparent and modifiable.

It is an inescapable truth that human behavior is patterned and that these patterns are often underlying or latent, and persistent. Virtually all explanatory methods and theories are about recognizing and understanding or explaining patterns. Not unlike other explanatory methods, classic grounded theory was designed to identify behavioral patterns found in data, some latent, some not. A primary difference between classic grounded theory and other methods is that it does this inductively without incorporating intentional constructivist elements and it conceptually names concepts directly from the data.

There are other clear indicators in Glaser and Strauss' and Glaser's subsequent writings that are counter to the idea that classic grounded theory assumes an underlying, obdurate reality. For example, in *Theoretical Sensitivity*, Glaser wrote, "The goal of grounded theory is to generate a theory that accounts for a pattern of behavior which is relevant and problematic for those involved," i.e. multiple subjective realities (p. 93). Another example is Glaser's (2002) criticism of the role of "worrisome accuracy" in qualitative data analysis (p.1). It is fair to say that objectivists are high on accuracy. Interestingly, in Puddephatt's (2006) interview of Charmaz she recognized the contradiction between objectivism and Glaser's concern with worrisome accuracy, when she says, "That sort of flies in the face of usual positivism, which aims towards accuracy." However rather than seeing this as an indication that Glaser might not be as much of an objectivist as she and others claim, she sees it merely as "odd" (p. 10).

If one's primary interest is in using sociological theory as a basis for practical action, at some point the rhetorical wrestle between objectivists and constructivists and who is which and to what extent becomes largely immaterial and even a waste of time—a bit like counting arguing angels on the head of a pin. From a practical perspective, the objectivist/constructivist debate is an age old esoteric argument that will never be settled. The strength of one position or the other in the social/behavioral sciences is probably more related to professional politics than epistemology and ontology. If we hold out for ontological/epistemological purity, we'll have a long wait. Let us be careful that we don't throw out the baby with the bathwater.

In an action scenario, the real test of a theory is the extent to which it works for action. Actors in the social world who want to bring about constructive change have no investment in and usually no awareness of whether or not theories were developed with an underpinning of constructivist or objectivist ontological/epistemological assumptions. They only care if the theory is relevant to and will be useful in their efforts. They just want it to work.

If a theory is grounded in and generated from action scenes around a particular problem or issue on which participants are working, it is much more likely to be useful as theoretical foothold for planning and bringing about desired changes (a subjective issue), with a minimum of unintended consequences. One could certainly argue that the extent to which it is constructivist or objectivist may ultimately be related epistemologically to how well it works, but the proof is in the pudding.

As I suggested above, adopting either a staunch constructivist or objectivist posture may make for an interesting philosophical debate, but in the end it has little impact on the real world. Classic grounded theory stakes out a middle territory by adopting reasonable and limited rather than absolute features of both and offering an alternative that is of greatest practical use because it solidly grounds explanatory theory in data and provides theoretical foothold for effective actions and change initiatives. Let the diehard constructivists and objectivists continue their rhetorical wrestle; for others, as Glaser tells his students, "just do the work!"

References

Bryant, A. (2002a). Re-grounding grounded theory. *Journal of Information Technology Theory and Application.*

Bryant, A. (2002b). Bryant responds: Urquhart offers credence to positivism *Journal of Information Technology Theory and Application*, 4(3). Art. 6.

Charmaz, K. (2000). Grounded theory: Objectivist and constructivist methods.

In N.Denzin & Y. Lincoln (Eds.), *Handbook of qualitative research*, 2nd edition (pp.509-535). Thousand Oaks, CA: Sage Publications.

Charmaz, K. (2006). *Constructing grounded theory: A practical guide through qualitative analysis.* Los Angeles, CA: Sage Publications.

Durkheim, E. (1938). *The rules of sociological method.* (Ed. By G. E. G. Catlin) (S.A. Solovay & J.H. Mueller, Trans.). Chicago, IL: University of Chicago Press.

Glaser, B. G. (1978). *Theoretical sensitivity.* Mill Valley, CA: Sociology Press.

Glaser, B. G. (1992). *Basics of grounded theory analysis: Emergence vs. forcing.* Mill Valley, CA: Sociology Press.

Glaser, B. G. (1993). (Ed.). *Examples of grounded theory: A reader.* Mill Valley, CA: Sociology Press.

Glaser, B. G. (1994) (Ed.). *More grounded theory methodology: A reader.* Mill Valley, CA: Sociology Press.

Glaser, B. G. (1995). (Ed.). *Grounded Theory: 1984-1994.* Mill Valley, CA: Sociology Press.

Glaser, B. G. (1996). (Ed.). *Gerund grounded theory: The basic social process dissertation.* Mill Valley, CA: Sociology Press.

Glaser, B. G. (1998). *Doing grounded theory: Issues and discussions.* Mill Valley, CA: Sociology Press.

Glaser, B. G. (2001). *The grounded theory perspective: Conceptualization contrasted with description.* Mill Valley, CA: Sociology Press.

Glaser, B. G. (2003). *The grounded theory perspective II: Description's remodeling of grounded theory methodology.* Mill Valley, CA: Sociology Press.

Glaser, B. G. (2005). *The grounded theory perspective III: Theoretical coding.* Mill Valley, CA: Sociology Press.

Glaser, B. G. (2006). *Doing formal grounded theory: A proposal.* Mill Valley, CA: Sociology Press.

Glaser, B. G. (2008). *Doing quantitative grounded theory.* Mill Valley, CA: Sociology Press.

Glaser, B. G. (2009). *Jargonizing: Using the grounded theory vocabulary.* Mill Valley, CA: Sociology Press.

Glaser, B. G. & Strauss, A. L. (1965). *Awareness of dying.* Chicago IL: Aldine Publishing Company.

Glaser, B. G. & Strauss, A. L. (1967). *The discovery of grounded theory.* Chicago IL: Aldine Publishing Company.

Glaser, B. G., & Holton, J. A. (2005). Basic Social Processes. *The Grounded Theory Review,*), 1-21.

Glaser, B. G., & Holton, J. A. (2007). (Eds.).*The grounded theory seminar reader.* Mill Valley, CA: Sociology Press.

Gregory, T. A., & Raffanti, M. A. (Guest Eds.). (2006). Grounded action: An evolutionary systems methodology [special issue]. *World Futures: the Journal of General Evolution.* 62(7).

Haig, B. D. (1995). Grounded theory as scientific method, *Philosophy of Education,* Retrieved from http://www.ed.uiuc.edu/eps/pes-yearbook/95_docs/haig.html

Kools, S., McCarthy, M., Durham, R., & Robrecht, L. (1996). Dimensional analysis: Broadening the conception of grounded theory. *Qualitative health research.* 6(3), 312-330.

Lee, J. (1993). Doing time. In B.G. Glaser (Ed.), *Examples of grounded theory: A reader*(pp. 283-308). Mill Valley, CA: Sociology Press.

Maddy, M. D. (2007). Maximizing potential: A grounded theory study. *Dissertation Abtracts International,* 69/01 (UMI No. 3300310)

Martin, V. B. (2007). The postmodern turn: Shall classic grounded theory take that detour? A review essay. *The Grounded Theory Review,* 5(2/3), 119-129.

Mills, J., Bonner, A., & Francis, K. (2006). The development of Constructive grounded theory. *International Journal of Qualitative Methods,* 5(1), 25-35.

Patnode, G. R. (2004). Shoring up: A grounded theory of process in political leadership. *Dissertation Abstracts International,* 66/01, 250. (UMI No. 3160516)

Puddephatt, A. J. (2006). Special: An interview with Kathy Charmaz: on constructing grounded theory. *Qualitative Sociology Review*, 2(3), 5-20.

Raffanti, M. A. (2005). Weathering change: A grounded theory study of organizations in flux. *Dissertation Abstracts International*, 66/03, 853.(UMI No.3166425)

Simmons, O. E. (1993). The milkman and his customer: A cultivated relationship. In B.G. Glaser (Ed.), *Examples of grounded theory: A reader* (pp. 4-31). Mill Valley, CA: Sociology Press. (Reprinted from Bigus, O.E. (1972, July). *Urban Life and Culture*, I, 131-165).

Simmons, O. E. (1994). Grounded therapy. In B. G. Glaser (Ed.), *More grounded theory methodology: A reader* (pp. 4-37). Mill Valley, CA: Sociology Press.

Simmons, O. E. & Gregory, T. A. (2003). Grounded action: Achieving optimal and sustainable change. *Forum: Qualitative Social Research*, 4(3), Art. 27.

Spieker Slaughter, S. A. (2006). Leveraging: A grounded theory of collaboration. *Dissertation Abstracts International*, 67/02 (UMI No. 3208849)

Stillman, S. B. (2007). Working the system: Aligning to advantage: A grounded theory. *Dissertation Abstracts International*, 68/10. (UMI No. 3287694)

Strauss, A. L. (1987). *Qualitative analysis for social scientists*. Cambridge, England: Cambridge University Press.

Strauss, A. & Corbin, J. (1990). *Basics of qualitative research: Grounded theory procedures and techniques*. Newbury Park, CA: Sage Publications.

Strauss, A. & Corbin, J. (1998) *Basics of Qualitative Research: Techniques and procedures for developing grounded theory*. (2nd ed.) Newbury Park, CA: Sage Publications.

Urquhart, C.(2002). Regrounding grounded theory-or reinforcing old prejudices? A brief reply to Bryant. *Journal of Information Technology Theory and Application*, 4(3), Art,5.

Vander Linden, K. (2005). Navigating a new experience: A grounded theory. *Dissertation Abstracts International*, 67/02. (UMI No. 3208852)

Vander Linden, K. (2008) Navigating new experiences: A basic social process. The *Grounded Theory Review*, 7(3), 39-62).

Endnotes

[1] Glaser devotes Chapter 2 of *Doing Grounded Theory* (1998) to the topic of how he transformed ideas he gained from his Columbia University graduate school professors into grounded theory. For further discussion of this topic see Holton, in the present volume.

[2] Differences can be seen in Strauss' (1987) book *Qualitative Analysis For Social Scientists*, although they are less pronounced in part because, as he acknowledged, he borrowed heavily from Glaser's (1978) *Theoretical Sensitivity* in his introductory chapter. Strauss saw these differences as minor. As he stated, "There are some differences in his [Glaser's] in his specific teaching tactics and perhaps in his actual carrying out of research, but the differences are minor" (p. xiv).

[3] I developed the core ideas for using grounded theory as a basis for action when I was still a graduate student with Glaser and Strauss. I was holding a weekly seminar in my house for graduate students from other universities in the San Francisco area who were unable to enroll in Glaser and Strauss seminars because of limited space. They referred these students to me. My ideas were a response to a member of my seminar who said, "Glaser and Strauss said that grounded theories would give theo-

retical foothold for actions, but they didn't say how." Over the next week I wrote memos on how this could be done. I've used these ideas in my various professional capacities ever since, as grounded therapy and grounded action. For a series of articles on grounded action, see the special edition of *World Futures: the Journal of General Evolution*, devoted to the action method (Guest editors: Gregory & Raffanti, 2006).

[4] "Theoretical sampling is the process of data collection for generating theory whereby the analyst jointly collects, codes, and analyzes his data and decides what data to collect next and where to find them, in order to develop his theory as it emerges" (Glaser & Strauss, 1967, p. 45).

[5] Although qualitative data, particularly open-ended interviewing, is used most commonly in classic grounded theory studies, as Glaser maintains, "all is data" (Glaser, 1998, p. 8). "Grounded theory does well with qualitative data, but it has rightfully no part in the wrestle between quantitative and qualitative.... Grounded theory was not discovered to foster a qualitative ideology" (1998, p. 43). For a discussion of quantitative data in classic grounded theory see Glaser (2008).

[6] It isn't entirely clear if Glaser intended for "latent" and "underlying" to represent two different phenomena or was just using a common stylistic device—that of using two terms to affirm the same meaning. Regardless, although "underlying" may suggest it more than "latent," neither term is synonymous with "objective reality." However, latent and underlying are cross referenced as synonyms in Roget's International Thesaurus 4th Edition and other thesauruses. Like most words, both words have multiple meanings, some that are more suggestive of objective reality than others. In general, which meaning variations an individual selects or reads into words being used by others is likely to be related to their particular ontological view and often reveals constructivist assumptions they are making about the user of the word.

2

Atmosphering for Conceptual Discovery

Astrid Gynnild
University of Bergen

In the international workshop arena for researchers at PhD-level and above, the grounded theory seminars have found a place of their own. While there is a general tendency to structure up PhD tracks according to existing ontology, these seminars provide a supplementary route for developing systematized theorizing proficiency. Elsewhere, novice researchers are trained in theory-testing, theory-to-theory comparisons or problem-solving methods, but seldom in theory generation. An argument often used is that a theorizing approach might be too challenging for novices, and even for established researchers. GT troubleshooting seminars are therefore one of few face-to-face arenas where novice researchers are trained in specified procedures for generating new theory.

In this chapter I propose that *atmosphering*, used as a gerund, is what sets the stage for conceptual discovery in a grounded theory seminar context. Atmosphering is of concern to teachers and supervisors as well as PhD candidates and researchers. The very specific term will be thoroughly discussed throughout the chapter; initially, the aim is to bring to attention the potential impact of the opening spirit, feel, character, flavour or "air" of a learning milieu. The term was developed by Barney Glaser at an early stage of his teaching career and is mentioned intentionally during his opening lecture at GT seminars but not further explicated. I will first introduce the atmosphering process and provide some more background information about grounded theory seminars as a unit of study. I will briefly define and discuss the two central terms of the chapter, atmosphering and conceptual discovery, and from there, the theoretical proposition will be discussed at length. I will finalize the chapter with a more general discussion of atmosphering for learning in the light of educative aspects of Carl Rogers' person-centered theory.

Proposition in brief

In this study, I propose that in a GT seminar context, atmosphering for conceptual discovery is a conscious teaching act aimed at escalating participants' learning curve as much as possible within the given time frame. As such, atmosphering refers to a multiple set of deliberate, sequentially spread actions before, during and after grounded theory seminars. The actions manifest both structurally and psychosocially, where the structuring of seminars conditions the emergent individual and collective learning processes. Within the particular learning environment, researchers are invited in as peers in a multi-disciplinary, multinational and loosely organized network where participants gradually learn to apply the method theoretically and practically.

The proposition is based on a comparison of data from nearly a dozen troubleshooting seminars over a five-year period in the first decade of the 21st century. A brief comparison with established learning theory adheres to its robustness. It suggests that atmosphering for conceptual discovery refers to a holistic, experiental, exploratory, and yet grounded mentoring approach to the generation of new theory. As the approach is firmly grounded in previous and current data about participants' learning curves and moment-to-moment needs, the atmosphering approach is about the application of theory in practice. As such, it demonstrates how the believability and relevance of a grounded theory might be tested on its own substantive field.

Good vibes through playfulness

Before we analyze central elements that fuels the enhancing of theorizing skills in more depth, I will sensitize the reader to the chapter's topic by including an example of what might be identified as "moment capturing" through a humorous approach.

The scene is a restaurant in Covent Garden, London, and about a dozen researchers from different disciplines and different countries are just about to get seated. It is quite noisy and crowded around the long table. Suddenly, in a broad American accent, the man next to me exclaims, "Hey, what's going on here?" He has a big grin on his face. His eyes look like a pair of question marks. While the words are being uttered, the man's mouth is moving slowly, exaggerating every moment, as if something really dramatic has just happened. And it almost did. In fact, he was just about to fall through the seat of his chair, which appeared to be loose. In the next moment, the man turns around, grabs the seat, blows on it and puts it back into place, still exaggerating the movements and still with a teasing expression on his face. Then he makes a quick gaze around the restaurant table as if to see whether anyone else in the group observed his behavior. I still wasn't sure who the man was, but his gestures were so funny that the both of us burst out laughing. In the next moment I found myself engaged in a conversation with Barney Glaser, the co-founder of grounded theory.

Later, by constantly comparing behavioral patterns that Glaser engages in, it became evident that during these few minutes in an informal setting in London, Glaser practiced moment capturing. I was being atmosphered in a specific manner and for specific purposes, without, at that time, being aware of Glaser's atmosphering patterns or of the terminology.

The episode is illuminating for several reasons. Anecdotal evidence suggests that the potentially lasting effect of a first encounter – be it with a person, a book, a piece of art, or a theory – should not be underestimated. An experience shared by many is that even after human encounters of profound influence, people tend to recall only fragments of what has been said or done. What they do tend to remember in the long run, though, is the atmosphere surrounding the situation. When people are at ease with an initial encounter, they tend to be more open the next time. A process of building trust through a feeling of authentic communication might have begun. By contrast, if bad feelings are aroused by a first encounter, it is less likely that the persons involved in the situation are going to build a trustful relationship later.

The sequence also illustrates that creating good vibes is not something that happens by accident. It demands deliberate, initial action from at least one of the two parties involved. In this case, Glaser's deliberately humorous approach demonstrated a personal detachment to the situation, as if he said "don't worry." Thus, the approach foreshadowed playfulness as a more general pattern that Glaser often engages in while teaching the method. When issues or situations are so evidently hopeless that they cannot be taken seriously, he frequently turns to what in his own terminology is called reversal humor. This particular form of playfulness might best be explained as a warm variant of irony, in the sense that it turns the apparently obvious upside-down so the situation or issue becomes quite funny.

Researchers who get engaged in grounded theorizing often refer to their first troubleshooting seminar as a distinctive, mind-opening experience. One participant says: "I came home with GT in my thinking all the time"; another one says, "It's a brilliant way of learning and mixing with others"; and a third says, "The true combination of Barney's talks was to set a mode of 'competence' among the researchers, something that is not easily developed without interpersonal mentorship." A fourth person emphasizes that "it is an entirely different exercise to situate inductive theory within the theory than it is to bend the prior literature into a theory of the substantive area."

As I attended several grounded theory seminars and was repeatedly exposed to the spiralling flow of analytic skills that evolved in a group within three days, I became curious about the troubleshooting method as a way of learning grounded conceptualization. Why did seminar participants get so energized that the spirit created in the group lasted for months afterwards? Why did so many experience the seminars to be so intellectually challenging, rewarding and yet so fun compared to other academic seminar settings?

Researching the seminar setting

And so I started observing what was going on during seminars and taking notes of the process itself; I wanted to find out how the seminars help PhD candidates and mature scholars across disciplines, nationalities and varying levels of scholarly skills transcend personal boundaries and move from descriptive to conceptual levels of research. The resulting proposition, *atmosphering for conceptual discovery*, explains how the grounded theory seminar model supports and escalates participants' development of conceptual research competence. As will be further discussed later in the chapter, the study might have implications for other learning situations in higher education and elsewhere.

In the context of GT-seminars, atmosphering for conceptual discovery is a deliberate, pedagogical process which is aimed at facilitating a new learning awareness, as well as productive training of conceptual theorizing in groups of 15-20 academics. The theory brings to focus that when people relate, there is an interactive process of toning constantly going on. It brings to awareness that people might tone each other consciously or unconsciously. More specifically, atmosphering for conceptual discovery refers to a multivariate set of deliberate toning actions that are carried out before, during, and after seminars. At a structural level, these actions include five distinct, framing principles: *across-ism*, *fly-in-fly-out*, *ambience*, *sense-orchestrating, dressing down, and group individualism.* These structural principles facilitate what we might call *initial contact, focused contact,* and *closing contact* within and between participants, including the supervisor. This contact, in turn, might create within each researcher an intrinsic closeness to his or her own data that helps ensure a productive outcome. The contact stages are thus a cyclic process within which a range of psychosocial atmosphering actions are carried out. The actions are initiated by Barney Glaser as a teacher to implement the openness and awareness that breeds the ground for raising conceptual levels. In this process, Glaser serves as a mentoring role model for the partcipants.

Included in the psychosocial atmosphering is a full range of human behaviors intended to stimulate participants to think "out of the box." Within lectures as well as during the so-called troubleshooting sessions participants are taught and trained also in human aspects of teaching grounded theory, in particular *authentic presence, explicitness, full acceptance, and playfulness.* Implicitly, properties of outspoken curiosity and active problem-solving are encouraged, in particular during sessions where "troubleshootees" present their ongoing work and articulate needs for help at their stage of learning grounded theory. During the run of a seminar, the variety of interactional approaches is mirrored by participants who gradually grow into roles as sharing peers and *reciprocal inspirators.*

The specific conceptual environment of grounded theory seminars requires authentic presence and self expression relevant to the situation. As demonstrated by Barney Glaser, a teacher's self-awareness and self-acceptance influences his or her situational sensitivity. His behavioral patterns and the switches between approaches are guided by what comes up in the group from moment to moment. According to Glaser, "workshops cannot be preplanned except in the broadest of outlines" (Glaser 1998), and "its details are planned on the spot according to the ability level of participants as they emerge" (Glaser 1998, p. 231). To bring the group right to the edge of current problems, Glaser explains, he usually travels with several lectures.

Thus no wonder, probably, that at grounded seminars, many participants are taken aback by Glaser's multiple ways of "breaking up" mindsets. The letting go of preconceptions, and "story talk" for its own sake, are but two issues that, according to Glaser's applied use of theory, must be dealt with continuously in order to transcend from a descriptive to a conceptual level of thought. The task at hand guides the character of the communication. It follows that atmosphering for conceptual discovery concerns contact with one's self, personal grounding, and contact between oneself and one's surroundings, relational grounding. Above all, the proposition explicates ways of building trust with and between people in a minimum of time.

I will now briefly elaborate on GT seminars as a unit of study, followed by an introduction to the key theoretical concepts *atmosphering* and *conceptual discovery*. Next the implications of atmosphering for conceptual discovery are presented in breadth, and will finally be discussed in a broader perspective of relevant learning theory.

Background and purpose of GT seminars

The grounded theory seminars are run in Europe, North America, as well as in Asia several times a year. They are conducted by Dr. Barney Glaser and fellows of The Grounded Theory Institute and attract participants from all continents and a multitude of disciplines. The several dozens of dissertations that have come out of the seminars indicate that the model serves its purpose. The seminars were initiated by Glaser and the Grounded Theory Institute around the turn of the millennium. By this time, Glaser was an internationally established scholar who had supervised dozens of PhD-candidates during his many years at the University of California in San Francisco. The seminars were based on the one-year long grounded theory seminar he used to teach at UCSF. They were later reduced to a three- day work shop, developed as a means to support PhD candidates who were minus-mentored in grounded theory, even though they had appointed supervisors at their institutions. Throughout the years, Glaser got many questions from researchers around the globe who grappled with learning grounded theory from his books. This material provided data for Glaser to apply his own method when designing the seminar approach.

A dilemma for Glaser, when designing the seminars, was that learning the basics of grounded theory method takes about one and a half years of intensive work (Glaser, 1998). It is not something that can be learned in a few days of intensive teaching, no matter how efficient the seminar design is. The ability to integrate the specific procedures of theory development is what Glaser terms a "delayed action process," where the relatively slow pace hangs together with the processing speed of the subconscious mind. It presumes more than a shift in research focus from deductive to inductive thinking. Implicit in the process is the need for a letting go, or the de-construction of, a descriptive mindset in order to integrate a conceptual approach. Thus, grounded seminars mainly serve as *inspirational loops* (Gynnild 2007) for participants as they move forward with grounded theory projects that normally take two to four years of work. Inspirational loops refer to a variety of breaks and shifts that fuels energy and intrinsic motivation during creative processing of research problems.

From the beginning, the seminars have been arranged independently of any university schedule or curriculum. They are sought by researchers from all continents, often particularly independent individuals who navigate multiple pathways during their PhD period. Each seminar is open to a dozen "troubleshootees" and six to eight "observers" who do not submit any work in progress but take part in the solution-oriented discussions of other people's work. Often, observers are selected researchers already experienced in doing grounded theory. Since the first European seminar in Paris 2002, several dozen dissertations worldwide have been produced as a direct result of seminar inputs. The model has later been adopted by several fellows of the Grounded Theory Institute who were trained in Glaser's early seminars.

The main purpose of grounded theory seminars is to support and accelerate each participant's development of conceptual research skills and successive research autonomy. Glaser makes it explicit that he wants to bring each participant one step further, wherever they are in their research, and the overall goal is to help people get their PhD degrees. Furthermore, in order to get participants' theorizing skills as far as possible within the limited amount of time, it appears that seminars are designed to increase their awareness of "what is really going on" in a broad sense.

The what-is-going-on phrase refers to two aspects of awareness. The first aspect is data sensitivity, learning how to identify layers of data previously identified by a grounded theory: vague, interpreted, properline and baseline data respectively (Glaser 1998). The second aspect follows from the first and concerns purposive presence. Being fully present and fully focused on what's going on within a group setting as well as in one's own mind requires much practice. The larger a group gets, the easier it is for individuals to drop out of focused discussions and let thoughts wander off in their own ways. This multifaceted challenge is resolved through a number of atmosphering approaches aimed at opening up for the discovery of new ways of seeing, hearing, and experiencing what might be in the data.

The concepts of atmosphering and conceptual discovery

Atmosphering is a gerund constructed from the noun atmosphere. In writing, the concept is hardly mentioned anywhere else other than in Glaser's books (Glaser 1998) and seminars. At least it is not yet a recognized term in dictionaries. The noun "atmosphere," on the other hand, refers according to the *Oxford Dictionary* to a feeling, for instance of good or bad, which the mind receives from a place or conditions. In the Visual Thesaurus, atmosphere is defined as "a particular environment or surrounding influence."

To begin with, using atmosphere as a gerund might seem a bit strange. From a social action point of view, however, atmosphering is an eye-opener.

Whereas the noun "atmosphere" suggests that an environment or surrounding influence of something is outside of individual control, the switch to gerund implies that atmosphering is an act. The activity is carried out by someone towards someone, and there is an interaction between two or more parties taking place. Atmosphering suggests that when people relate, they also tone each other, consciously or unconsciously. When people are aware of the toning process, they can deliberately choose between proactive and reactive responses and interplay. However, if people are not aware that atmosphering is constantly going on, they might feel helpless in the situation and become passive recipients of other people's actions. A main challenge in a learning environment is for the teacher to evoke autonomous engagement among students. In other words, the degree of atmosphering awareness influences the degree of personal control of the situation, and also personal responsibility. As a preliminary conclusion, atmosphering concerns awareness of individual opportunities to create a particular environment or surrounding in order to influence others.

The other significant term of the main category in this study, *conceptual discovery*, is also rarely found in the existing literature. The term is associated with studies within information science, where researchers set up data mining programmes to search for specific concepts in digital texts (Butcher, Bhushan, and Sumner, 2007), and educational approaches online and offline (Alessi, 2000; Furth, 1963). The well-known noun, discovery, is, simply put, an act of detecting something new. With references to academic disciplines, *The Oxford Dictionary* defines the term discovery as "the observation of new phenomena, new actions, or new events and providing new reasoning to explain the knowledge gathered through such observations with previously acquired knowledge from abstract thought and everyday experiences." Terms related to discovery are exploration, knowledge, actions, learning and sensing – and questioning. Questions play a leading role in any discovery; without questions, no new observations, reasoning or unveiling of patterns of behavior would occur. Discovery is thus an act, a process of finding out what is going on in a field of interest and the outcome is awareness. The concept of conceptual discovery emphasizes the direction of the detection; it concerns imagination, or "conceptions" to bear on facts (Rennie and Phillips, 1988). As such, conceptual discovery concerns the creation of new perspectives on perspectives. It is a developmental trajectory where the atmosphering is a necessary first step to get open for abstract conceptualization (Kolb 1984, 1999).

The outcome of this particular context is a feeling of psychological safety for most participants. Creating a psychologically safe space for participants frees energy for productive work.

In Glaser's books, the same strategy of atmosphering is used to create that safe psychological space in his readers. His refusal to copy-edit his work is designed to demonstrate that ideas are ascendant over presentation. It is similar to his dressing down in seminars. The focus on ideas creates a dynamism in thinking, movement is always forward, conceptual progression is not halted by looking backwards at presentation. Whereas Glaser's books create a textual approach to atmosphering for conceptual discovery, GT seminars provide a multidimensional human approach.

Structuring principles

GT-seminars take place in specific virtual and physical arenas. The latter refers to the actual localities where seminars are held. The virtual arenas are initiated and run by seminar facilitators and include wikis and other internet forums where participants are expected to connect and contribute to group learning before and after seminars. Glaser never engages in these discussions as he wants to know as little as possible about the concrete troubleshooting issues beforehand, but they provide an opportunity for participants to connect before they physically meet.

Within these virtual and physical frames, the first structuring principle is across-ism. This concept identifies diversity and multiplicity as an important premise for conceptual discovery. Across-ism refers to an opening for, in theory, indefinite mix of research disciplines, substantive areas and GT skill levels represented at a seminar. Since a stated goal is to help any candidate exactly where he or she is in their study, a basic idea is one of diversity; participants will be better at helping each other when competencies and levels of skill are complementary. Across-ism also points to a diversity of cultural and national backgrounds. As a structural approach, across-ism takes the edge off unproductive competition between participants in favor of productive, collaborative sharing. As one participant said in an email to the author of this chapter, "Haven't been to anything similar. Its informal atmosphere was great however I was nervous when it came to me to present, but I knew that people were there to support you not to criticise you which makes it better."

Second, there is the fly-in-fly-out principle. This dimension gives a direction to selection of physical seminar sites. Obviously there are practical reasons why grounded theory seminars, just like many other academic seminars and conferences, are typically set up close to huge international hubs such as Paris, London, New York, and San Francisco. Big cities are easily accessed and thus, participants save time and money on travel. However, in an abstract grounded theory perspective, the fly-in-fly-out contributes to a temporary de-contextualization on behalf of participants. When heading for an inspirational loop, in practice, a grounded theory seminar, participants literally have to let go of their everyday life routines and environment. A participant says: "Seminar locations influence me in several ways. By going global this way, you are suddenly in the midst of the creative connectivity of big cities. Detached from colleagues, and yet this feeling of learning so much through being with people from other disciplines. It creates much excitement."

Physical de-contextualization from routine environments speaks to the subconscious mind as well, and accelerates the process of opening up for new discoveries (May, 1994; Gynnild, 2007). In that sense, participants are out of familiar space. Getting to know people mostly by their first names adds to the experience of the fly-in-fly-out principle as a kind of temporary time, place, and people upheaval.

A third, structuring dimension of atmosphering for conceptual discovery is what I have chosen to call sense-orchestrating. This dimension concerns the facilitation of purposeful interaction between, and stimulation of, senses. In many academic settings, the auditory sense is privileged. Presentations of lectures and papers are often utterly focused on detached intellectual reasoning through listening and reading, whereas other sensory information is out of focus. At grounded theory seminars, it appears that appeals to all main sensory systems are part of the atmosphering package; hearing and vision, as well as touch and taste. As very sensitive receptors of information, these sensory systems transform information from the physical world to the realm of the mind. In any setting where there are people, there are multifaceted processes of sensory information going on in the visual, the auditory, and the somato-sensory systems. To what extent such messages are identified and further processed, depends on receptors' degree of awareness and alertness in the situation.

Informality is key to enabling a warm atmosphere first at the meet-and-greet events, which are staged prior to the seminar. At the meet-and-greets, participants are introduced to one another and invited to share a meal together. The informality helps reduce anxiety while at the same time increases participants' anticipation of the experience to come. Glaser sees these events as a means of binary deconstruction, that is, a levelling of relationships which implies that any interaction is between equals.

Another aspect of sense-orchestrating is purposive ambiance. This concept refers to the choice of locations within the frames of the fly-in-fly-out. Purposive ambiance points to the vibes that people pick up from a specific physical environment such as a building or an area of a city. Often, locales are found in classic buildings, which in themselves convey an aura of distinction, or rather, upheaval, from common trivialities of everyday life. With few exceptions, seminar participation in uplifted, historical contexts nourishes the feeling of well-being and creative flow in a world of good taste, abundance, and other signs of energy surplus.

It further supports the deconstruction of held beliefs; the detachment from context-bound environments helps participants to move from the descriptive to the abstract by cutting across time, space, and people. It is about liberation from unnecessary disturbances that would otherwise take energy away from the conceptual discovery process. It is about selecting what is purposeful in the situation and at the same time insisting upon the highest standards with the result that participants feel valued and respected and respond to the stimulus. Implicitly, sense-orchestrating includes the facilitation of switching fluently between environments that, in different ways, might influence general sensitizing. For instance, seminar rooms might be of such a small size that people more or less run into each other physically during breaks; appeals to the kinaesthetic sense is but one way of informally breaking the ice and actually "get in touch" with other people.

The fourth structuring principle of atmosphering is dressing down, a well-known pattern among academics to emphasize informality and thereby helping other people relax. As an integrated part of Glaser's appearance at GT seminars, dressing down signalizes a peer-to-peer approach.

The fifth and last structural principle is that of group individualism. This principle points out that the GT seminar model takes care of participants' binary needs for plenary and individual sessions in one. Within a group of fifteen to twenty people, space is given for up to twelve individual sessions. During these sessions the whole group engages in a plenary troubleshooting process lead by the supervisor. The troubleshooting focuses on what research problem needs to be resolved for the participant to move to the next step. Participants also have the option to be observers throughout the whole span of a PhD dissertation. In practice, status as observers provides an opportunity to avoid presenting one's own work but a subsequent obligation to participate actively in helping others. Implicit in group individualism is the option, on the third day, for one-on-one or two-on-one discussions with the seminar leader, in this context Barney Glaser. Group individualism supports the idea of multiple experientiality; by engaging in conceptualizing across disciplines and stages of theory development, participants take a shortcut.

So far, we have theorized the main *structuring* features of atmosphering GT seminars; these are all frames and physically observable facts that are easily identified by participants. The structuring has been discussed as a premise for creating the ultimate setting for conceptual troubleshooting of a multidisciplinary group of evolving grounded theorists of mixed nationalities. The structuring helps *setting a tone* for productive theorizing.

Interacting principles

At a psychosocial level, the multiple atmosphering actions before, during, and after grounded theory seminars are identified as aspects of a cyclic interacting process: initial contact, focused contact and closing contact.

Initial contact refers to the introductory stages of interaction that takes place during a seminar, more specifically the informal meet-and-greet in the evening before the seminar starts, and also at the opening lecture on the first day. Focused contact refers to the very concentrated stages of interaction with and within the group as lecturing and troubleshooting unfolds. Closing contact refers to the finalizing sequences at the end, the summing up session followed by an informal meal at a location different from that of the seminar. It should be noted that the initial and finalizing contacts stages can be even more sequentially spread in those cases where online discussion groups are set up prior to a seminar, so that participants are able to start the process of getting to know each other beforehand.

Before we move on, it should be kept in mind that the theorizing in this chapter is focused on an empirically grounded facilitation of a particularly fruitful environment of learning. The interest lies in analyzing how Glaser's theorizing around teaching grounded theory unfolds in practice. The intensity and duration of the first step of getting acquainted with another person varies, however. Sometimes the ice is broken very quickly, as exemplified by the meet-and-greet encounter between Glaser and me early in the chapter. By humorously exaggerating the drama with the broken chair, Glaser managed to move from initial contact to focused contact with me in a matter of seconds. This was a practical demonstration of impression management (Goffman, 1959) and doing it quickly but naturally. Glaser could, of course, have chosen other approaches to the same situation. The analysis of the interacting principles of atmosphering indicates that the psychological space established during the first hours of a grounded theory seminar sets a norm for the days to come.

At a psychosocial level, a term that might be used synonymously with atmosphering is toning. In music, toning implies the specific pitch, quality and duration of a note, the character of a sound. In speech and writing, it refers to the manner in which something is expressed; the firmness, pitch or quality of something (dictionary.com). In his opening lecture at seminars, Barney Glaser repeatedly applies the same concept to explain what atmosphering is: "It's all about toning. The first thing you do is set tones and you ride the tone. It's tone riding whether the tone is open, friendly or rude." With that said, Glaser explicitly states that he constantly uses himself as a multi-stringed instrument to achieve the intended impact on participants. In the following, I will discuss the main principles of psychosocial atmosphering approaches within a seminar context, as applied by Barney Glaser as a teacher. All the principles relate to awareness.

I have called the first *focused presence*. Included in focused presence are the teacher's sensitivity and direct initiatives and responses to what is going on. Initially, focused presence is established through the making of eye contact with individuals as well as with a group as a whole. Eye contact is succeeded by mindful switching between the posing of questions, active listening, the posing of follow-up questions, the asking for clarifications, reflective pausing, role playing, and joking. All approaches are soaked with abstract concepts, well known, unknown or exploratory. When carrying out the conceptual overloading, Glaser as a teacher might go back and forth between more refined aspects of the above mentioned approaches, all depending on context. The pace of speaking is slow, which lends gravitas to each word. The reflective pausing of Glaser, as well as participants, reduces pace even more. Some pauses might last for minutes. This pace adds to the experience of time upheaval – time might be fast or slow. Reflective pausing makes the switching between active listening, as well as questioning and humorizing, more pronounced. As one participant explained when I commented that he appeared to be thinking very hard: "Yes, I'm really listening, you know. I'm listening so hard that I'm afraid others can hear it." The variation in approaches contributes to keeping energy and engagement up.

Focused presence is both about being present to oneself, self-awareness, and about being present to others, empathetic awareness. Facial expressions and body language emphasize focused presence. The switches between the aspects of focused presence appear frequently and with ease, in a manner that indicates authenticity. When a teacher's interacting patterns appear to be honest, participants tend to feel connected to what he is and what he stands for. As an approach to teaching, purposive presence embraces intuition for participants' needs for breaks and shifts to keep energies up, in short mindfulness to uncover subtle signs.

Another principle is explicitness. As repeatedly pointed out by Glaser in books and seminars, as a method, grounded theory is based on trust in the data (Glaser 1967, 1978). However, to achieve the required sensitivity and closeness to the data, aspiring grounded theory researchers usually need to unlearn conjecture, which in GT amounts to not trusting the data. Thus, for a seminar leader, it appears crucial to be aware that, in order to encourage researchers to have confidence in the data, he needs to ensure that they trust the method. Such trust implies the conviction or belief that the method works and is relevant, which implicitly challenges the seminar leader to demonstrate expertise skills in the field. Being relaxed, flexible, and in control of the situation is part of building confidence through explicitness.

In his seminars, Glaser builds trust in ways similar to that of his books, by being explicit and staying close to the data when he theorizes. Professional authority is established by stating clearly and explicitly what grounded theory is and is not; by bringing details from his own career trajectory and successful use of the method, by conceptualizing seminar goals and how they are going to be achieved; and advising participants on what they will experience professionally and personally during and as a result of the seminar. The list includes a full range of personal and professional discoveries, and emotions and questions that any researcher tends to grapple with throughout a career. According to Glaser's analysis, these include issues of confusion, contribution, originality, autonomy, depression, confidence building, empowerment, inspiration, networking, and "a sense of youthification through the joy of it all." Glaser also makes clear that:

> We're going to do perspectives on perspectives. People will see data and I'll expect you all to chime in with a potential concept, for the data. I want you to start getting abstract. So leave the data and get on a conceptual level which is abstract of time, place and people, and start talking

> about the general patterns of life. The one thing I can't stand is tiny topics.

Another aspect of explicitness manifests through an extensive use of illustrations that support the conceptualizing. An example often used is from Dallas, Texas. Glaser had just finished a seminar and wanted to know the quickest way to the airport.

> There were ten of them{seminar participants} who developed ten different theories on how to get to the airport. It's like running a commando corps; you just don't drive to the airport. They all had their theories based on several indicators they had accumulated over a long time. You have to be in this lane and that lane. I followed one of the theories and I made the airplane. See, one of the most natural theories is how to get from here to there. You base it on different indicators based on traffic patterns. You do it all the time based on interchangeable indices. And the concept is routing.

A third principle of psychosocial atmosphering is full acceptance. Many participants at grounded theory seminars are surprised by the degree of openness created in a very short time within the existing group. Moreover, they talk of feeling included and valued as productive, autonomous contributors to the discussion even if they are new to grounded theory. Full acceptance is demonstrated through a genuine curiosity in participants' work, and a helpful attitude towards research problems that are brought to the fore – without being judgmental. Asking for help is also a way of credentializing others.

By emphasizing that each participant is an expert on his own data, Glaser creates a peer-to-peer route to conceptual discovery. This fosters a spiraling flow of idea development in the group and within the mind of the individual. Included in the full acceptance approach is a non-judgemental openness towards persons, ideas, and research problems. However, the full acceptance approach requires the deliberate detachment of personal, emotional, political as well as other presuppositions in the situation. Such detachment means to let go of something old in order to replace it with something new (May 1994).

Applying grounded theory in practice, in other words, implies training in non-judgmental attitudes. The approach ultimately requires modes of deliberate detachment. As Glaser responded to a presenter who was afraid of not doing well enough in front of the group: "To me, every time you speak, I'm just thinking of the data. I'm trying to transcend it conceptually." He was focussing totally on the ideas.

Yet another principle is vigor. As a co-discoverer of his own method, Glaser is not only devoted to teaching grounded theory. He is passionate about it. In troubleshooting seminars, his "embodiment" serves as a walking proof of the energy which is built into theory generation based on GT procedures. He is also walking proof of the empowerment and autonomy that evolves from developing grounded theory expertise. As emphasized by Glaser, one of the virtues of the seminars is that:

> Grounded theory is youthifying. If you look around the table you'll see that people who attempt to do grounded theories tend to be older – they have come to a stage in life where they know it's OK to wonder and that you don't know everything constantly. Did I tell you about Tara? She was 80 when she started and ran around like a 30 years old when she found her core category.

The power and durability of grounded theories is repeatedly exposed through providing examples of generated theories that have made a difference in their fields.

The "youthfying effect" might hang together with the explorative opportunities that are an extant feature of grounded theory. It is fun to explore and it is fun to be the first one to discover new, latent patterns. This fact brings us to the fifth interaction principle of atmosphering for conceptual discovery, which is playfulness. As the most experienced grounded theorist worldwide, Barney Glaser is at an intuitive expertise level (Gynnild, 2007). Intuitive expertise implies that focused presence, explicitness, full acceptance, and vigor are executed with ease and fluency, and rapid switching between different states of mind. It is a state where thinking and doing converge, like a theorizing tango where time is pulled together and dragged out according to the situation.

As such, Barney Glaser typically demonstrates how a teacher might be in full control of what he is doing and how he is acting. As a troubleshooter, Glaser is constantly juggling a repertoire of human approaches to help participants "get to the next step." The playfulness also demonstrates detachment to the situation; nothing is ever so serious that you cannot comment on it humorously. Humor helps keep energy up, and it is a generous way of telling people to keep on track. Moreover, playfulness helps create psychological space for exploration and for breaking out within a group. It takes courage to break off or interrupt a discussion, thus playfulness also fosters risk-taking; a humorous approach might take the edge off otherwise embarrassing situations. Glaser frequently demarcates the difference between grounded theory and qualitative data analysis by humorously telling people to "stop story talking," which in a grounded theory perspective brings the researcher nowhere as stories might go on and on without any conceptual outcome.

When Barney Glaser sums up the first session, he might point out "we're creating chance-taking, creating mutual confessions, permitting, we're laughing a lot and some can cry." By this statement, he makes clear to participants that an open space for explorative theorizing is facilitated. The group has "atmosphered and toned." It is time to start doing.

Similarities with person-centered theory

As a preliminary conclusion, the analysis of atmosphering for conceptual discovery suggests that Glaser's ways of teaching grounded theory are also applying grounded theory. A question that is still awaiting an answer is this: In what ways does atmosphering for conceptual discovery resemble existing theories about teaching and learning in intelligent careers? In the following I will briefly refer to literature that explicitly mentions the importance of creating a good atmosphere for learning. From there I will narrow the scope to focus on main aspects of humanistic experiential psychology, represented by Rogers' person-centered theory (Rogers, 1951, 1969). This theory appears to speak most directly to the grounded atmosphering approach discussed in this chapter. Other theories that appear to be of relevance include, but are not exclusive to, experiential learning (Kolb, 1984, 1999), and a more general heritage of social learning derived from psychotherapy.

Creating a good atmosphere for learning has puzzled teachers and researchers throughout history, and yet there is still a lack of clear definitions and explications. According to Gabbrielli, the term atmosphere has "become a slippery 'industry' buzzword, with teachers and researchers alluding to an impressionistic feel-good factor devoid of an inherent definition" (Gabbrielli, 2009). The current study, therefore, suggests that the gerund atmosphering fills a void in the literature of learning.

According to several studies (Hatwell, 1992; Maley, 1994; Gabbrielli, 2009), teachers intuitively tend to feel that a classroom atmosphere is important in order to create a good learning milieu. Young (1991) indirectly points out that student feelings of affect and low anxiety are frequently linked with vague perceptions of atmosphere. It might be expressed as the absence of activities or surroundings that might lead to unnecessarily unpleasant, uncomfortable, or frustrating experiences for the learner. Several researchers (Hadfield, 1992; Maley, 1994) claim the importance of atmosphere and chemistry in a group or class. According to Maley, classrooms are humanistic arenas of psychosocial and emotional encounters, and what students do in a classroom is mainly in accordance with that which they are allowed by the teachers to do (Maley, 1994).

The term atmosphere, however, does not occupy much space in the literature of learning.

In a more general learning context, psychosocial interaction principles are more extensively analyzed than physical and philosophical frames, which apparently are considered more in the periphery of teaching and learning. This fact brings to the fore some core issues presented and discussed in humanistic, educational psychology, in particular the person-centered theory developed by the American psychologist Carl Rogers. The best known presentation of his theory is found in the three editions of Freedom to learn (Rogers 1969, 1983; Rogers & Freiberg 1994), which are all based on Rogers' Client-Centered Therapy from 1951, and further developed for the education field. Similar to grounded theory, Rogers' person-centered theory represents a holistic approach to learning and human growth. As a main learning theory, it has had profound influence on experiental teaching and learning designs in the western world.

Whereas Glaser's mentoring approach is concerned with the skilled development of researchers as theorizers across disciplines and geographical borders, Carl Rogers is concerned with education helping individuals become optimally functioning persons, citizens, and leaders in a democratic society. Whereas Glaser stresses detachment from time, place, and people in order to develop expertise in abstract conceptualization (Kolb, 1984), Rogers stresses the more general aspects of being a well-functioning participant in society. At the same time, Rogers emphasizes that significant learning is most effectively promoted in environments in which threat to the self of the learner is reduced to a minimum, and a differentiated perception of experience is facilitated. These propositions are congruent with the grounded atmosphering approaches that are carried out at grounded theory seminars.

Furthermore, in their approaches to learning, both Glaser and Rogers are concerned with opening up to, and theorizing from, experience. Rogers (1969) insists that specific attitudinal qualities in the facilitator-learner relationship yield significant learning; these attitudes and qualities all relate to the facilitator's approach and can be summed up in three ideas, realness or genuineness, acceptance, and empathy.

Realness or genuineness refers to a teacher or facilitator's ways of being and presenting himself. When values and expressions are in congruence, he or she might enter a relationship with the learner without presenting a front or a façade. He or she is aware of own feelings as they are experienced and communicates what is appropriate, which opens up or real or genuine person-to-person encounters. The second quality that stands out is what Rogers terms "prizing, acceptance, trust", which is a kind of non-possessive caring. By valuing and accepting the other person as she or he is, the facilitator demonstrates trust in the situation and trust in the other person. The same attitudes are found in Glaser's focused presence and full acceptance approach when teaching grounded theory.

The third quality that helps create a climate for self-initiated, experiental learning, as proposed by Rogers, is empathetic understanding. This aspect concerns sensitivity and awareness on behalf of the facilitator or teacher. By empathy, Rogers refers to an attitude where a supervisor is standing in the other's shoes and is able to view the world through the eyes of the students. When students feel understood, not judged, the likelihood of significant learning is increased. In a grounded theory perspective, however, the "empathetic understanding" that participants tend to experience in grounded theory seminars is not primarily based on the teacher's, that is Glaser's, individual skills in human understanding. Rather, the equivalents to "empathetic understanding" are grounded in empirical data collected in the substantive area of teaching grounded theory. Thus, the theorizing that resulted from a thorough analysis of these data provided the fundamentals when creating the seminar design. For each seminar, the theory is adjusted and modified according to new data, which in turn ensures that the design is relevant and fits participants' needs.

Another similarity between grounded theory seminars and a person-centered education approach is a stated flexibility in teaching methods and a transparency of teaching design. Included in the design is usually the resolution of relevant and real problems, instruction intended to meet individual needs, and peer tutoring. A hallmark of students and facilitators of both approaches is, furthermore, a willingness and openness for personal change and growth. Again, according to Rogers, it all starts with an initial, genuine trust in learners by the facilitator, followed by the creation of an acceptant and empathetic climate. He also emphasizes the importance of creating an atmosphere of permissiveness, acceptance, and student reliability within given limitations or circumstances, an issue that again speaks directly to the explicitness and full acceptance dimensions of atmosphering for conceptual discovery. To sum up, learning tends to be enhanced in contexts where learners have supportive relationships, a sense of ownership of the learning process, and can learn with and from each other in environments where they feel valued and trustworthy (Thorne, 1992; McCombs, 2004a), in short, where a good atmosphere or climate is created.

From this brief comparison it becomes evident that there are some very interesting psychosocial teaching similarities between grounded theory and person-centered learning. And yet the differences are probably just as many. The strength of both approaches lies in the focus on presence and relational qualities in a learning situation, and the supervisor or facilitator's function as a role model. What Rogers terms realness, genuineness, or congruency, is by Glaser, more informally, referred to as "walking the talk." This issue also speaks directly to double loop theory (Argyris, 1976), which in turn is based upon a "theory of action" perspective outlined by Argyris and Schön (1974). In theory of action, an important aspect is the distinction between an individual's espoused theory, what he says he is going to do, and his theory-in-use, what he actually does. Thus, a main concern in double loop learning, just as in applied grounded theory and person-centered education, is to bring about a congruency between talking and walking. They all have their chosen approaches that surely work.

When supervising or teaching others in grounded theory, though, the staying-close-to-the-data-dictum implies that teaching methods are applied according to grounded theories of what is relevant and important to participants. Thus, atmosphering for conceptual discovery demonstrates that from a teaching perspective, the grounded theory requirement of getting open, and staying open, concerns more than a constant awareness of oneself and one's surroundings. Atmosphering for conceptual discovery concerns the development and implementation of a specific, conceptual mindset among researchers - across disciplines, and across social, cultural, and psychological borders. As such, the atmosphering approach at grounded theory seminars supports research fundamentals, that is, the general relevancy, credibility, and believability of abstract conceptualization and academic theorizing.

References

Alessi, S. (2000). Designing educational support in system dynamics based interaction learning environments. *Simulation Gaming, Volume 31(2)*, pp. 178-196.

Argyris, C. and Schön, D. (1974, 1992). *Theory in practice. Increasing professional effectiveness.* San Francisco: Jossey-Bass Inc.

Butcher, K.R., Bhushan, S. and Sumner, T. (2007): Multimedia display for conceptual discovery: Information seeking and strand maps. *Multimedia Systems 11(3)*, pp. 236-248.

Furth, H.G. (1963). Conceptual discovery and control on a pictorial part-whole task as a function of age, intelligence and language. *Journal of Educational Psychology 54(4)*, pp. 191-196.

Glaser, B. G., & Strauss, A. L. (1967). *The discovery of grounded theory: Strategies for qualitative research.* Mill Valley, CA: Sociology Press.

Glaser, B.G. (1978). *Theoretical Sensitivity: Advances in the methodology of grounded theory.* Mill Valley, CA: Sociology Press.

Glaser, B. G. (1998). *Doing grounded theory: Issues and discussions.* Mill Valley, CA: Sociology Press.

Gabbrielli R. R. P. (2009): Demystifying classroom atmosphere. *The ELJ Journal, Volume 1(2).*

Goffman, E. (1959). The presentation of self in everyday life. New York, NY: Anchor books.

Gynnild, A. (2007): Creative cycling of news professionals. *Grounded Theory Review, 6(2)*, pp. 67-94 .

Gynnild, A. (2006). Growing open: The transition from QDA to grounded theory. *Grounded Theory Review 6(1)*, pp. 61-78.

Hadfield, J. (1991). *Classroom dynamics.* Oxford: Oxford University Press

Kolb D. (1984). *Experiental learning: Experience as the source of learning and development.* Englewood Cliffs, New Jersey: Prentice Hall.

Kolb D. (1999). *The Kolb learning style inventory, Version 3.* Boston: Hay Group.

Lei, P. W., & Wu, Q. (2007). Introduction to structural equation modeling: Issues and practical considerations. *Instructional Topics in Education Measurement, 26(3)*, pp. 33-43.

McCombs, B. L. (2004). *The learner-centered psychological principles: A framework for balancing a focus on academic achievement with a focus on social and emotional learning needs*, In J. E. Zins, R. P. (New York: Teachers.)

Maley, A. (1994). The anatomy of atmosphere, P*ractical English Teaching, 14(3)*, pp. 67-68.

May, R. (1994). *Courage to Create.* New York: W.W. Norton.

Rennie, D. L., Phillips, J.R & Quartaro, G. K. (1988). Grounded theory, a promising approach to conceptualization in psychology. *Canadian Psychology 29(2)*, pp. 139-150.

Rogers, C. (1951). *Client-centered therapy: Its current practice, implications and theory*. London: Constable.

Rogers, C. (1969). *Freedom to learn: A view of what education might become.* Columbus, Ohio: Charles Merill.

Rogers, C. (1983). *Freedom to learn for the 80s.* Columbus, OH: Charles Merrill.

Rogers, C. with Freiberg, H. J (1994). *Freedom to learn,* third edition. New Jersey: Prentice Hall Inc.

Thorne, B. (1992). *Carl Rogers.* London: Sage.

Young, D. J. (1991). *Affect in foreign language and second language learning. A practical guide to creating a low-anxiety classroom atmosphere.* New York: Mc Graw-Hill.

3

GETTING THROUGH THE PHD PROCESS USING GT: A SUPERVISOR-RESEARCHER PERSPECTIVE

Wendy Guthrie
Loughborough University

Andy Lowe
Glyndwr University

This chapter is a collaboration between Andy Lowe, an experienced GT PhD supervisor, and Wendy Guthrie, one of his researchers, whose doctoral thesis of *Keeping Clients in Line* (Guthrie, 2000) was externally examined by Dr Barney Glaser. The main inspiration of the chapter is to help both PhD supervisors and researchers to navigate their way through some of the obstacles they are likely to confront while doing a PhD using the classic grounded theory method in a conventional university setting. The chapter identifies five main issues which have to be confronted and dealt with in order to ensure that the PhD, using a classic GT research methodology, is achieved:

(i) Establishing a consensus on the purpose of the PhD
(ii) Finding and evaluating the most appropriate supervisor
(iii) Understanding how the specific university regulations might impact on the research process
(iv) How to manage your committee
(v) Getting published

The future is grounded

Sometimes being ahead of your time is difficult. Barney Glaser can vouch for this. Experienced GT researchers will identify with this too. How many times have you had to explain the whys and wherefores of doing GT? Of course we should not be surprised by the need to painstakingly unpack what GT has to offer. Anything which is complex and unfamiliar takes time to be adopted, and, indeed some things take longer than others! We remember Barney saying a number of years ago that GT was some 20 - 30 years ahead of its time.

We aim to raise awareness of the multi-faceted challenges likely to be faced by those wishing to gain a PhD using GT and for their supervisors tasked with navigating the route to the successful completion and award of the degree. The advice is designed for fledgling GT researchers, supervisors new to GT and those seeking comfort in the knowledge that they are not alone. It is intended that the advice offered deflects some of the pitfalls that ordinarily would be discovered through experience.

We hope that the information offered smoothes the path ahead in two fundamental ways. Firstly our writing should raise your levels of awareness about critical issues which may impinge upon progress and, in so doing, enable you to anticipate what challenges might be encountered. At the very least you should be better placed to take appropriate decisions with the benefit of other's hindsight. We are the test pilots, so there is no need for you to crash and burn. We invite you to latch on to the opportunity to make swift progress based on our learning and above all enjoy the Grounded Theory journey.

(i) Establishing a consensus on the purpose of the PhD

Western European universities from the Middle Ages onwards have been issuing academic degrees of various types. Throughout Europe universities were dominated by the Christian church. The church sought to be the sole custodian of both knowledge and power. The first doctoral degrees to be awarded were all doctor of divinity degrees, then came Master of Arts and Science degrees. The PhD was the most recent newcomer of all higher degrees. The contemporary PhD has its academic roots in Germany in the 1850s, shifting soon after to the USA. Later it caught on in the UK and then the rest of the world. By tradition the holder of a PhD degree is a person who is qualified to practice high quality research without supervision. Around the world today there is a vast variation of both the purposes and objectives of the PhD degree. This variation extends to different universities within the same country. At one extreme the PhD process is completely formularistic while at the other extreme the PhD is vague and ambiguous. Even the way in which the PhD is evaluated is extremely varied. The PhD candidate does not have the opportunity to defend the thesis in person in most Australian universities which award PhDs. Instead the PhD researcher simply submits the thesis and awaits the written evaluation of the committee. In most other countries there is the opportunity for the PhD candidate to personally defend the thesis but there are hundreds of different ways in which the process is conducted. The issue of importance for the PhD researcher and supervisor involved in the PhD is to have an in-depth understanding long before the process begins.

What a PhD should be

It should be a process of empowerment to achieve intellectual autonomy and creativity. The PhD should be an important vehicle for the development of personal self-confidence. It is a process of becoming a PhD. In addition to gaining intellectual independence the researcher is also becoming a member of a social grouping; the PhD community.

What a PhD should not be

A PhD should not be done just to fulfill other people's desires and requirements. This type of "surrogate PhD" will neither bestow intellectual autonomy nor achieve personal self-confidence. Often parental expectations, possibly driven by their own lack of opportunities or achievements, can pressurize some researchers to commit to doing a PhD against their better judgment. It must be the researcher's PhD not someone else's.

Why a Grounded Theory PhD is different

Making sure you really want to discover how to use GT is important. After all there are many alternatives available. Question your motivation to study for a PhD in the first instance and reasons for wishing to use classic GT methodology in particular. Once you have verified these fundamentals, seek out good support from others. Early on, get to grips with the pertinent texts which articulate the theory of using the chosen methodology. The principal sources of authentic material are generally to be found through the Sociology Press (Glaser, 1978, 1992), with the exception of the original text by Glaser and Strauss, *The Discovery of Grounded Theory*, published in 1967. Read the texts which are examples of grounded theory in action in order to become familiar with the real thing (Glaser, 1993; Glaser (Ed.), 1995; Glaser & Strauss, 1971, 1972).

We urge you to study examples of good GTs to observe how experts tackle the challenges. Participate in workshops when possible to examine how more experienced researchers practice their art. Cultivate dialogue with your peers and read diverse examples of well-constructed literature to become sensitive to the constituents and how they are integrated. If possible establish a small writing group and convene regularly. If this option is not physically viable consider utilizing the virtual environment. Building in a group dimension is helpful because it avoids the potential negative impact of feeling isolated. It has the added bonus of building a disciplined approach to writing activity. The purpose of this suggestion is to encourage writing to become routine and in this way to enhance forward momentum. It is easy to become over-committed to the process of data collection at the expense of writing and analysis.

Beginning to analyze the data in concert with generating the data is essential. Stockpiling superb data to be analyzed later is at best unhelpful. Worse still this type of counterproductive activity risks unnecessarily prolonging the time taken in the field by failing to use the theoretical sampling process to best effect. It may be somewhat intoxicating to generate more and more fascinating data from the field, adhering to GT procedure, however, prevents this happening and diverting energy.

Classic GT states we must curb our inclination for indiscriminate data collection, however captivating. Furthermore it encourages us to begin analysis immediately and consequently the overall data collection process becomes superbly focused because it is directly guided by what is emerging from the ongoing analysis. Activity is thus purposeful and well directed.

It is normal to feel some degree of apprehension regarding just how to begin to analyze your data. Imagine how terrible it would be to begin the PhD process and discover that you have no analytic ability or aptitude for the task! Convince yourself that this is unlikely to be the case and get going with initial attempts to make sense of the data. Do this by strictly disciplining yourself to the business of fragmenting the data into open codes.

It is human nature to avoid exposing oneself as possibly incompetent and this may explain the underlying reluctance to leap headlong into analysis. Simply recognize this phenomenon. Rationalize it as a way of dealing with the risk of potential embarrassment and the threat of public ridicule. Acknowledge that these concerns are normal and legitimate. Consciously choose to begin analysis at the earliest point. This is a learning process. These are the first attempts at a difficult task and not a finished product. There should be no expectation of perfection at the outset. Beginning the analytic process is only that and is energizing in itself.

There are particular difficulties associated with doing GT for the first time. Be aware of these but don't get unnecessarily concerned. To a newcomer some of the key texts may appear difficult at first sight. Have you ever read a book which aims to teach you how to ski, surf, ride a horse? None, no matter how well written, can mimic what it is *really* like to feel the full range of these real experiences as they are lived. Using GT is no different. Indeed reading and re-reading the texts, such as *The Discovery of Grounded Theory* (Glaser & Strauss, 1967) and *Theoretical Sensitivity* (Glaser, 1978), which describe the methodology, explain its use and reveal how to do classic GT, is essential. Many years after beginning to discover GT, revisiting these resources throws up different perspectives. Perspectives which, based on experience, and, with the passage of time reveal, previously imperceptible, nuances about what is *really* involved in doing classic GT at an accomplished level.

Skills become honed through the practice of doing GT. As skills develop, greater depth of understanding is possible and the texts take on new meanings. With growing knowledge finesse in doing grounded theory increases. Equally we can look back and visualize how our abilities to generate theories from data have evolved. This is empowering. We can also begin to pinpoint how to develop further as a grounded theorist as a fuller grasp of the subtlety of doing GT research becomes possible. This is incremental. Many who purport to be using GT do not, in fact. There are many reasons why they fail to deliver an authentic rendition of the methodology in practice. Among these are:

- Those who want the legitimization GT offers. These types cannot let go of their pre-understanding; they demand to be in control and consequently deliver "fraudulent" theory which is usually grounded in extant knowledge and does not reflect the reality of what is in the data. They are likely to be experts in their field; they think they know the answers already and want to pretend they have discovered them systematically. Professionals cannot admit they do not have all the answers. It would be unprofessional, wouldn't it? In this sense classic GT challenges the socially structured myth of the omniscient professional. It may therefore be seen as anti-establishment and its exponents risk being labeled as unorthodox.

- GT is so often misused that this in itself breeds further misuse. For example, "so and so used GT like this so, I will too – that must be how it is done…"
- People fail to take the time and effort to understand correctly what doing GT entails. These users latch on to it because they think it will be quick and easy.
- Too few people are well qualified to judge what a good GT is. Often those who are tasked with such evaluation impose inappropriate criteria on such work, once again perpetuating misuse.

One key dimension of grounded theory is the highly effective focusing that happens as a consequence of the requirement to simultaneously generate and analyze data. This integration stimulates directed activity where relevant to the emerging theory. It does not waste effort elsewhere. It is advisable not to latch on to any apparently obvious discoveries too soon. For relatively inexperienced GT researchers there is a risk of prematurely misinterpreting where the action is. Counter this inclination by continuing the process of coding the data and in this way ensure you are seeing what really matters.

The Grounded Theory research method demands that the researcher allows the latent patterns embedded in the data to emerge. It is predicated on the belief that human behavior can be understood by latent pattern analysis. The GT method is not based on the preconceptions of others. To some university systems this will be intolerable.

The GT PhD is not for everyone. It is best suited to those researchers who have an affinity for conceptualization coupled with a high tolerance for uncertainty. It is neither contextually based nor focused on any existing body of knowledge. It will not be preceded by a discursive literature review. Neither will it be possible at the outset to give very specific research objectives. This does not mean that it is impossible to write both a literature review or give very specific research objectives in order to comply with some institutionalized prerequisite when doing a classic GT PhD. Rather, it means that the novice GT PhD researcher needs to learn the importance of fulfilling institutional requirements without violating the basic tenets of the classic GT research method. Achieving these twin goals may at first appear unlikely especially if you find yourself in an apparently hostile, to GT, institutional environment. Do not despair. Although any individual PhD researcher will always be out flanked by the powerful institutional forces of a university, there are ways in which this imbalance can be successfully managed.

The important issue for the novice GT PhD researcher to understand is that it is in fact possible to both satisfy the institutional requirements of a university without violating the classic GT method. The way to achieve this is to treat both issues quite separately. Initially you must fulfill the basic institutional requirements of the university because without doing so you will neither be registered nor will you be awarded the PhD. If the regulations state that any PhD research proposal must be accompanied by a literature review, then do a literature. If the regulations state that a literature review must become the first chapter of the PhD then again give them a literature review. However the most important thing for the novice GT PhD researcher to understand is to do the GT research in the classic manner *first* rather than attempting to retro fit a contrived literature review into the PhD thesis. The literature review is written after the core variable has emerged in the classic GT method. In reality what will happen is that once your core variable has emerged the external examiners will be far more interested in that than in any literature review you did before the research evolved.

Although the GT research method is now an internationally recognized as a robust and legitimate research methodology its power has been diminished by several academics (Strauss & Corbin, 1990) who have sought to label their own descriptive research as GT research. There are even authors who have hijacked the label to present it as something entirely different. From the outset the aspiring GT PhD researcher has to be able recognize classic GT research from the pretenders. This is quite easily done however. The only legitimate source of the classic approach to the GT research method is to be found in the publications of Dr Barney Glaser and at the Grounded Theory Institute website (www.groundedtheory.com). Researchers using any other adaptation of GT will be deluding themselves and misleading others. The classic grounded theory research method is a very specific methodology with each step of the process very specifically delineated. Those who adapt and amend the process should not label the research method they have used as GT. Instead they should say their research was "influenced" or "inspired by GT" and then go on to create a new label for the research process they have used. At a minimum, supervisors should insist on methodological clarity. This practice is worth cultivating because it establishes as routine the need to be clear about how research processes are conducted and acts as an indicator of intellectual competence.

(ii) Finding and evaluating the most appropriate supervisor

Historically most GT PhD researchers have used the "minus-mentoring" approach. This means that the researcher does the GT research in the absence of a knowledgeable and experienced GT research supervisor. The researcher often has an institution based PhD supervisor who knows little of the GT process but does have knowledge of the institutional requirements of the PhD process. "Minus-mentees" then seek help and guidance from experienced GT researchers and practitioners who become their unofficial mentors. These GT mentors are to be found via the Grounded Theory Institute's own website (www.groundedtheory.com) as well as through the usual academic channels of networking, via conferences, academic journals and on-line research discussion groups.

It is still quite rare for the GT PhD researcher's formal supervisor to be an experienced GT practitioner, but it is happening more every year. This follows a traditional supervision model where the supervisor and supervisee enter into a mutually beneficial contract with the end game of the completion of a GT PhD. In recent times it is becoming more possible to have a supervisory contractual agreement between a geographically remote, internet based supervisor and a researcher. For example, Andy is based in Bangkok, Thailand and his most recent GT PhD researcher (Christiansen, 2007) was based in the Faroe Islands some 7,000 miles away. This achieved a satisfactory conclusion because it was stipulated that the supervisor had to have a face to face meeting with the researcher at least once a year. The site Grounded Theory Online (www.groundedtheoryonline.com) provides an invaluable facility dedicated to supporting researchers and classic GT where researchers may easily choose appropriately qualified supervision or sign up for internet based methodology workshop sessions.

Andy has found that from the supervisor's perspective there is much to commend GT PhD supervision by internet. The main advantage for both the supervisor and PhD researcher is the record of emails sent and received. In conventional university-based supervision not all supervisors diligently make detailed and pertinent records of all supervisor researcher interactions. Another benefit of communication by email is that it allows both parties to consider their responses to issues raised. With the global introduction of broadband services and the greater availability of video conferencing these types of interactions are now much more feasible. Finally when the annual face to face meeting does happen between internet-based doctoral student and supervisor, the encounter is highly focused and very productive.

How to ensure that you have the most appropriate supervisor

Many novice PhD researchers may feel rather uncomfortable about the notion of doing background checks on their potential supervisors. They should not be concerned about doing this because one of the main reasons for PhD researchers' failure to complete is poor supervision. In many universities there is an absence of systematic PhD supervisor training. Because of this it is important for the researcher to find out if their chosen institution has a PhD supervision "mentoring" system. This type of system is an on the job training program for PhD supervisors where other colleagues and faculty members meet with the supervisor on a regular basis to assist them with their PhD supervision tasks. These kinds of systems are necessary because it does not always follow that a successful career academic is also a competent PhD supervisor.

The job of a career academic can be classified into four main types of tasks; administration, teaching, research and publication. The fast track career academic has to make sure that he gives an appropriate priority to these four types of activities. The type of PhD supervisor to avoid is the PRAT. By this we mean a person who first prioritizes his own publication, then his research, then administration and finally teaching and supervision. The reason why it is dangerous for the novice PhD researcher to get involved with a PRAT is because they would categorize PhD supervision as a low priority and see it as a combination of administration and teaching. Excellence in either activity is not valued in terms of their personal career advancement. This is a consequence of the structural reward system existing in academic institutions. The novice PhD researcher should try and look for the career academic that is caught in the TRAP. This type of academic prioritizes his time with students and researchers because to this person the order of priorities are as follows; teaching, research, administration and publication.

The researcher must ascertain the detailed prior supervision experience of his potential supervisor. Avoid institutions that arbitrarily allocate researchers to supervisors as an administrative convenience. The researcher must first research all potential supervisor candidates carefully. This can be done by first establishing an accurate assessment of the candidate's track record in supervision. There are at least three ways to discover more about the supervisor's competence. Firstly visit the library of the institute where you are considering registering to a PhD. Ask the library officials to point you in the right direction where you can view all the PhDs awarded in your area of interest selected by both researcher and supervisor. Secondly, track down all those researchers who have been supervised by your potential supervisors and arrange to meet them in person. If this is not possible get hold of their e-mail addresses and ask them the following questions:

- How long did the PhD process take?
- Were you encouraged to meet the other researchers this person was supervising?
- What happened at your first PhD supervision meeting?
- How many meetings a year were scheduled?
- What goals and objectives were agreed at these meetings?
- What were the best aspects of your supervisor's supervision regime?
- What were the worst aspects of the supervision regime?

- Would you recommend your supervisor as an appropriate person for my proposed PhD?
- Did your supervisor give you an overview of the whole PhD process during your first meeting?
- At which point in the PhD process did your supervisor explain in detail the formal assessment criteria for the PhD?

Some consequences of having an inappropriate supervisor
One consequence is that you may have to find a surrogate supervisor. A surrogate supervisor is a person who volunteers to offer the PhD researcher continued assistance throughout the PhD process. The surrogate supervisor neither has any official standing as supervisor nor receives any financial reward. Why do surrogate supervisors agree to do this? When Andy was doing his PhD he had an inexperienced and inept supervisor and his PhD was saved by the timely intervention of a surrogate supervisor. In his case the surrogate supervisor was a professor from a different university than the home institution where he was registered as a PhD researcher. The surrogate heard him give a presentation at an international conference and generously made several positive comments about his work and indicated how it could be further developed. Surrogate supervisors exist because there are some generous people in the academic world who are willing to offer help to those in need. However, it is an unwise strategy to do a PhD with the aim of finding a surrogate supervisor. It's much wiser to make sure that your own supervisor is really up to the job.

Another consequence is that you may have to dismiss your PhD supervisor. This may seem to be rather drastic action, but there are instances when this is the only course open to the PhD researcher. Andy recalls being approached at a conference by a young PhD researcher from a different university from where he worked and long after he had gained his own PhD. The PhD researcher explained that he had been allocated a very lazy and incompetent supervisor and was in danger of being de-registered as a PhD researcher because of his allegedly poor work. In actual fact it was the lack of feedback and general absence of supervision which was the cause of the problem. The PhD researcher was from Asia and it was totally alien to him to openly criticize any teacher especially his PhD supervisor. Andy advised him that he should make an official complaint for "mal-supervision" and contact the student's union lawyers to start legal proceedings to reclaim the supervision fees and other living costs he had incurred during the research process. The advice worked perfectly. The university made an investigation into the lazy PhD supervisor's behavior and found that other researchers also complained about his conduct. The university apologized to the PhD researcher and allocated a much more experienced PhD supervisor. The researcher received his PhD without any further problems in 2008.

(iii) Understanding how university regulations might impact the research process

All universities however diverse have regulations. All require the PhD researcher to write a formal research proposal. There are several different ways to accomplish this and it is probably better to write three different research proposals, one for each type of audience, for the same research project in order to obtain the most efficient results. The type of research proposal you use depends on the audience it is addressing. There are three broad categories of research proposal audiences; your supervisor, the institution where you are going to register your PhD and the external funding constituencies.

The informal research proposal for your supervisor

The research proposal for your supervisor should focus on the type of supervision relationship you are hoping to establish. This means that you should clearly indicate your preferred pattern of supervision that will ensure the attainment of the PhD in a reasonable time frame. To do this the researcher has to think very carefully about both the nature and frequency of supervision meetings. It is usually more effective to have goal oriented supervision meetings rather than calendar based ones. There must be a built in mechanism for the researcher to receive frequent written feedback from the supervisor. The feedback mechanism is a safety device for the researcher so that when the PhD is being evaluated patterns of inadequate supervision can be distinguished from shortcomings in the researcher's own skills.

The formal research proposal for the university

The novice GT PhD researcher must follow the detailed guidelines from the university where the PhD is to be registered. If the university does not have a detailed and inflexible research proposal procedure and structure, the researcher is advised to take full advantage of this and write a research proposal for GT PhD as follows:

- Pre-understanding: containing a summary of the author's previous exposure to the subject area. This includes an overview of the author's own subjective influences which are likely to affect the generation and interpretation of the research data.
- Pilot Study: a primary data generation phase to obtain a better understanding of the main issues of concern amongst those being studied.
- Refinement of research objectives: dialogue between the pilot study indicators and the author's own subjective understanding which leads to a more focused research agenda.

- Outline of the chosen research design: description of the chosen research method, including explanation of the research procedures to be used; and justification of the chosen research method.
- Explanation of generated data: analysis and synthesis of the data generated.
- Data Interpretation and Comparative Literature Review: a succinct interpretation of the data is followed by a comparative literature review.
- Recommendations: policy guidelines or indicators for the different constituencies of the research community, policy makers and others.

If the university does have very specific and inflexible, detailed regulations governing the structure, format and process of the PhD there is no need for the classic PhD researcher or supervisor to panic. You must comply fully with the university regulations. But it should be done in a skilful manner. All formal research proposals should be submitted exactly as specified. This may mean it is necessary to start the research with a formal literature review. There are at least two different creative ways to approach any inflexible regulations. The first way can be called "pre-GT" and the second way is called "post-GT".

The "pre-GT" research proposal

In order to fully comply with the university regulations write a logically plausible (but quite irrelevant) literature review.

The "post-GT" research proposal

Here the researcher first completes the classic GT research in the correct manner, and then subsequently re-writes the thesis in order to fully comply with the university regulations. Once the PhD has been awarded the researcher reverts to the classic GT approach and publishes.

The research proposal written to obtain external sources of funding

Here are some general guidelines to help novice GT PhD researchers be more effective in communicating their ideas to those in control of resources. The research proposal should contain a statement of the main research objectives, including:

- An indication of the area of study and an explanation of the context in which the research is to be set.
- A clarification of the underlying assumptions of the research proposal.
- An indication of the possible values of the research outcomes to different constituencies.

- A detailed research design which includes what the intent of the research is, how it can be operationalized, and why the chosen or recommended method is especially appropriate for the type of problem selected. The main criteria by which classic GT research should be judged are clearly articulated in Glaser (1978).
- An indication of a detailed resource plan which includes the financial, technical and human resources needed for successful completion.
- The production of a detailed time plan indicating the key points in the research project, especially when feedback will be given to the sponsor/s of the research project.
- An explanation of the type of dissemination strategy to be adopted once the research has been completed.

In addition to the above, novice researchers should ask the following questions:

- What is the story I am telling? All research projects are written by people with very different backgrounds and interests. This influences the perspective that will be given to each research project. By giving the story in the context of the issues shown above, this may help would-be sponsors of the research to become more personally interested in a given proposal. It will also force the researcher to be more honest, transparent and open about motivations for interest in a specific research topic.
- Who are the audiences to which the research proposal is being addressed? The researcher has to decide whether the research proposal is a plea for finance, recognition, moral or intellectual support, or rather more straightforwardly, simply gaining the approval of his or her research supervisor or supervisory committee. Occasionally the need may arise to write different research proposals for the same project, as it may be aimed at different audiences. For instance, academic funding bodies will place more emphasis on the academic credibility of the research design, whereas practitioners will more interested in the utility of the research outcomes.
- Why does it matter? Readers and evaluators of research proposals are likely to be very busy and the last thing they want to do is read a rather dull and ordinary proposal, or alternatively read a proposal which is plainly lacking in the basic elements of desirable form and structure. It can be argued that the researcher has a duty to readers, and indeed themselves, to make the proposal interesting from both a readability and structural perspective.

- Why now and why me? By the time the research proposal is written there is already a sense of knowledgeability in relation to the chosen subject area. This can be demonstrated by injecting a sense of urgency into the research proposal, by explaining consequences of the research not taking place, especially when applying for research funding. In particular, researchers are advised to make a point of communicating any unique qualities or skills possessed which will enable the research outcomes to be achieved.

The researcher should also bear the following in mind:

- Perfect the research proposal
- Research proposals should never be written in isolation. Often, there are many other highly experienced people whose advice, assistance or critical comments may enhance the probability of successful research progression.
- Allow sufficient time: constructing a good research proposal is time consuming. Time spent in reconnaissance is rarely wasted. Before beginning detailed development of a proposal it may be worthwhile to investigate other successful proposals, where these are available in the public domain or otherwise through personal ingenuity. Scanning previously acceptable proposals may assist in generating funds, where necessary, or submitting a successful proposal. Learn from the effort of others before submitting anything. This has now become much easier because all the governmental research bodies have their own web sites. Where dissertation or thesis proposals need to be written, it is crucial to access the relevant literature, either internal or external to the appropriate research domain. Nothing is more frustrating than having to rewrite a proposal for academic research over and over again; notably, for externally funded research, there is no second chance.
- Consider all opportunities for financial help: in addition to the usual governmental research bodies there are a number of specialist charitable organizations who have research foundations who may well be able to offer financial resources. Obtaining money from the corporate sector is possible but it is wise not to oversell any possible research outcomes and to be honest about time scales. One of the most useful techniques in obtaining funds is called 'snowballing'. This happens when an initial modest amount of money is gradually increased by attracting more funds from other bodies who can be persuaded to collaborate. Often, firms related to your topic area may be interested in sponsoring some, or all, of the project. The question

of 'what's in it for them' must be thought through carefully before approaching any firm.
- Discuss the proposal: wherever possible enter into dialogue with experienced research professionals, including colleagues, in advance of submitting anything. It is especially important when seeking external funds that the research agendas of these organizations are understood in relation to your own project.
- Justify all financial requirements of the proposal: it is not sufficient merely to give an indication of financial requirements relating to a given project. Expenditures need to be justified. This is necessary for all kinds of reasons. One of the less obvious reasons is that the more meticulous one is in explaining requirements, the more it gives a number of positive messages to would-be sponsors. Firstly, it demonstrates that the researcher is not profligate and can be trusted. Secondly, it indicates that the researcher has a firm intellectual grasp of the nature of the project to be embarked upon. Finally, it is simply more professional.

It is a mistake to get derailed by the circular arguments as to which research method is the best one to use. Instead, focus on two other aspects: firstly, the nature of the problem; and, secondly, the extent of the researcher's abilities, talents, skills and temperament. Certain types of problems are best suited to deductive research and others to inductive research. Since all research contains elements of both methodological approaches, purists may miss the point entirely. To do good research demands a firm intellectual grip of the nature of the research problem and a deep understanding of one's own emotional and intellectual capabilities. If researchers remember that the journey is more important than the destination then not only should some interesting research be earned out, but also one's own potentially blinkered mindset may be altered or expanded.

(iv) Managing your committee

If the university allows it, it is important for the both PhD researcher and supervisor to select the committee. If neither of these two parties has an influential role in the selection process there is scope for many problems in the evaluation process. Classic GT research is still not fully understood by many leading academics. At one extreme are the simply ignorant, at the other are the prejudiced. There will be problems having a committee comprised of either of these extremes. One way to ensure that this does not happen is to regularly publish your research and present conference papers. This will increase the span of awareness of your work as well as widen your own personal network of potential PhD committee members.

How to survive the formal PhD evaluation process

In Scandinavian countries the evaluation process takes on a more transparent format. Often the process is in two parts. The first part only involves the PhD committee and the second part is a ritualized procedure to which the general public is invited to participate. In reality, although the decision to award the PhD is never announced until after the public participation, the decision is made exclusively by the PhD committee. One of the PhD researchers Andy supervised to a satisfactory conclusion submitted his thesis to a Danish university. His committee was comprised of three professors, two of whom were openly hostile to the classic GT method. Prior to his formal evaluation the research candidate received a very detailed written report about his thesis from the PhD committee. The PhD researcher wanted to respond with an equally detailed written rebuttal. Here is an extract of the e-mail Andy sent him:

> The most important aspect about how to respond to the written comments of the committee is to establish what is the correct protocol for that university? Are you as a PhD candidate expected to make a formal written response prior to the date of your evaluation? Or are they only expecting you to do that on the day? I am asking you these questions because it is of special importance not to unnecessarily alienate the committee before the due date. If there are neither any formal requirements nor expectations for you to make a formal written response prior to the evaluation then I would definitely NOT write to them with a detailed rebuttal. Instead I would write to them thanking them for their very detailed comments which you are looking forward to discussing with them on the due date. There are five reasons why communicating to the committee in writing with a detailed rebuttal is a very bad idea. Firstly, if members of the committee do not agree with your perspective now they never will. Writing to them will only reinforce their prejudices. Whereas talking to them face to face in an open forum will put peer pressure on them to reconsider their own perspectives. Secondly, whenever we receive communications in writing which we do not wholly endorse it tends to reinforce our prejudices and appears to be confrontational and hostile. It is always easier to embrace new perspectives verbally. Thirdly, the committee could well open up new lines of criticism if you chose to send them a detailed written rebuttal prior to your evaluation. Fourthly, if you do not give a detailed written rebuttal before the PhD evaluation date the committee are disadvantaged because they are unsure as to how you will respond to their comments. This gives you an advantage during the evaluation process. Finally, by withholding your response it demonstrates the development of your own intellectual maturity. By holding your fire until the evaluation you are showing that you are able to acknowledge that research, as in the rest of life, has multiple per-

> spectives. By resisting the temptation to send a written rebuttal it demonstrates your ability to discuss why different perspectives exist. If you do it in a relaxed and non-confrontational manner you are more likely to persuade them of the power of your line of argument. No PhD candidate, however brilliant, should be arrogant. If you did send a detailed rebuttal to your committee, prior to the formal evaluation, they are very likely to consider you to be rather arrogant. Relax, smile and be humble when the committee interrogates you on your evaluation day. Calmly explain your own perspective without denying the possibility that they too have legitimate perspectives even if they are quite different.
>
> Remember that you are in a very privileged position having been sent very detailed comments about your work by the PhD committee prior to your formal evaluation. If you are skilful you can transform a potentially stressful confrontation on the day of your evaluation into the opportunity of a lifetime to clearly express yourself with confidence about your research. After all no one knows more about your own research than you! Getting a PhD is the final part of the basic sociological process of becoming a PhD. You have to demonstrate to the committee, in a respectful and diplomatic way, that you have already emerged to become part of the wider community of PhDs. In other words, you have to show that you are now "one of them". This means that when you hear some of their comments that either appear to you as "ignorant" or "irrelevant" pause before you reply with an instinctive rebuttal. After the pause be skilful and thank them for their comments and politely explain your own line of argument. To become a PhD it is essential that you have a perspective and are able to clearly articulate it. It may well be that some of the apparently "negative" or "misinformed" comments made by the committee have deliberately been made polemical in order to gauge your own reaction and to give you the opportunity to demonstrate that you are worthy of joining the PhD community. From the committee's perspective getting a PhD is not just about doing robust, sound, research it is also about you being able demonstrate your own intellectual autonomy.
>
> I have gone to great pains to explain why you must resist the temptation to reply with a detailed written rebuttal prior to your evaluation. However should protocol demand that you reply in writing before the formal evaluation do so in a very concise, polite and enigmatic manner. (A. Lowe, personal communication, 1st September 2007)

We are pleased to report that this PhD candidate got his PhD in 2007.

(v) Getting published

All classic PhD theses have the distinct advantage that they are dealing with issues of substance and interest. If the thesis is appropriately packaged for the journal or publisher they have a good prospect of being published. Be aware that your research might need to be re-packaged to meet the needs of different journals. Before this is explained we would like to share the experience of an unsuccessful attempt in publishing some research. Fifteen years ago Andy asked Dr. Barney Glaser if he would agree to co-authoring a paper on how classic GT impacts on theory of marketing in business. They wrote the paper and then submitted it to a very prestigious academic journal in this field. The three "blind" reviewers unanimously rejected it. The first reviewer said that whoever wrote this paper clearly did not understand grounded theory. The second reviewer stated that the author of this paper was someone from a country where English was not their mother tongue and they should hire a more skilled interpreter. It was subsequently discovered that this reviewer was from Norway. The final reviewer said that there were insufficient citations.

There are several teachings from these comments. First prejudice and ignorance are alive and well in the academic publishing world. Secondly, there are 100s of academic journals out there and the choice of which journal to position the research is often more important than the research itself. Finally, the editorial boards of each journal must be carefully scrutinized so that the extent and nature of the prejudice and ignorance can be better assessed prior to submitting the research for publication.

However, the PhD that has used classic GT will have revealed authentic latent patterns of human behavior that are transcendent of their original context. This has three important benefits which should enable publishing in academic journals. Firstly, the range of academic journals to which your research can be disseminated is vast. A single piece of classic GT research has the potential to be published in methodological journals, contextual journals as well as research strategy journals. Secondly, the GT author can also rewrite the grounded theory to higher levels of theoretical abstraction giving yet more publication options. Finally on rare occasions the GT author can also generate a formal theory, which has the potential to be published in an even wider range of titles.

References

Glaser, B. G., & Strauss, A. L. (1967). *The Discovery of Grounded Theory: Strategies for Qualitative Research.* Chicago: Aldine.

Glaser, B. G., & Strauss, A. L. (1971). Status passage. London: Routledge and Kegan Paul.

Glaser, B. G. (1972). *Experts versus laymen: A study of the patsy and the subcontractor.* Mill Valley, CA: Sociology Press.

Glaser, B. G. (1978). *Theoretical sensitivity: Advances in the methodology of grounded theory.* Mill Valley, CA: Sociology Press.

Glaser, B. G. (1992). *Basics of Grounded theory analysis: Emergence vs forcing.* Mill Valley, CA: Sociology Press.

Glaser, B. G. (1993). *Examples of Grounded theory: A Reader.* Mill Valley, CA: Sociology Press.

Glaser, B. G. (Ed.). *Grounded theory: 1984-1994: A Reader (Vols. 1-2).* Mill Valley, CA: Sociology Press.

Guthrie, W. (2000). Keeping clients in line: A grounded theory explaining how veterinary surgeons control their clients (Doctoral dissertation, unpublished). University of Strathclyde, Glasgow.

Lowe, A. (1998). Managing the post merger aftermath by default remodeling. *Management Decision*, 36(2), pp. 102-110.

Strauss, A., & Corbin, J. (1990). *Basics of qualitative research: Grounded theory procedures and techniques.* Newbury Park, CA & London: Sage.

4

Learning Methodology Minus Mentorship

Antoinette McCallin
AUT University, Auckland, New Zealand

Alvita Nathaniel
West Virginia University School of Nursing

Tom Andrews
School of Nursing and Midwifery
University College Cork, Ireland

The aim of this chapter is to examine how learning grounded theory without a mentor impacts methodological development and contributes to the remodeling that is so common. It is argued that because many researchers have learned grounded theory without the supervision of an experienced grounded theory researcher, methods and methodology have been compromised by epistemological debate that has had little impact on the end product of grounded theory, except to make the product more descriptive. Background issues related to learning grounded theory with mentoring support are outlined. Next, methodological misunderstandings and understanding are clarified. This is followed with an analysis of the epistemological debate that is illustrated with particular reference to the constructivist challenge, which has had an inordinate influence on grounded theory development. In the final section of the chapter the moral implications of methodological remodeling on disciplinary knowledge development are discussed. The chapter concludes with a reflection on mentorship and how this influences scholarship when mentoring is not available.

Background

Grounded theory is a popular research methodology that has made a significant contribution to knowledge development in many disciplines. Part of the attraction for researchers is the methodological emphasis on understanding behavior and action in practical situations (Glaser & Strauss, 1967). It is not surprising that novice researchers from all sorts of disciplines are attracted to a methodology that is practically focused and aims to explain behaviors and meaning in a particular setting. The expectation that participants identify the research problem and explain action in the area of interest has been embraced enthusiastically by many. Enthusiasm, however , has undermined methodological rigor, as so many new researchers have learned grounded theory by trial and error learning, i.e., minus mentorship. Like any research methodology grounded theory has its own procedures, which are understood more easily when the novice is mentored by an experienced researcher.

Indeed, Glaser (1998) recognizes the importance of mentoring grounded theory research students and suggests the methodology is learned in apprenticeship with an experienced grounded theory researcher. Those who do not have this privilege are left interpreting methodology and using methods in "ways that make sense to them" (Corbin, 2009, p. 52). Others have traveled to Mill Valley, London and New York to attend The Grounded Theory Institute Troubleshooting Seminars where novice researchers learn from Barney Glaser. Learning from the master mentor and from experienced grounded theory researchers in what has become an international community of scholars is, for many, a momentous learning experience. The learning environment is such that teachable moments abound with "aha" learning experiences, bringing the methodology to life. Non-competitive class discussions and one-to-one conversations with the master mentor give the student a glimpse of potential theoretical development. Thinking is challenged and students are constantly exhorted to: "Conceptualize! Conceptualize! Conceptualize!" Gentle guidance and reminders to "trust in emergence" and "just do it!" encourage students to become more confident users of the method. Students learn from each other, clarifying methodological thinking in a supportive learning environment. Without doubt, mentoring is vital to the passing on of the grounded theory method.

Although mentoring is seldom discussed as an essential component of grounded theory research training, mentorship advances understanding. Mentoring occurs when a person who is wise, knowledgeable and highly respected in an area works with a novice, deliberately encouraging and inspiring the mentee to extend the self and develop full potential in the area of interest (Carroll, 2004). Indeed, Fuller (2000) argued that "mentorship plays a critical role in setting a standard … for those individuals who want to be involved in research and ultimately, for the preparation of the next generation of [scholars]" (p. 2). Mentorship is invaluable to guide and motivate the learner to engage in excellence and scholarship. Mentoring not only supports methodological engagement, but orients and socializes new members into a community of scholars (Schumacher, Risco, & Conway, 2008). For many years, Barney Glaser has offered himself willingly as a mentor to research students all over the world. Seminars are a meeting place for learners to network with others in a similar situation. Conversations abound, while students listen and reflect, gaining new understandings. Student vulnerability quietens, as the master responds sensitively to student problems. Unfortunately, many students learn methodology in less supportive circumstances, which is where remodeling begins.

Because many researchers learn grounded theory minus mentored supervision individuals have interpreted methodology in all sorts of ways to the extent that today there are several different versions of the method (Morse, Stern, Corbin, Bowers, & Clarke, 2009). Indeed, Charmaz (2009) believes that grounded theory is an over-arching term that refers to a variety of methods. Part of the problem is that novice researchers seldom appreciate the fine distinctions of the method until they are engaged in the research process. If a researcher is under supervision from a qualitative researcher who is not familiar with grounded theory at all, the learner is exposed to qualitative generalizations that are at odds with what is a specialist methodology. Although research-learners seek understanding from the literature, much that is written is methodologically inaccurate. All too often grounded theory research published in well-respected international journals bears little resemblance to the original methodology.

Many of the problems began when the original authors of the methodology (Glaser & Strauss, 1967) explained the approach in more detail (Corbin & Strauss, 2008; Glaser, 1978; Strauss, 1987; Strauss & Corbin, 1990; 1998). Two versions of grounded theory emerged: the original approach that is now known as classic grounded theory, and the Strauss and Corbin model. The latter version stimulated a debate that has had a significant impact on methodological development since (Glaser, 1992). Strauss and Corbin grounded theory shares commonality with classic grounded theory in that both approaches include comparative analysis, theoretical sampling, memo-writing, and saturation. The purpose of theory development, however, may differ. Some researchers focus on description, while others believe conceptualization, integration and theoretical explanation are fundamental to knowledge development in practice-based disciplines (Corbin, 2009). Descriptive conceptual models certainly provide some explanation of practice but are very different to the integrated theoretical accounts typical of classic grounded theory (Glaser, 2001). The varied interpretations of grounded theory (Morse et al., 2009) may have occurred because so many researchers learned methodology minus mentoring. This has caused all sorts of misunderstandings.

From misunderstanding to understanding

Located in accounting, business studies, education, engineering, information management, journalism, management studies, media studies, medicine, nursing, psychotherapy, psychology, social work, sociology, sports sciences, and vetinary science, grounded theory researchers have wide-ranging professional backgrounds informed by different disciplinary ideologies, all of which affect understanding of methodology. Although disciplinary diversity is not necessarily of issue, professional values and disciplinary ideology are intertwined, and in some disciplines more than others influence how ontology, epistemology and methodology are understood. Every discipline has a unique emphasis that affects knowledge value, knowledge generation and teaching, and the rules and regulations that govern scientific rigor. Methods, the techniques for data collection, are defined according to methodology, the plan of action that is influenced by the philosophical perspective, which is informed by epistemology (Crotty, 1998). However, misunderstandings about methods, methodology and epistemology have created much confusion in some disciplines (Harding, 1987). The knowledge claims of individual professions have become intertwined with individualistic views of knowledge that have undermined disciplinary knowledge development (Gergen, 1994).

Disciplinary interpretations of what methodology is or should be have limited understandings, and contributed to the remodeling of grounded theory. The newer professions for example have taken methodology and interpreted it according to a disciplinary agenda. Innovative approaches to knowledge generation were welcomed (Cutcliffe, 2005), as methodology was adapted in a bid to move a discipline forward, so that members could be recognized as legitimate professionals. This occurred commonly in nursing (Morse et al., 2009), and also sociology to some extent (Charmaz, 2006; Clarke, 2005), where theory development was used as a strategy to advance disciplinary knowledge development (Meleis, 1997). The strategy supported the belief that professional knowledge was theoretical abstract knowledge (Abbott, 1988).

The issue of abstract knowledge draws attention to the conceptualization-description debate that is the basis of grounded theory remodeling. Although some researchers produced theoretical explanations about practice, others rejected theory on the grounds that abstractions about practice could not possibly explain individual everyday activities. This belief about the value of knowledge may account for those grounded theory researchers that favored qualitative descriptions as legitimate knowledge. Gergen (1994) argues that descriptions about the world are problematic in that descriptive language raises questions of representation that are contrary to theory development and theoretical language, which aim to predict real-world events. It seems that researchers studying methodology minus mentorship did not understand the subtle distinctions underpinning classic grounded theory. When this occurred in a discipline where some, not all, rejected theory development, theoretical integration was replaced with descriptive problem solving frameworks about practice, and the issue of methodological rigor was not addressed at all.

Another methodological misunderstanding concerns philosophy. Glaser (1998) states that grounded theory is a-philosophical. Nonetheless, many researchers link grounded theory to symbolic interactionism, usually on the grounds that Strauss had strong interactionist and pragmatist connections with the University of Chicago, where the sociological tradition was well developed (Strauss & Corbin, 1998; Layder, 2000). Grounded theory though is not underpinned by symbolic interactionism, as some authors claim (MacDonald & Schreiber, 2001; Milliken & Schreiber, 2001). Glaser (2005) clarifies the situation, arguing that symbolic interactionism is a potential theoretical code that may be used in theory development if it emerges in the data. While Alvesson and Skoldberg (2000) realize that grounded theory and symbolic interactionsim are not connected automatically, Morse (2009) claims the symbolic interactionist connection promotes innovation. Similarly, Charmaz (2000) identified a philosophical void and situated grounded theory within constructivism, a position that has been accepted by many (Charmaz, 2004; Freeman, McWilliam, MacKinnon, DeLuca, & Rappolt, 2009; Jones, Torres, & Arminio, 2006; Kushner & Morrow, 2003; MacDonald & Schreiber, 2001; Morse et al., 2009). Philosophical alignment with constructivism, which focuses on what is real in the minds of individuals (Lincoln & Guba, 1985), is incorrect in terms of original grounded theory. This philosophical positioning has contributed to the remodeling of grounded theory, compromising scientific rigor, as research learners misunderstand how such a stance impacts theoretical development.

Methodological innovation has confused understanding in that it is celebrated by some and frustrates others on the grounds that innovative superficiality does little to advance understanding or knowledge development in a discipline (Travers, 2009). Methodological innovations are evident in many guises. Passed off as "rhetorical manoeuvrings" (Alvesson & Skoldberg, 2000, p. 3), "tensions, tussles, and resolutions" (Morse, 2009, p. 13), flexible progression (Layder, 2000), cutting edge discussion (Gubrium & Holstein, 2003), innovation is linked to modernization and the assumption that newness improves what already exists (Travers, 2009). More often than not innovation challenges tradition and is couched in terms of critical review of methodology, which is necessary to uncover the truth (Denzin & Lincoln, 2005). Along these lines the concept of emergence, integral to classic grounded theory, is seen as a passport to methodological innovation. Emergence has opened up broad interpretations that researchers across disciplines have embraced with misplaced enthusiasm that has revitalized qualitative research (Kushner & Morrow, 2003). The end result of this highly creative modification of the methodology under-mines methodological rigor to the extent that theory is little more than an exploratory analysis. Part of the problem is the poor understanding of methodology.

Similarly, another misunderstanding concerns the notion that the aim of grounded theory study is to discover a basic social process (Charmaz, 2008), and if this does not occur theoretical development is limited (Cutcliffe, 2000). The fact that theory development is not restricted to the generation of basic social processes that may or may not be relevant in the research seems to have been missed altogether. Interpretation about the place of social processes in grounded theory research is not a recent disavowal of methodology (Charmaz, 2008), or a change in stance (Bryant & Charmaz, 2007), although it certainly led to a re-interpretation about the place of the basic social process (Reed & Runquist, 2007), so that it was congruent with the disciplinary knowledge generation agenda in nursing in particular. The debate added little to methodological discussion, simply increasing confusion.

Methodological confusion may have stimulated Bryant and Charmaz (2007) to view grounded theory as a family of methods. Such a position implied that researchers were free to interpret methodology individually. Broad interpretations resulted, supposedly legitimizing qualitative research in general (Glaser 2003). The end product however was very different to classic grounded theory. While it is difficult to account for the wholesale remodeling of grounded theory, poor scholarship arising from a partial reading and understanding of the method provides one explanation. All too often authors who have never used grounded theory write authoritatively about something that can only be learned from close study and application in research practice (Glaser & Holton, 2007).

Interestingly, Charmaz (2009) appreciates that most researchers use few if any of the grounded theory strategies and those that do often have a limited understanding and alter them beyond recognition. So, what is grounded theory? Classic grounded theory is a simple inductive methodology for systematically generating theory about a conceptualized latent pattern (Glaser, 1998; Glaser, 2003). Classic grounded theory assumes that the social organization of life is such that individuals are always in the process of resolving relevant problems (Glaser, 1978). The purpose of grounded theory is to provide a theoretical explanation of how problems are managed. Theory is based on conceptualization that transcends description and interpretation. Theory is abstract of time, place and people (Glaser 2002a). Analysis is not a description of the "voice" of participants, rather an abstract theoretical explanation (Glaser, 2002b).

Glaser (2008) has promoted understanding reminding researchers that grounded theory is a general method (Glaser & Strauss, 1967) that can be used with qualitative or quantitative data. The unit of analysis is individual behavior although the research purpose is to generate a theory that explains group patterns of behavior. The explanation is conceptual not interactional (Glaser, 2005) and differs from qualitative research or simple inductive qualitative analysis that focuses on interpretation of individual experiences. Grounded theory contrasts with qualitative data analysis (QDA), where the emphasis is on interpretation rather than conceptualization and generalization. The aim of QDA is to produce accurate representation of the participant's voice, using reflexivity to reduce the bias that occurs as data is co-constructed. Thus the two approaches are quite different, with QDA emphasizing the lived experience of participants (Patton, 2002). In contrast, grounded theory examines the main concern of people and how that is continually resolved or processed (Glaser, 1978). The two approaches have their origins in a wider epistemological debate.

Epistemological Debate

Known for her constructivist version of grounded theory Charmaz (2000, 2006) has opened up the epistemological debate and "moved the methodology further into interpretive social science" (Charmaz, 2009, p. 136). Charmaz argues that classic grounded theory is positivist with objectivist underpinnings, in that it assumes an objective, external reality. Crotty (1998) notes that constructivism is about unique individual experiences while constructionism is about the influence culture has on worldviews. Constructivism is useful to explain what is in the minds of individuals, while constructionsim focuses on understanding social relationships. Constructionism as a research paradigm denies the existence of an objective reality, assuming instead a relativist ontological position (Mills, Bonner, & Francis, 2006a). Applied to grounded theory, the constructivist perspective recognizes multiple realities and rejects Glaser's assumption that truth emerges from a neutral, objective external reality in which the researcher is quite separate from participants in data collection and analysis.

Charmaz (2009) recognizes that epistemologically, constructivist grounded theory challenges classic grounded theory, arguing that this version is a "contemporary revision" of the original method (Charmaz, 2009, p. 129). According to Charmaz (2009), constructivism offers an alternative to the classic objective stance by acknowledging the subjectivities inherent in qualitative research. Charmaz suggests that constructivist grounded theory is essentially interactive. Mutual construction of the data takes place through interaction between the researcher and the participant, which ensures that the individual's voice is at the for-front of data collection and analysis (Charmaz, 2004). A constructivist grounded theorist aims to get close to participants, "getting inside the experience and taking it apart" (Charmaz, 2009, p. 142). The constructivist grounded theorist ensures that interpretations are such that they resonate with and are validated according to participants. Despite the use of reflexivity, active participant involvement in qualitative research is contested, as is the claim that constructivism is a way of integrating the ontological and epistemological divergence that was inherent in the original methodology (Charmaz, 2009).

Charmaz's (2000) view of constructivism, which assumes multiple realities exist, contrasts sharply with objectivist grounded theory, which presumes there is an external reality (Charmaz, 2009). Charmaz's assumptions conflict with those of Berger and Luckmann (1991), who were influential developers of social constructionism. Berger and Luckman (1967) view society as existing both as objective and subjective reality. As people interact within their social world, they influence others, while they become familiar with the "common taken-for-granted knowledge that ordinary people use to manage everyday life" (Cheek, Shoebridge, Willis, & Zadoroznyj, 1996, p. 150). This knowledge consists of socially accepted behaviors, roles, and beliefs that are legitimized within society.

Berger and Luckmann (1991) view society as an objective reality that is humanly produced, but is nevertheless real on a subjective level to people. Gergen (1994) suggests that individuals actively create reality according to the actions and meanings given to those actions. Society is an accumulation of actions, interactions and interpretations that are more significant than social structures. Thus reality is socially defined, although subjective and objective understandings are different. Objective reality is about the rules that structure society. Subjective reality refers to concepts such as conversation that are readily shared with others. Shared meaning and understanding are intertwined and taken for granted, so much so that concepts do not need to be defined each time they are used (Berger & Luckmann, 1967). This view of social constructionism is consistent with classic grounded theory that seeks to conceptualize patterns of behavior, as people resolve or process their main concern. It is though, at odds with constructivism that is more relativist. Relativism maintains that because there are multiple realities, there are multiple interpretations of those realities (Bury, 1986).

Relativism complicates the epistemological debate, because there is no way of judging one account of reality as being better than another (Burr 1995). Charmaz (2009) recognizes the problem but believes that qualitative inquiry and grounded theory in particular have shifted over the years. Constructivist grounded theorists pick up data in particular situations and therefore data are problematic, relativistic, and partial - depending on conditions at the time. This perspective is different from the views of both Berger and Luckmann (1967) and Glaser (2001). The position though is consistent with the increasing tendency within qualitative research to adopt the relativist position, despite the fact that multiple perspectives undermine the legitimacy of knowledge generated within the qualitative paradigm. If all perspectives are valid how do we judge truth? If qualitative knowledge is uncertain, research relevance becomes debatable, and its usefulness may be questioned, particularly in relation to health care research (Murphy et al. 1998).

Charmaz (2000) addresses questions of relevance by ensuring that participants have a voice. Voice is seen as an individual creative expression that ensures participant subjective experiences come through in full, rich descriptions, which will be verified by individuals (Morse & Field, 1996). Creative expression of data is designed to mediate the effect of the researcher on data collection and ensure participant subjective experiences come through in full, rich descriptions, which will be verified according to individuals (Morse & Field, 1996). Despite claims that this leads to constructivist theory generation, more often than not the end result is multiple descriptions that are far removed from theory generation. Clearly, epistemological issues have caused methodological tensions to the degree that methodology has been remodeled even though Glaser (1998) maintains that the social world exists and the role of the researcher is to find out what is going on in a substantive area. Multiple realities exist and realities are dependent on the different perspectives of people (Glaser & Strauss, 1967). The notion that researchers should focus on problems that are of concern to participants (Mullen & Reynolds, 1994) seems to have been lost in the epistemological debate.

Overall, it is evident that realism and relativism represent two polarized perspectives on a continuum between objective reality at one end and multiple realities on the other. Both positions are problematic for qualitative research. The former position is incompatible with the fundamental principle of classic grounded theory, while the latter leads to questions regarding the function and utility of research. Adopting a realist position ignores the way the researcher interprets findings and assumes that what is reported is a true and faithful interpretation of a knowable and independent reality. Relativism leads to the conclusion that nothing can ever be known absolutely, and that there are multiple realities, none having precedence over the other in terms of claims to represent the truth about social phenomena.

Hammersley (1992), however, has proposed a potential solution to the realist-relativism debate, suggesting a midway position between objective and multiple realities in which researchers might position themselves as a subtle realist. This position acknowledges independent reality, but denies that there can be direct access to that reality, emphasizing theoretical representation. In this middle course common-sense knowledge is considered useful, although the validity of all types of knowledge is rejected. Central to this is a rejection of the view that knowledge is independent of the researcher, whose reality can be known with certainty. Both realism and relativism share the view of knowledge in that both define it in this way as the starting point of their stances. This results in the current dichotomy in qualitative research. The contention is that by avoiding such a definition, the negative implications for research associated with both philosophical perspectives can be avoided.

Blaikie (2000) though challenges the philosophical emphasis, suggesting a pragmatic approach to research is more important. (Murphy et al., 1998) observe that qualitative research tends to draw inferences about meaning. Danermark (2002) argues that current trends in qualitative research have moved beyond the realism-relativism debate (Danermark et al., 2002; Denzin & Lincoln, 1994). Avis (2003) nonetheless supports a disentanglement of methodology and epistemology replacing concerns of truth with a pragmatic epistemology, which depends on internal consistency of arguments presented to support a knowledge claim. This is based on the fit between evidence, social theory and existing knowledge, which introduce moral concerns into the discussion.

Moral Concerns

Moral issues deal with important social values such as respect for life, freedom, and love; issues that provoke the conscience of such feelings as guilt, shame, self-esteem, courage, or hope; issues to which we respond with words like *ought, should, right, wrong, good, bad;* and issues that are uncommonly complicated, frustrating, irresolvable, or difficult in some indefinable way (Jameton, 1984). Since "doing good" is the underlying purpose of the disciplines, every facet of professional life - practice and research - is imbued with moral concerns in that every action has a moral basis. The purpose of research within a discipline is to add to what is known, to improve the practical application of a discipline's knowledge, and to enhance some aspect of life. Therefore, regardless of the broader methodological, epistemological, or ontological issues discussed so far in this chapter, knowledge generation and research in a discipline carries moral weight.

Research is vital to a discipline's overall goals. Sometimes studies are important on a chemical or molecular level and sometimes on the social/behavioral level. Grounded theories focus on this social/behavioral realm, the domain in which humans live their lives. Classic grounded theory is unique among methods in that it seeks to understand people's experiences from this point of view. If humans deserve a certain degree of respect and autonomy, as most professional codes of ethics would suggest, the starting point of any inquiry begins with seeking to understand individuals' problems from their own perspectives. Rather than imposing biased, preconceived, irrelevant, or contrived notions upon the research, classic grounded theorists begin with problems that are important to the people involved. Grounded theorists seek to depict, on a conceptual level, how people deal with their most basic problems. Hence, grounded theory provides the best method of uncovering concepts that are central to people's lives. The grounded theorist values the person and strives with humility to understand his/her life circumstance—a morally relevant approach.

The theorist who utilizes the classic grounded theory method allows theory to emerge naturally from grounded data (Glaser, 1978). Thus, although abstract and conceptual, classic grounded theories based upon qualitative data maintain the shape and substance of reality as perceived by study participants. This is why classic grounded theories have grab (Glaser, 1978). People who have experienced the same phenomenon as described in a theory immediately recognize their truth. They understand the theory and recognize the explanations of reality that are presented.

Truth, in the context of grounded theory, is subjective and relative. It involves the informant's perception of reality. If grounded theory methodology is based upon American pragmatism, as many suggest, we might argue with Peirce that truth as an ideal is the "final opinion" of investigators (Peirce 1868, CP 5.311, Peirce 1878, CP 5.407). Thus, as theorists gather grounded data (i.e., truth from the point of view of various sample populations) modifiable conceptual theories emerge. Subsequent studies modify earlier theories and move toward a better depiction of the ideal reality, perhaps identifying a final opinion.

The moral foundation of discipline-specific research turns on the ultimate goal of the profession. Classic grounded theory seeks truth in that its goal is to uncover important problems and patterns of social behavior as experienced, understood, and communicated by individuals absent of bias, value judgments, and interpretations of the theorist. It is a unique theory-generating approach to understanding human experience. The moral imperative of research in the social sciences is to produce the best possible knowledge that can be used to positively affect those who require the services of a professional. So, there seems to be valid moral justification for adherence to the tenets of classic grounded theory in disciplinary research. Furthermore, inadequate, skewed, misinformed, biased, or capriciously interpreted data and thoughtless, preconceived analysis of research data fails to attain the moral imperative central to disciplinary development.

The suggestion here is that there are moral implications involved with remodeling of the original classic method. Various post-modern tangential methods in research have produced derivative self-proclaimed grounded theory methods that stray from the original ontological, epistemological, and methodological stance. Some have little methodological similarity to the method as first described by Glaser (Glaser, 1965, 1978, 1992, 1993, 1998, 1999, 2001, 2005; Glaser & Strauss, 1967; Glaser & Tarozzi, 2007). Methods that deviate illogically from the basic tenets of the original "classic" grounded theory create confusion among the scholars as they blur distinctions between grounded theory and other methods (Baker, Wuest, & Stern, 1992) and call into question the trustworthiness of grounded theory studies. As pointed out by scholars over the years, there continue to be conflicting opinions and unresolved issues regarding the nature and process of grounded theory (Anderson, 1991; Annells, 2006; Backman & Kyngas, 1998; Baker et al., 1992; Benoliel, 1996; Boychuk Duchscher & Morgan, 2004; Bunch, 2004; Cutcliffe, 2000, 2005; Eaves, 2001; Elliott & Lazenbatt, 2005; Fleming & Moloney, 1996; Glaser, 2002; Greckhamer & Koro-Ljungberg, 2005; Heath & Cowley, 2004; Katada, 1990; Kushner & Morrow, 2003; Lingard, Albert, & Levinson, 2008; Lomborg & Kirkevold, 2003; Mills, Bonner, & Francis, 2006a, 2006b; Mills, Chapman, Bonner, & Francis, 2007; Reed & Runquist, 2007; Teram, Schachter, & Stalker, 2005; Wilson & Hutchinson, 1996; Wolanin, 1977). Revisionist methods that are the product of remodeling misinterpret the ontological and epistemological stance of the original classic method. Scholars who are unfamiliar with the method may legitimately question the trustworthiness of grounded theories because of this methodological confusion. Even the best grounded theory studies might be discounted. Therefore, research that produces theories that fail to uncover truth as recognized by those whose lives are examined are inherently immoral, because they fail at both their immediate and ultimate goals.

Grounded theory is one of the most popular research methods used with qualitative data. Discussions about classic versus revisionist versions of the method are all too common. Remodeling of the method presents many implications for theory development. It blurs the boundaries between methods, calls into question validity and trustworthiness of theories, confuses students and scholars, and slows disciplinary progression and development. Although many of the remodeling issues discussed may be due to the fact that so many researchers have learned grounded theory minus mentoring it is surprising that scholars within disciplines have not challenged the lack of methodological rigor and have accepted the associated sloppy scholarship. Problem identification though can become a catalyst for change. The change that is required if grounded theory researchers must be supervised by non-grounded theory researchers, is that supervisors must seek the methodological mentorship that is now well established internationally. Methodological mentorship is essential to guide researchers, as they engage in the research process. Today, there are all sorts of resources for novice grounded theory researchers. There is no need for a research-learner to go through the research process without expert mentors. Fortunately, support systems are available in person, on-line, via books written by the original author, and through world-wide seminars. The next step in fostering rigorous scholarship is for supervisors unfamiliar with grounded theory to recognize their limitations and refer students to the international grounded theory community of scholars where research rigor is promoted.

Conclusion

It is difficult not to conclude that current understandings of grounded theory are due to sloppy scholarship based on a partial reading and understanding of the method. In the social science literature there is no discussion about the strategy of conceptualization and little understanding as to what it means. This has led directly to an uncritical determination to re-conceptualize grounded theory in terms of constructivism, resulting in an end product that is descriptive in nature rather than conceptual. Attempts, however well intentioned, to re-write grounded theory in a way that is consistent with current understandings of ontology and epistemology have resulted in a misrepresentation of the methodology and an undermining of its power to generate theory. This has made a straightforward methodology overtly complex, despite attempts to keep it simple. While this may have occurred because so many students learn the methodology minus mentorship, working with qualitative researchers who do not understand grounded theory, it is imperative that grounded theory is re-emphasized as a conceptual methodology that leads to theory generation rather than tolerate the current situation in which grounded theory is used in ways that endanger the intellectual reputation of social science research. At the very least there is a moral imperative and responsibility to take note of and respond to grounded theory interpretations that are not recognizable to the co-originator of the methodology. There is much work to be done.

References

Abbott, A. (1988). *The system of the professions.* London: University of Chicago Press.
Alvesson, M., & Skoldberg, K. (2000). *Reflexive methodology: New issues for qualitative research.* London: Sage.
Anderson, M. A. (1991). Use of grounded theory methodology in a descriptive research design. *Abnf Journal, 2*(2), 28-32.
Annells, M. (2006). Triangulation of qualitative approaches: Hermeneutical phenomenology and grounded theory. *Journal of Advanced Nursing, 56*(1), 55-61.
Avis, M. (2003). Do we need methodological theory to do qualitative research? *Qualitative Health Research, 13*(7), 995-1004.
Backman, K., & Kyngas, H. (1998). Challenges of the grounded theory approach to a novice researcher. *Hoitotiede, 10*(5), 263-270.
Baker, C., Wuest, J., & Stern, P. N. (1992). Method slurring: The grounded theory/phenomenology example. *Journal of Advanced Nursing, 17*(11), 1355-1360.
Benoliel, J. Q. (1996). Grounded theory and nursing knowledge. *Qualitative Health Research, 6*(3), 406-427.
Berger, P., & Luckman, T. (1967). *The social construction of reality.* New York: Penguin.
Blaikie, N. W. H. (2000). *Designing social research: The logic of anticipation.* Malden, MA: Polity Press.
Boychuk Duchscher, J. E., & Morgan, D. (2004). Grounded theory: Reflections on the emergence vs. forcing debate. *Journal of Advanced Nursing, 48*(6), 605-612.
Bryant, A., & Charmaz, K. (2007). Grounded theory in historical perspective: An epistemological account. In A. Bryant & K. Charmaz (Eds.), *The handbook of grounded theory* (pp. 31-57). London: Sage.
Bunch, E. H. (2004). Commentary on the application of grounded theory and symbolic interactionism. *Scandinavian Journal of Caring Science, 18*(4), 441.
Burr, V. (1995). *An introduction to social constructionism.* London, United Kingdom: Routledge.
Bury, M. (1986). Social constructionism and the development of medical sociology. *Sociology of Health and Illness, 8*(2), 137-169.
Carroll, K. (2004). Mentoring: A human becoming perspective. *Nursing Science Quarterly, 17*(4), 318-322.
Charmaz, K. (2000). Grounded theory: Objectivist and constructivist methods. In N. Denzin & Y. S. Lincoln (Eds.), *Handbook of qualitative research* (2nd ed., pp. 509-535). London : Sage.
Charmaz, K. (2004). Grounded theory. In S. Nagy Hesse-Biber, & P. Levy (Eds.), *Approaches to qualitative research: A reader on theory and practice* (pp. 496-521). New York: Oxford University Press.
Charmaz, K. (2006). *Constructing grounded theory. A practical guide through qualitative analysis.* London: Sage.
Charmaz K. (2008). Reconstructing grounded theory. In L. Bickman, P. Alasuutari, & J Brannen (Eds.), *Handbook of social research* (pp. 461-478). London: Sage.
Charmaz, K. (2009). Shifting the grounds: Constructivist grounded theory methods. In J. M. Morse, P. N. Stern, J. M. Corbin, B. Bowers, & A. E. Clarke (Eds.), *Developing grounded theory: The second generation* (pp. 127-154). Walnut Creek, CA: University of Arizona Press.
Cheek, J., Shoebridge, J., Willis, E., & Zadoroznyj, M. (1996). *Society and health: Social theory for health workers.* Melbourne, Australia: Longman.

Clarke, A. E. (2005). *Situational analysis: Grounded theory after the postmodern turn.* Thousand Oaks, CA: Sage.

Corbin, J. M. (2009). Taking an analytic journey. In J. M. Morse, P. N. Stern, J. M. Corbin, B. Bowers, & A. E. Clarke (Eds.), *Developing grounded theory: The second generation* (pp. 35-53). Walnut Creek, CA: University of Arizona Press.

Corbin, J. A., & Strauss, A. L. (2008). *Basics of qualitative research: Techniques and procedures for developing grounded theory* (3rd ed.). Newbury Park, CA: Sage.

Crotty, M. (1998). *The foundations of social research: Meaning and perspective in the research process.* Sydney, Australia: Allen and Unwin.

Cutcliffe, J. R. (2000). Methodological issues in grounded theory. *Journal of Advanced Nursing, 31*(6), 1476-1484.

Cutcliffe, J. R. (2005). Adapt or adopt: Developing and transgressing the methodological boundaries of grounded theory. *Journal of Advanced Nursing, 51*(4), 421-428.

Danermark, B. (2002). *Explaining society: An introduction to critical realism in the social sciences.* London: Routledge.

Denzin, N., & Lincoln, Y. S. (2005). *The Sage handbook of qualitative research.* London: Sage.

Eaves, Y. D. (2001). A synthesis technique for grounded theory data analysis. *Journal of Advanced Nursing, 35*(5), 654-663.

Elliott, N., & Lazenbatt, A. (2005). How to recognizes a 'quality' grounded theory research study. *Australian Journal of Advanced Nursing, 22*(3), 48-52.

Fleming, V. E., & Moloney, J. A. (1996). Critical social theory as a grounded process. *International Journal of Nursing Practice, 2*(3), 118-121.

Freeman, A. R., McWilliam, C. L., MacKinnon, J. R., DeLuca, S., & Rappolt, S. G. (2009). Health professionals' enactment of their accountability options: Doing the best they can. *Social Sciences and Medicine, 69*(7), 1063-1071.

Fuller, S. S. (2000). Enabling, empowering, inspiring: Research and mentorship through the years. *Bulletin of the Medical Library Association, 88*(1), 1-10.

Gergen, K. J. (1994). *Realities and relationships: Soundings in social constructionism.* Cambridge, MA: Harvard University Press.

Glaser, B. G. (1965). The constant comparative method of qualitative analysis. *Social Problems, 12*, 10.

Glaser, B. G. (1978). *Theoretical sensitivity: Advances in the methodology of grounded theory.* Mill Valley, CA: Sociology Press.

Glaser, B. G. (1992). *Emergence vs forcing: Basics of grounded theory analysis.* Mill Valley, CA: Sociology Press.

Glaser, B. G. (1993). *Examples of grounded theory: A reader.* Mill Valley, CA: Sociology Press.

Glaser, B. G. (1998). *Doing grounded theory: Issues and discussion.* Mill Valley, CA: Sociology Press.

Glaser, B. G. (1999). The future of grounded theory. *Qualitative Health Research, 9*(6), 10.

Glaser, B. G. (2001). *The grounded theory perspective: Conceptualization contrasted with description.* Mill Valley, CA: Sociology Press.

Glaser, B. G. (2002). Constructivist grounded theory? *Forum Qualitative Social Research, 3*(3).

Glaser, B. G. (2003). *The grounded theory perspective II: Description's remodeling of grounded theory.* Mill Valley, CA: Sociology Press.

Glaser, B. G. (2005). *The grounded theory perspective 3: Theoretical coding.* Mill Valley, CA: Sociology Press.

Glaser, B. G. (2008). *Doing formal grounded theory.* Mill Valley, CA: Sociology Press.

Glaser, B. G., & Holton, J. A. (2007). *The grounded theory seminar reader.* Mill Valley, CA: Sociology Press.

Glaser, B. G., & Strauss, A. L. (1967). *The discovery of grounded theory: Strategies for qualitative research.* Chicago, IL: Aldine Publishing.

Glaser, B. G., & Tarozzi, M. (2007). Forty years after discovery: Grounded theory worldwide. *The Grounded Theory Review: An International Journal, Special Issue*, 21-41.

Greckhamer, T., & Koro-Ljungberg, M. (2005). The erosion of a method: Examples from grounded theory. *International Journal of Qualitative Studies in Education, 18*(6), 729-750.

Gubrium, J., & Holstein, J. (Eds.). (2003). *Postmodern interviewing.* Thousand Oaks, CA: Sage.

Hammersley, M. (1992). *What's wrong with ethnography? Methodological explorations.* London: Routledge.

Harding, S. (1987). *Feminism and methodology.* Bloomington, IND: Indiana University Press.

Heath, H., & Cowley, S. (2004). Developing a grounded theory approach: A comparison of Glaser and Strauss. *International Journal of Nursing Studies, 41*(2), 141-150.

Jameton, A. (1984). *Nursing practice: The ethical issues.* Englewood Cliffs, NJ: Prentice-Hall.

Jones, S. R., Torres, V., & Arminio, J. (2006). *Negotiating complexities of qualitative research in higher education.* New York: Routledge.

Katada, N. (1990). Grounded theory approach: Trends in grounded theory in nursing research. *Kango Kenkyu, 23*(3), 283-289.

Kushner, K. E., & Morrow, R. (2003). Grounded theory, feminist theory, critical theory: Toward theoretical triangulation. *Advances in Nursing Science, 26*(1), 30-43.

Layder, D. (2000). *New strategies in social research.* Cambridge, MA: Polity Press.

Lincoln, Y. S., & Guba, E. G. (1985). *Naturalistic inquiry.* Beverly Hills, CA: Sage.

Lingard, L., Albert, M., & Levinson, W. (2008). Grounded theory, mixed methods, and action research. *British Medical Journal, 337*, 567.

Lomborg, K., & Kirkevold, M. (2003). Truth and validity in grounded theory - a reconsidered realist interpretation of the criteria: Fit, work, relevance and modifiability. *Nursing Philosophy, 4*(3), 189-200.

McDonald, M., & Schreiber, R. (2001). Constructing and deconstructing: Grounded theory in a postmodern world. In R. S. Schreiber & P. N. Stern (Eds.), *Using grounded theory in nursing* (pp. 17-34). New York: Springer.

Meleis, A. I. (1997). *Theoretical nursing: Development and progress* (3rd ed.). Philadelphia, PA: Lippincott.

Milliken, P. J., & Schreiber, R. S. (2001). Can you "do" grounded theory without symbolic interactionism? In R. S. Schreiber & P. N. Stern (Eds.), *Using grounded theory in nursing* (pp.177-190). New York: Springer.

Mills, J., Bonner, A., & Francis, K. (2006a). Adopting a constructivist approach to grounded theory: Implications for research design. *International Journal of Nursing Practice, 12*(1), 8-13.

Mills, J., Bonner, A., & Francis, K. (2006b). The development of constructivist grounded theory. *International Journal of Qualitative Methods, 5*(1), 1-10.

Mills, J., Chapman, Y., Bonner, A., & Francis, K. (2007). Grounded theory: A methodological spiral from positivism to postmodernism. *Journal of Advanced Nursing, 58*(1), 72-79.

Morse, J. M. (2009). Tussles, tensions, and resolutions. In J. M. Morse, P.N. Stern, J. M. Corbin, B. Bowers, & A. E. Clarke (Eds.), *Developing grounded theory: The second generation* (pp. 13-22). Walnut Creek, CA: University of Arizona Press.

Morse, J. M., & Field, P. A. (1996). *Nursing research: The application of qualitative approaches.* London: Chapman and Hall.

Morse, J.M., Stern, P.N., Corbin, J.M., Bowers, B., & Clarke, A. E. (2009). *Developing grounded theory: The second generation.* Walnut Creek, CA: University of Arizona Press.

Mullen, P. D., & Reynolds, R. (1994). The potential of grounded theory for health education research: Linking theory to practice. In B. G. Glaser (Ed.), *More grounded theory methodology* (pp. 127-145). Mill Valley, CA: Sociology Press.

Murphy, E., Dingwall, R., Greatbatch, D., Parker, S., & Watson, P. et al. (1998). Qualitative research methods in health technology assessment: A review of the literature. *Health Technology Assessment, 2* (16).

Patton, M. Q. (2002). *Qualitative evaluation and research methods.* Thousand Oaks, CA: Sage.

Peirce, C. S. (1901). Truth and falsity and error" (in part). In J. M. Baldwin (Ed.), *Dictionary of Philosophy and Psychology*, vol. 2. Reprinted (CP 5.565–573), (pp. 718-720).

Peirce, C. S. (1868). Some consequences of four incapacities. *Journal of Speculative Philosophy, 2*(1868), 140–157. Reprinted (CP 5.264–317), (CE 2, 211–242), (EP 1, 28–55). Eprint. NB. Misprints in CP and Eprint copy. Accessed July 11, 2008 at http://www.cspeirce.com/menu/library/bycsp/conseq/cn-frame.htm

Peirce, C. S. (1878). How to make our ideas clear. *Popular Science Monthly, 12* (1878), 286–302. Reprinted (CP 5.388–410), (CE 3, 257–276)), (EP 1, 124–141). Eprint. Accessed July 11, 2008, at http://en.wikisource.org/wiki/How_to_Make_Our_Ideas_Clear

Reed, P. G., & Runquist, J. J. (2007). Reformulation of a methodological concept in grounded theory. *Nursing Science Quarterly, 20*(2), 118-122.

Schumacher, G., Risco, K., & Conway, A. (2008). The Schumacher model: Fostering scholarship and excellence in nursing and for recruiting and grooming new faculty. *Journal of Nursing Education, 47*(12), 571-575.

Strauss, A. L. (1987). *Qualitative analysis for social scientists.* New York: Cambridge University Press.

Strauss, A. L., & Corbin, J. M. (1990). *Basics of qualitative research: Grounded theory procedures and techniques.* Newbury Park, CA: Sage.

Strauss, A. L., & Corbin, J. A. (1998). *Basics of qualitative research: Techniques and procedures for developing grounded theory* (2nd. ed.). Thousand Oaks, CA: Sage.

Teram, E., Schachter, C. L., & Stalker, C. A. (2005). The case for integrating grounded theory and participatory action research: Empowering clients to inform professional practice. *Qualitative Health Research, 15*(8), 1129-1140.

Travers, M. (2009). New methods, old problems: A skeptical view of innovation in qualitative research. *Qualitative Research, 9*(2), 161-179.

Wilson, H. S., & Hutchinson, S. A. (1996). Methodological mistakes in grounded theory. *Nursing Research, 45*(2), 122-124.

Wolanin, M. O. (1977). Confusion study: Use of grounded theory as methodology. *Community Nursing Research, 8*, 68-75.

PART II

DOING GROUNDED THEORY

5

Conducting Grounded Theory Interviews Online

Helen Scott
groundedtheoryonline.com

I use video conferencing software to connect with people from all over the world. I am continually delighted that people from every continent can gather in the same meeting room, see each other's screens, exchange complex ideas and carry out demonstrations in real time. Communicating this way is an entirely different experience than the one-dimensional nature of chatting by text only, yet both of these communication channels can be used very successfully when collecting data online. When I first started data collection in 2003, online interviewing was still a very new technique and was text-based. Audio communications were not yet "commonly available for Internet-accessible computers" (Mann & Stewart, 2002, p.3), video conferencing was a distant dream and online interviewing was unheard of as a means of collecting data for a grounded theory (GT) study. My experience is that the technique lends itself very well to the research method and that the theoretical fit is good. There are two important points: the first is to make sure that the researcher has considered the basics of interview design from a grounded theory perspective; and the second is to use the technology with which both researcher and participant are comfortable and which their combined connection speed can support. This chapter will address both issues as it discusses the techniques of interviewing online, ethical matters and issues of data quality.

Interviewing as a conversation

Grounded theorists use the gentlest definition of the word *interview*, defining it as a conversation between equals which is led by the participant (Glaser, 2003 b). The researcher's aim is to create the conditions under which the participant feels able to talk about what matters most to him or her: to "instill a spill" (Glaser, 1998, p111). A good way to start is to ask broad 'grand tour' questions (Nathanial, 2008; Simmons, 2010). Glaser suggests being as broad as "How are you today?" or "What's up?" (Glaser, 2003 b). It takes courage to ask open questions; the fear is that the participant may not talk about the research topic. For some novice researchers there are enormous consequences of "getting it wrong" and the need to control is great. Understanding the method will help novice interviewers to let go of control and it is therefore important to remember that the method is exploratory. It is also critical to realize that a grounded theory will conceptualize the main issue of the substantive population and how that issue is resolved or processed (Glaser, 1992 p22). For the grounded theorist, therefore, it is essential not to know beforehand, not to impose an issue on the participants but rather to listen, to hear, what the participants themselves consider to be the main issue. Thus, in the opening stages of a study, the researcher is advised to ask the broadest questions she dares and then follow the conversation where the participant leads. How this is achieved online is discussed below.

The ways in which we can communicate over the internet include text, audio and audio-visual singly or in combination. We can communicate one-to-one or in groups of three or more. Communications can be carried out synchronously where the parties meet online at the same time, or asynchronously when messages are left for the different parties to retrieve at their convenience and which may or may not be at the same time. Table 1 lists some of the current software which researchers might like to investigate using, and indicates which can be used synchronously and which asynchronously.

Table 1: Online communication software

	One-to-one interviews	**Group interviews**
Text: email (asynchronous)	Many. Facebook (http://www.facebook.com)	Yahoo Groups (http://groups.yahoo.com) Windows Live Groups (http://groups.live.com) Google Groups (http://groups.google.com)
Text: chat (synchronous)	Yahoo Messenger (http://messenger.yahoo.com/features/chatrooms) Windows Live Messenger 2011 (http://explore.live.com/windows-live-messenger?os=other) Google Talk (http://www.google.com/talk) Skype (http://www.skype.com) Facebook (http://www.facebook.com)	
Audio (asynchronous)	Ipadio (http://www.ipadio.com) Wimba (http://www.wimba.com)	
Audio and Audio-visual (synchronous)	Nefsis (http://www.nefsis.com) Skype (http://www.skype.com) ooVoo, http://www.oovoo.com) Webex (http://www.webex.com) Adobe Connect (http://www.adobe.com/products/adobeconnect.html)	

For interviewing purposes the choice as to which style is used is best left to the participant who understands his/her individual characteristics, competencies, constraints and preferences. Asynchronous is often preferred when people lead busy lives and responding to inquiry can be fitted into an individual's schedule more easily. Synchronous sessions may be preferred where there are issues of literacy, as an opportunity to practice language skills, or when something is difficult to communicate and synchronous communication enables continuous clarification of meaning. Or a synchronous conversation can take place just because researcher and participant happen to be online at the same moment, using the same technology and the moment works.

A first objective for interviewing online is to minimize the actual and perceived risk to the participant. It is important to create the conditions under which the participant feels technically and emotionally comfortable, and is able to share his or her experiences online. The processes and behaviors required by ethical review bodies can help to establish trust on the part of the recipient (for explanation of the development of an ethical approach to online data collection, see Scott 2007 a). Information sheets which, for example, carefully detail how the interview data will be used, stored and disposed of; the participant's right to withdraw at any time; and the right to confidentiality all convey an attitude of respect for the participant. The researcher thus works to earn the participant's trust right from the beginning of the interview process. *Atmosphering* and *toning* are two techniques that Glaser uses in seminars to create the emotional context in which it is safe for people to risk sharing ideas (Glaser, 2003 b). In face-to-face meetings, atmosphering includes selecting conducive physical surroundings considering; accessibility, light, temperature, noise, cleanliness, space and privacy. Online meetings flourish where the software is familiar or intuitive to use, reliable, easy to access and affords privacy. Toning is more to do with communication style and content; where what is said and done is designed to show the participant that he or she is safe from harm, is respected and will not be judged. In a research context it is ethics in practice.

Regarding the participant's technical comfort, an important principle is to use a communication tool commensurate with the participant's online experience and the bandwidth available. Avoid suggesting using Skype (http://www.skype.com) if the participant does not already have a good quality internet connection and a working headset. Practice using the software beforehand and test any hardware (e.g. headphones and microphone) that will be used. This will improve the interviewer's confidence in the method of communicating and may also enable him to troubleshoot basic problems that the participant may experience. Skilled use of the communication software avoids wasting participants' time and causing problems for those who are technically insecure, for whom problems can undermine already fragile confidence levels. Thus, also avoid asking users to download and install new software in order to take part in an interview unless the corresponding gain is significant and the participant is known to be a competent user. Even then it is a risky strategy, given the need to ascertain whether or not the bandwidth is sufficient, and whether or not the participant has the appropriate hardware and is comfortable using it. This is a lot to ask at the beginning of a relationship when the generosity of the participant, in agreeing to be interviewed, has already been called upon. However, when confident of the user, his hardware and internet connection, the relative simplicity of downloading and installing good quality video conferencing software is worth the effort, for the richness of communication. Asking the technically unaware however, to wrestle with digital matters wastes their time and causes frustration which breaches my ethical rule of "do no harm".

Individual interviews: email and chat

When I first started using email for interviews, my concern was whether people would tell me what they thought in enough depth. I often started the email by asking 'how are you?', and in the face-to-face world it is easy to carry on a conversation from this point. In the focused world of email with its lack of immediacy, however, this question is not sufficient on its own to elicit a useful response. Given the relative paucity of text-based communications, I consider that a direct approach is most likely to elicit useful responses and so mention the substantive area I am researching. Since I am asking about participants' experiences, I ask how they feel and what they think. For my research study I framed my questions as follows:

- What do you feel about your experiences of online learning?
- What are the issues for you when working/learning online?
- How did you/do you resolve those issues?
- What else should I know, that I haven't asked?

The last two questions elicited very little useful information and were quickly dropped. A better way to get that information was to ask for more details in a follow up email. Participants did, however, expend a great deal of effort giving full replies to the first two. This is one participant's response to the question asking how he felt (text reproduced here as it was received):

> I am into my third unit come January and so far I would not say that I have had a bad experience. It is a change from classroom learning and I am still trying to establish a routine that would not have me under stress come assignment and exam time. I have failed miserably. The challenge I have with online learning is that it is very easy to fall behind because I still have to fulfil my job requirements and my domestic duties. I think that online learning although convenient to people who are working, it

> can be very difficult if one does not put aside time every day to get some studying in. (participant A, personal communication, July, 2005)

And these are the issues that he experienced:

> The major issue I have at this point is the time zone between here and the UK We are four or five hours depending on the time of year, behind the UK and this causes major headaches in relation to the Virtual Class. All the classes are either held when I am sleeping or when I am at work. .... [In one online unit] we were placed in Virtual Groups. In my group there were three guys from the UK and two of us [from this time zone] Of course we met very sparingly and I don't think we were able to maximise the learning of being in a group. (participant A, personal communication, July, 2005)

The main concern of finding time and integrating study into his life, and how this person largely failed to process the concern could not have been expressed more clearly, and this concern was voiced over and over again by other participants. When I receive an initial response, if I am unclear or need more information about what has been written, I quote the participant and request that he/she 'tell me more about' the issue raised. I emailed one participant: "You write: 'with a busy working and family life, planning my own hours of study was also the advantage of online learning'. Please can you explore this a little more?" Quoting the participant and asking for more information is a very effective technique and works as well in the online world as it does when one uses the same technique in audio or face-to-face situations. As my study developed, I asked more specific questions. For example, I wrote to another participant:

> You describe a process where the work seems to become of less and less value to you. Despite this you completed the course and so must have satisfied the attendance requirements as evidenced by the number of postings. Did your way of working/reasons for working change over the course?

He replied:

> I completed the requirements because I had paid for the course and wanted the qualification. I lost heart and interest. I changed from wanting to do the course to really learn something and improve myself at the beginning to simply just wanting to finish by the end! I picked and selected bits of the material that did interest me and I did the readings and work for the assignments. I did the final paper – but for my own self

> pride and satisfaction. (participant D, personal communication, March 2005)

Pacing the exchange of emails matters, and again I take my lead from participants. If someone responds quickly, so do I; but if someone takes more time it seems unkind to rapidly return the burden of response to the participant, and so I also take more time. Thanking the participant for his or her help is obvious good manners, but I also ask if I may email again if I have further questions as analysis develops. Generally I have been humbled by the generosity of participants in their willingness to help.

I use the same type of questions when using 'chat' software. Chat can be a little more problematic and the main lessons I have gleaned are:

- Patience is essential because chatting by text is infuriatingly slow;
- do not interrupt;
- a webcam is helpful in that it shows the participant that their chat partner is alert and listening; and
- chat can be non-linear and so it is necessary to scan up and down the discussion so as not to lose threads.

It can sometimes be difficult to tell whether one's chat partner has finished typing or not. I find sitting on my hands very useful; if one starts to type too soon and interrupts, another thread is started, the participant has not said all he or she wants to about the earlier thread and the conversation gets very messy. An early chat session went very well despite me, when my chat partner completely ignored my interruptions and carried on with his points regardless. The joy of chat, however, is that it is more dynamic and the researcher can ask for more details during the conversation. The chat below occurred later in my study, when I had realized that time was the issue for my participants and I was seeking more information about how they process that concern:

Paul: other than my work, i'm also involved in church activities so I need to plan well in advance so i dont end up in a mess
Helen: (thinking)
Helen: This is difficult to phrase because I am not sure what I am asking.
Helen: I think it is to do with the thought processes that you go through in order to plan
Helen:?
Paul: do you mean, whts going thru my head when I make my timetable?
Helen: yeah i think so

Paul: let me see, at the start of the unit, I first see how the course is evaluated. If it's both coursework n assessment, I see whats the time gap between them (is this what u are asking, or am I completely off track?)
Helen: it's absolutely what I'd like to learn about.
Paul: that's good, then the next step (participant E, personal communication, September, 2005)

And so the conversation about how Paul integrates learning into his life continues in some detail. This is also a good example of when it is not helpful to literally code line by line: the detail is not relevant to my study, but rather I look for incidents which may run across many lines. Conversely, there may be many lines where I do not notice any incidents, which is ok.

Group discussion by email

Instilling the spill in a group discussion is a function of the interview design and the medium (software). Interview design will include issues such as the role of facilitator and the amount and type of facilitator intervention. This means, that as well as coaxing or challenging participants to share relevant information relating to their main concerns and how they resolve their main concerns, there might also be a need to coach participants in the process of online communicating.

Initially, at least, the issues for collaborative learners online and online group interviewees are likely to be similar and the aim for all is to develop online conferencing skills. Communicating and learning online can be a distressing experience when implemented badly (Hara & Kling, 2000), and it is therefore essential to design for success. Salmon (2000) describes the stages learners go through in becoming independent online learners and offers strategies for design and moderator behavior that are aimed at encouraging learners to participate and to develop their competence and comfort with the process. I recommend using the first three stages of Salmon's model: Access and motivation, online socialization and information exchange when designing the group interview process and in informing facilitator behavior.

Yahoo!® Groups is a useful communication tool offering both a web-based bulletin board which enables access to the discussion from any computer connected to the internet, and a group address which enables group discussions conducted by email. Using a group address is essential, as otherwise group members can get left out of conversations, for example, when email addresses are input incorrectly or missed out completely. A group address is particularly helpful where there are novice users who are not familiar with email *reply to* functions. I also suggest that, whichever software is chosen, that there is the facility to 'add' individuals to the group (as opposed to having to 'join' the group) so that inconvenience to the individual is minimized.

The process for setting up a group might be:

- Send an invitation to an individual to participate including a statement of ethical matters;
- Receive acceptance from the individual together with an informed consent statement;
- Add the individual to the group;
- Send a welcome message to the individual, asking them not to email the group until all members have been added;
- Send invitations to all group members inviting them to introduce themselves;
- Send first questions to all group members.

Where participants are inexperienced in online communication, the role of the researcher is to model behavior to begin with, to ease out of discussions as they develop, and to prompt should discussion lag. In practice:

- Excellent organization is need to make sure that participants are able to take part in the group in as trouble-free a manner as possible, since participants have no time to waste on the concerns of the researcher;
- Weaving posts and summarizing issues helps to create a conceptual platform from which the conversation can move forward; the researcher must also
- Manage the tension between the online facilitator's function of facilitating discussion, and the concern of the grounded theorist that interventions on the part of the researcher should not be allowed to inhibit or direct discussion.

I advise finding a group of patient friends to practice with before embarking on a group discussion in earnest.

Group interviews using chat

Group sessions using chat software can be anarchic, with comments from one participant running over two or more lines being interspersed by the comments of others, which to the unpracticed or the slow typist can prove overwhelming. I therefore recommend the chat guidelines of Melanie Heard-White (Heard-White, Saunders, & Pincas, 2004) and suggest the use of a seating plan as shown in Figure 1. I also recommend using basic chat conventions such as '#', to indicate when typing has stopped. Furthermore, I suggest breaking long messages over several lines, indicating that there is more to follow by typing and extended version of an ellipsis (......) and indicating the continuation of the message by starting the next line with it as well. Using this technique can help a participant carve out a space to 'speak' and improves the pace of the conversation for other participants.

Figure 1: Sample Chat Plan and Seating Plan for Group Chat Sessions

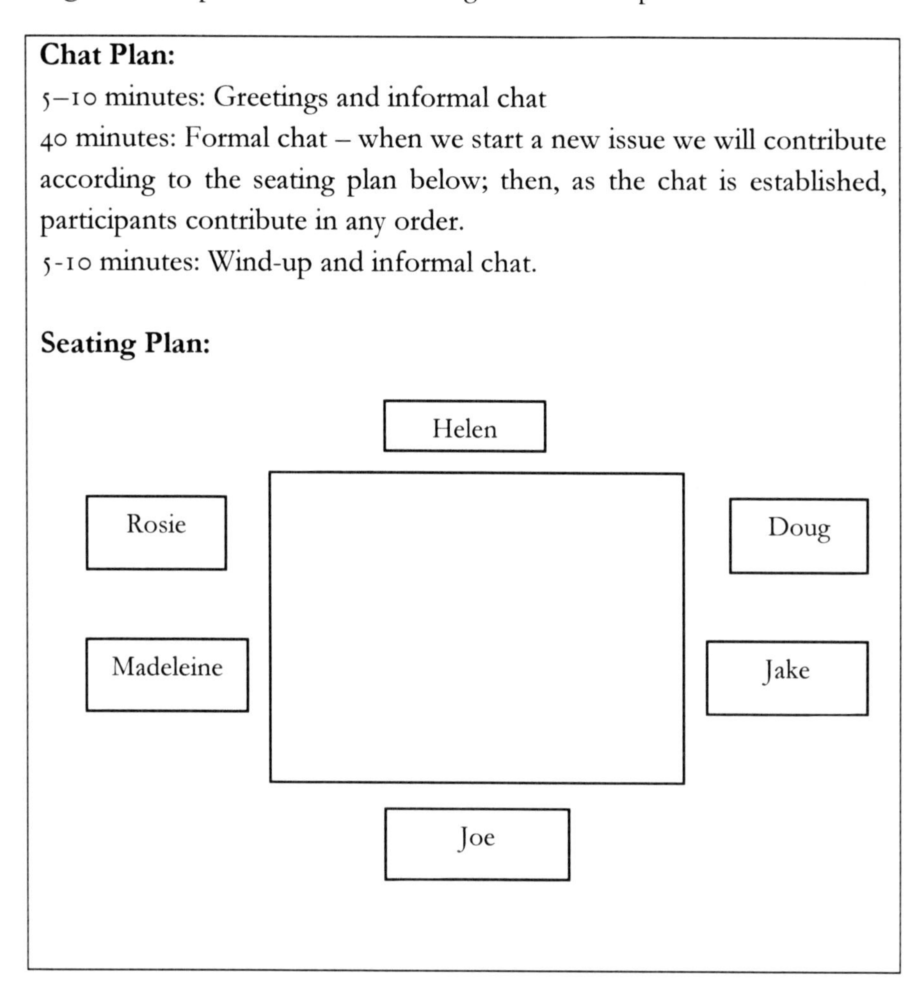

The twin joys of text-based communications are: First that the researcher does not have to make field notes, as the data is already captured digitally and second that they do not require much bandwidth. The twin problems with audio and audio/visual methods are that the researcher does have to make field notes and they do require a lot of bandwidth.

Audio and audio-visual interviews

Given the ubiquity of the mobile telephone, most of the world is practiced at remote audio conversations. Conducting a one-to-one interview on the internet using, for example, Skype is a very similar experience, particularly since many modern mobile telephones can be Skype-enabled. For the researcher the main issue is making sure, as much as possible, that both the researcher and the participant have protected time and space for the conversation. Since using the internet will require connecting to it and opening specific software, I ask for an appointment. An audio interview online can be conducted along similar lines as a face-to-face conversation. In both instances, once the researcher has asked the grand tour question, the researcher is advised to speak as little as possible and active listening is encouraged. Where the participant pauses, it is suggested that the researcher merely repeat the last word the research participant spoke. When necessary, further questions are limited to what has already been said, such as "Tell me more about ..."

Audio-visual conversations can be held in the same way; the additional benefit is that active listening can be displayed through movements: For example, nodding, smiling, seriously attending or leaning forward. When conducting audio or audio-visual group interviews, I adopt Melanie Heard-White's technique and combine it with an attempt to create a relaxed atmosphere by inviting participants to 'sit' round (an image of) my kitchen table. Participants are invited to contribute in turn according to the seating plan. It is helpful to agree beforehand whether interruptions are acceptable or whether comments should be made when invited.

Audio and audio-visual software are more demanding of both the participant's competence and of the researcher's and the participant's combined bandwidth. If a novice internet user requests that we use audio or audio-visual software, then I send very precise instructions with screenshots and suggest meeting a little earlier so that we can address any difficulties. If possible I will have other lines of communication open, for example chat, email or telephone, so that any connection issues can be addressed. In some cases my efforts will not be sufficient: Perhaps the screen shots are slightly different than those presented by the participant's internet browser, or a firewall prohibits access, or the hardware is incompatible. Reasons can be many and varied and in these cases it is best to try another medium. Where participants are able to connect, we start by using video but if insufficient bandwidth means that either the video or the audio is compromised, we turn the video off. Sometimes the bandwidth is simply inadequate and the connection is frequently dropped. In these cases the disruption and frustration often means that it is best to reschedule or try another medium.

Data capture is by field note. With one-to-one interviews I make notes as I listen; with group discussions it is easier to make brief notes and to write up field notes after the event. Some grounded theorists advocate only making field notes after the event as noting may detract from the conversation: I suspect it is a matter for each researcher and participant to decide. For instance, I find my note taking relaxes participants and allows them to pace the interview; sometimes a participant will pause to think and there is an easy silence while I continue to write to catch up. I have not previously recorded interviews since I consider that the amount of time taken to analyze the volume of data would outweigh any benefit gained from it. Many researchers are required to record interviews for the purposes of monitoring for audit purposes or to evidence ethical behavior. In my opinion not only is it unethical to risk the privacy of participants for the sake of project concerns but also recording interviews subverts the grounded theory process.

Ethical issues

My approach to ethically gathering data online is the product of a careful review of the literature (Scott, 2007a) and has been developed in praxis. Planned interactions for the purpose of research always require informed consent. It is therefore necessary to gain informed consent when a researcher plans to collect data from a private or semi-private meeting since the researcher's presence is likely to impact and field notes are likely to contain identifying information. For the same reason, informed consent is appropriate should a researcher interact privately with participants at a public event. There are also ethical questions as to whether one should collect data from public online chat rooms and other online forums and how one should go about it if one does. I subscribe to Sharf's (1999) guideline:

> The researcher should strive to maintain and demonstrate a respectful sensitivity toward the psychological boundaries, purposes, vulnerabilities, and privacy of the individual members of a self-defined virtual community, even though its discourse is publicly accessible. (p. 255)

However, because of the unique characteristics of grounded theory, I consider that one does not require informed consent when using publicly available information or unplanned observations, including those collected in the course of one's occupation (the National Health Service of England, Wales and Northern Ireland does not agree with me). Producing a grounded theory entails collecting data and fracturing it, abstracting concepts from the fragments and re-integrating the concepts into a theoretical whole. There is no risk that an individual's contribution can be recognized in the exposition of a grounded theory unless a direct quote is made to illustrate a concept, and therefore my approach is to gain consent for quotes. The caveats to using unplanned observations are: Firstly, that the individual cannot be identified from any field notes taken, and secondly that the notes are benign and therefore that the loss or misuse of the data would not cause the individual harm. The way to avoid both of these issues would be to code these observations mentally and memo on the codes; thus the abstraction of time, person and place protects the individual observed. A respected colleague once observed that collecting data and not using it could be regarded as unethical; in a similar vein, one could argue that for the grounded theorist, not using valuable, observed data, collected intentionally or unintentionally, as part of one's work or during specific research activities, could also be unethical.

Collecting "good" grounded theory data online

But what if the participants are not ethical; it is easy to lie online, what if they lie? Collecting 'good' grounded theory data to enable the production of a reliable and valid study is a concern of every grounded theorist. But what comprises 'good' quality data and how is this effectively argued? Phillips and Pugh address this concern of data quality when they write of 'data theory', which they describe as being "concerned with the appropriateness and reliability of your data sources" (2000, p. 61). We know that data must relate to the substantive area or population being studied, and can be collected by observation of an occurring event and from any public or private record irrespective of form (Glaser, 1998 p. 8). If the data collected is collected directly from the substantive area or substantive population, or is about the substantive population or substantive area, it can therefore be considered to be appropriate data.

But is it reliable data? If, by this, Phillips and Pugh wish to ascertain that participants are telling the truth, then a grounded theorist would argue that this interpretation of reliable is inappropriate to a grounded theory. For grounded theorists data types include:

- Baseline data; "the best description a participant can offer";
- properline data; that "which is what the participant thinks it is proper to tell the researcher";
- interpreted data; that "which is what is told by a trained professional whose job it is to make sure that others see the data his professional way" and;
- vaguing out; where "there is no stake in the participant in telling the researcher anything, so he just vagues out" (Glaser, 1998, p. 9).

If a researcher is told a deliberate 'lie' by an individual and if the researcher accepts the 'lie' as baseline data but does not recognize the 'lie'; either:

- The 'lie' will not pattern out, being an isolated incident; or
- it will be compared as a 'truth' with other 'truths' and be subsumed in a larger pattern; or
- it will be compared to other similar 'lies' and will pattern out and form part of the theory.

In the second case, an angle will be missed. In the latter case it is unlikely that the theory will explain the behavior in the substantive area and thus will be a failed grounded theory. The grounded theory method is a process: If the researcher adheres to the strictures of the method, he or she will produce a grounded theory, the quality of which will depend on the skill and competence of the researcher. In a case where a researcher is systematically lied to, is unable to recognize the 'lies', and thus fails to analyze the data accordingly, the competence of the researched is called into question. Researcher competence is, however, a different issue to that raised over whether the participant is telling the truth, and is the responsibility of the researcher, not of the method. Grounded theorists do, however, rely on data to pattern out for categories to be saturated. Therefore data needs to be sufficient. Herein lay my main concern as a data collector: Would participants tell me in sufficient detail what mattered to them most, online, where a property of online conversation is that every word is digitally encoded and potentially recordable?

Knowing whether the data collected is sufficient, whether the study is theoretically complete or not, is apparent at all points of a GT study. Principally the analyst will be able to recognize whether or not concepts are saturated. A concept is saturated when "comparing more incidents yield no more properties of the category and it is verified, saturation is achieved by the interchangeability of indices" (Glaser, 1998, p.141).The interchangeability of indicators means that, for any one concept, the next indicator found could be swapped with any of the current indicators with no effect, without further developing the concept, its name, its properties, dimensions or degrees. (Glaser 2003b, 1998, p.183). An unfinished study with unsaturated concepts will produce a thin theory offering incomplete explanations and missing relationships, and the researcher will know. Thus Glaser writes of data quality:

> In sum, no matter what type of data is obtained, the data is the data even if the researcher does not particularly care for it. It is his or her job to let the data emerge in its own right and induce its meaning. (Glaser, 1998, p. 9)

Therefore, it only matters that sufficient data is collected. It does not matter whether or not one of the participants lies. It does not matter if all the participants lie – although in this case it matters that the researcher recognizes the lies and analyzes accordingly, can "induce the meaning of the data," Glaser argued in the same passage quoted above. Whether or not the researcher has adequately induced the meaning of the data will be found in the post hoc evaluation of whether the resultant theory, by its own criteria, is relevant, fits, works and is modifiable and explains how protagonists resolve or process their main concern (Glaser, 1998, p. 18). It therefore follows that good grounded theory data addresses the protagonist's main concern and, in this circular argument, the way in which that data might be elicited online is discussed above.

Complementary online data sources

Interviews are but one source of data for the grounded theorist and it can be helpful to search the World Wide Web for other data. Location tools and data sources are constantly changing and an early step is to use a search engine to identify current tools. This is as simple as typing ""finding images" or "finding chat rooms" into, for example, Google (http://www.google.co.uk) or Yahoo (http://uk.search.yahoo.com). Google Alert (http://google alert.com) scans that part of the web indexed by Google and offers email updates of new content as defined by the keywords selected. News clipping services like World News (http://wn.com) and Netvibes (http://www.netvibes.com) scan the viewable "published" web. Locating data on the veiled web (in the sense that one needs to register to access that area of the web) is usually a matter of finding an appropriate site, (for example, Facebook (http://www.facebook.com) or the UK's Financial Times (http://www.ft.com) and then registering and using the hosting site's search tools. I have previously found useful images from Google Images (http://images.google.com); quotes from Facebook and Twitter (http://twitter.com) and videos from Youtube (http://www.youtube.com/). Blogs can be useful sources of data., as can chat rooms like Paltalk http://www.paltalk.com) for synchronous conversations, and forums for asynchronous conversations on specific or general topics. Google Blog Search (http://blogsearch.google.com) and Social Mention (http://socialmention.com) are helpful tools to locate relevant data on social networks.

One of the joys of the internet is that text is but one communication medium and both public and private data sources can include video, audio and images. Some sites allow communications to be shared privately among one or a group of people, making it possible to develop data collection techniques to supplement interviewing which could, for example, involve asking participants to keep an online journal, video diary or audio diary, or otherwise communicating through or with video, audio or images. Although data was not collected online, Krekelwitz (2010) used participants' images sensitively and imaginatively as a technique to find a way of talking about abuses suffered by participants when they were children. Nilsson (2007) used video to research "how and if people with profound cognitive disabilities could benefit from training in a joystick-operated wheelchair," evidence that these types of data have produced excellent grounded theories. If it is appropriate, convenient or necessary to collect data online, then images and text, video and text and audio and text can be shared privately with one person, or a group of people, or publicly at flickr (http://www.flickr.com), YouTube (http://www.youtube.com) and Ipadio (http://www.ipadio.com) respectively. Given willing participants who are able to use a camera, video recorder and/or a microphone with a computer, these systems are very easy to use, with first-class help systems, demonstrations and tutorials.

Conclusion

The internet is a liberating tool offering the opportunity to collect data from geographically or temporally dispersed participants. The corresponding disadvantage is that all must be connected to the internet and must be competent internet users to the degree of the communication tool used. Given a population which connects to the internet, can you get better data from face-to-face contact as opposed to online communication? Probably. Can you get better data from face-to-face contact from a geographically or temporally dispersed substantive population within a given time period? Probably not. Can you get sufficient data from connected online sources? Yes, I think you can.

References

Bergman, M. K. (2001). The deep web: Surfacing hidden value. Retrieved from http://beta.brightplanet.com/deepcontent/tutorials/DeepWeb/index.asp

Glaser, B. G. (1978). *Theoretical sensitivity: Advances in the methodology of grounded theory* Mill Valley, CA: Sociology Press.

Glaser, B. (1992). *Basics of grounded theory analysis.* Mill Valley, C.A.: Sociology Press.

Glaser, B. (1998). *Doing grounded theory: Issues and discussions.* Mill Valley, CA: Sociology Press.

Glaser, B. (2003a). *The grounded theory perspective II: Description's remodeling of grounded theory methodology.* Mill Valley, CA: Sociology Press.

Glaser, B. (2003b, April). A *teachable moment.* Paper presented at the Grounded Theory Research Seminar, London, England

Glaser, B. (2005a). *The grounded theory perspective III: Theoretical coding.* Mill Valley, CA: Sociology Press.

Glaser, B. (2005b, May). *A teachable moment.* Paper presented at the Grounded Theory Research Seminar, Mill Valley, CA

Hara, N., & Kling, R. (2000). Student distress in a web-based distance education course. *Information, Communication & Society 3*(4), 557-559.

Heard-White, M., Saunders, G., & Pincas, A. (2004). *Report into the use of CHAT in education. Final report for project of effective use of CHAT in online learning*. Institute of Education, University of London.

Krekelwitz, C. (2010). *Self-care of incest survivor mothers.* (Doctoral dissertation, University of Manitoba). Retrieved from http://hdl.handle.net/1993/4203

Nilsson, L. (2007). *Driving to Learn. The process of growing consciousness of tool use – a grounded theory of de-plateauing.* (Doctoral dissertation). Lund University, Sweden.

Mann, C., & Stewart, F. (2002). *Internet communication and qualitative research: A handbook for researching online.* London: SAGE publications.

Phillips, E., & Pugh, D. (2000). *How to get a PhD* (2nd ed). Buckingham: Open University Press

Salmon, G. (2000). *E-Moderating the key to teaching and learning online.* London, Sterling VA: Kogan Page.

Scott, H. (2007a). The temporal integration of connected study into a structured life: A grounded theory. In B. G. Glaser & J. A. Holton (Eds.), *The grounded theory seminar reader 2007* (pp. 145-162). Mill Valley, CA: Sociology Press.

Scott, H. (2007b). The temporal integration of connected study into a structured life: A grounded *theory.* (Doctoral dissertation). University of Portsmouth, England.

Sharf, B.F. (1999). Beyond netiquette: The ethics of doing naturalistic discourse research on the internet. In S. Jones (Ed.), *Doing internet research; Critical issues and methods for examining the net.* London: Sage Publications

Walsh Y.S., (2009). *A qualitative exploration of cultic experience in relation to mental health difficulties.* (Doctoral dissertation, City University, London, England)

6

USING VIDEO METHODS IN GROUNDED THEORY RESEARCH

Lisbeth Nilsson
Lund University

I work with people who have profound cognitive disabilities. In this chapter I will recount the challenges I faced, and continue to face, as I seek to research my practice with my clients and patients. Working with people who have profound cognitive disabilities means having to learn a whole new system of communication where meaning is primarily conveyed through behaviors not words. Conducting research in this area places huge demands on the sensitivity of the research method to enable the collection and analysis of nuanced meanings expressed in blunt behaviors. I will explain why I chose grounded theory (GT) and how I resolved a potential conflict between using audio-visual technology and the tenets of grounded theory. I will discuss data collection, and the techniques of data capture using audio-video, and data analysis. I will address ethical matters and summarize the issues of using audio-video data from a grounded theory perspective.

My research is carried out in the field of rehabilitation and occupational therapy, and more specifically in the substantive field of people with profound cognitive disabilities. Each of my clients and patients has the potential to extend their severely limited physical and mental capabilities and, as a facilitator, I use powered wheelchairs as the tool through which the patient or client can learn and develop. Each person comes to understand his hand as a tool which he can use to activate the joystick which will move the chair; mentally connecting this series of movements with setting the chair in motion marks the start of our journey together. I call it Driving to Learn™ (Nilsson & Nyberg, 2003, Nilsson, 2007 and 2010).

The project started in 1992 with two participants and ended with a total of 126 participants, all of whom were engaged in powered wheelchair use at some point. The grounded theory developed explains how people may exceed others' or their own expectations of development or learning – de-plateau – if the right properties for their de-plateauing are available (Nilsson, 2005, 2007, 2010). This work informs the development of the "practice guidelines for power mobility for individuals with cognitive disabilities" (Wang, Durkin, Nilsson & Nisbet, 2010) of the Posture and Mobility Group, an international group based in the United Kingdom.

Grounded theory and audio-video data capture

I became interested in grounded theory methodology, which is often used in the health sciences where people's activity and concerns are the focus of professionals in the field. Hand in hand with my selection of research method went the decision of how to capture the data, i.e. how to record it. When conducting interviews and observations, the recommended method is to record the data in field notes as text using pen and paper or word processor. Since I research my own facilitating practice, it is not possible to adequately fill both roles of facilitator and researcher at once. My initial approach to a practice session focuses on one-to-one interaction with a participant, at which time I make decisions on appropriate responses and adaptations to the participant's reactions and outcomes. In some practice sessions others facilitate with me, so there are potentially multiple interactions with the physical environment and the persons attending going on at any one time. I am not able to adequately record the interactions and responses that occur at many different levels during a session by field note alone. A video recorded session, on the other hand, provides a multi-facetted data source which permits analysis from many different perspectives, by many people, on many separate occasions, and over multiple time periods (Gallup, 1994). The use of audio-video data in this research project is therefore primarily led by the need to collect sufficient data which enables the analyst to develop an understanding of the participants' concerns despite their profound cognitive disabilities.

Additionally, an early problem for me was that sometimes I was not able to understand what I was seeing such that I could write a field note: I needed a way to help me notice and recognize what I was seeing, to help me learn the language of behaviors. Riley and Manias (2004) comment that video recordings can help viewers to recognize behaviors that occur simultaneously or over time. For example, fast-forwarding a video recording provides the researcher with a fast-moving picture that enables a new perspective on the recorded activity. Conversely, recordings viewed in slow-motion or frame by frame, provide unique possibilities to analyze and compare events or processes that pass by very quickly; critical activity sequences can be the object for comparisons with new recordings over and over to discover change or processes that otherwise would go un-recognized (Nilsson & Nyberg, 2003). In my research the participants' additional physical limitations mean that they are not able to communicate and express their experiences and feelings using spoken language. Their body language and facial expressions as well as their vocalizations are often vague and difficult to interpret. Viewing and reviewing the audio-video data to evaluate and re-evaluate, validate or discard possible interpretations of participants' reactions and interactions when they practice in a powered wheelchair is and was essential.

The choice to use audio-video methods of data capture, however, appeared to be in contradiction of the grounded theory method. Specifically Glaser (1998) recommends that researchers avoid audio recording and transcribing of interviews: chapter seven of *Doing Grounded Theory: Issues and Discussions* (Glaser, 1998) is dedicated to the topic. Among the reasons for advocating the avoidance of audio recordings and connected procedures is that transcription is either time-consuming to the researcher or incurs financial cost where a transcriber is employed. Glaser (1998) also contends that it is important that the grounded theorist takes field notes on her own data and thereby receives the stimulus to latent creativity and the development of methodological skill (p. 112). Where the transcription is carried out by a person other than the researcher/grounded theorist, this opportunity to create field notes and access his or her latent creativity is lost.

One of Glaser's most critical reasons for avoiding coding from audio transcripts, however, is to prevent the researcher from becoming overwhelmed in detail. The sheer volume of transcribed data gives rise to many incidents, locking the researcher into a huge pile of potential codes such that the researcher is unable to identify indicators and patterns in the data. Audio recording for complete coverage, so that nothing will be missed, obstructs the grounded theory process. Part of the problem may come from the edict that 'all is data' (Glaser, 2001, p.145). Glaser (1998) contends that, irrespective of the type of data, whatever "may come the researcher's way in his substantive area of research is data for grounded theory" (p.8). This liberating perspective means that "the researcher does not need to buy into any particular data as sanctified, objective or valid. He need only see what incidents come his way as more "data" to constantly compare, to generate concepts and to induce the patterns involved" (Glaser, 1998, p.8). This is not to say that all data should be collected. Instead, Glaser (1998) proposes that grounded theorists make field notes and move quickly to delimiting data collection and analysis once the core category has been emerged.

By using video recordings the researcher creates the risk that too much data is collected, potentially causing 'data overwhelm' (Glaser, 2003, p.24). Video recordings provide both audio and visual data and can generate vast amounts of data, particularly when more than one camera is used to simultaneously capture more than one perspective of an activity scene. There are, therefore, many aspects on which to focus but the choice of what data to extract and analyze may not be obvious from the start of a research project.

Interestingly, Riley and Manias (2004) note the problems of 'data overwhelm' during analysis and state that the methods of analysis of video data are undeveloped and inadequately described in published articles. This stance is confirmed by several authors from different fields using video methods in their research (Janhonen & Sarja, 2000; Goldman, Pea, Barron, & Derry, 2007; Yu Zong, Smith, Park, & Huang, 2009; Durkin, 2009). These authors have described the difficulty of analyzing the data due to the richness of information captured by this medium. Their common feeling is that the greatest difficulty is to decide what to look for and what to focus on. There is also an awareness of how the selection of meaningful portions of the video stream is relevant to the analytical process (Parmeggiani, 2008). Some of these concerns may have stemmed from the field of anthropology, where researchers have learned how to handle and analyze the written word, but are not used to recognizing multiple forms of communication and behavior in a video recording of lived context (Hall, 1984).

Computer coding software is unlikely to solve the problem of too much data. In many cases using transcriptions and coding software leads to a proliferation of codes, which can obstruct the emergence and discovery of latent patterns and can also lead to the analysis being superficial and shallow. Glaser has argued that the fast technological categorization and sorting is considered to facilitate a descriptive approach and hamper the development of creativity and the skill of conceptualizing and doing grounded theory; the ability to find connections, latent concepts and to do the integration is a process which takes place in the researcher's own mind, not in the computer (Glaser, 2003, p.17). The use of such data collection and analysis techniques is therefore not recommended in GT methodology.

I therefore agree with Glaser's reasoning (1998, 2003) about the disadvantages of audio recording, a reasoning that can also be applied to the use of audio-visual recordings. However, there are situations, such as this one, where, if video recording is used thoughtfully, video data can operate as primary data. The exciting opportunity for video recording as primary data is in those situations where the researcher has to learn to notice what she sees and to sensitize herself to the data. In those situations video data can contribute to the emergence and generation of concepts by providing an audio-visual view of the action in a field of interest that can be repeatedly inspected by one or more theorists, first to notice what they see and second to enable comparison of incidents. In these cases, data is recorded not to enable descriptive completeness or verification, but to enable conceptualization – for the generation of concepts to be integrated into a hypothesis or theory (Glaser, 1998 p.107).

In 2003 when I attended my first trouble-shooting seminar with Barney Glaser I had used video data for more than a decade. I brought some video recordings to visualize my research as part of the trouble-shooting; in my experience people were not generally able to get an image of the participants and their activities in a powered wheelchair solely from my descriptions. When I tested the video-player before the seminar started, one of the other participants ventured that recordings should not be used in a GT study. And of course that was true, in respect of audio recording interviews, but I used the video recordings to capture the activity of people using non-verbal communication, their interactions with a tool (the joystick on a powered wheelchair) and their interaction with other people in the context of practicing powered mobility.

After viewing the video recording, Barney Glaser made the comment that in this field there is no other choice for collection of data (personal communication, September 23rd, 2003). Since the participants could not speak, the only original information explaining what was happening in the field of interest was their activities in the powered wheelchair. Interview data with people closely related to the participants would complement video data but would not be sufficient in itself: interview data alone could not give the information needed to understand the process that the participants underwent while practicing in powered wheelchairs. Moreover, people with profound cognitive disabilities develop at an extremely slow pace. Thus, for an observer it is almost impossible to remember and compare small shifts and nuances in behaviors and activities without the assistance of video data to refresh the memory. It is also difficult to develop the sensitivity needed to recognize and interpret very small changes and variations in those participants' non-verbal communications, activities and interactions.

Non-speaking participants is a new field for the application of GT methodology. Thus, Glaser (personal communication, September 23rd, 2003) modified his recommendation to avoid audio recording, since collecting audio-video data was the only method of data capture that could adequately enable the identification of incidents for analysis given the abilities of the participants acting in this specific field of interest.

Data collection and data capture

Data capture refers to the process and techniques of collecting the raw data from which incidents will be identified. Since a skilled grounded theorist aims to record incidents in field notes, data capture is an additional step in the data collection process. My preferred method of preparing field notes is to make notes immediately after the practice session and use these to guide which sections of the video to view first. I supplement the original field notes as I view the relevant sequences, before moving on to view the remaining video. I view and re-view the remainder of the tape seeking new incidents to record as further field notes.

Grounded theorists who plan to use video recording to capture data need to be clear of why they wish to do so and how to apply video methods in their chosen setting. They also need ideas about which equipment to use, how to carry out the recording and what scenes or events to capture in order to get useful data which can inform the researcher about the main concern of the actors in the field. General guidelines regarding use of video methods are very much concerned with pre-planning and pre-setting of conditions for recording (Derry, 2007). However, this level of detail is not the main concern when video methods are used in GT research. The primary concern is the ability to make constant comparisons of video data from many activity sequences under varying conditions such that the research can uncover latent social patterns in the substantive field of interest.

When planning and organizing the collection of video data for a grounded theory study it is important to have an initial plan. The researcher may wish to record different perspectives such as overviews of individual or group interactions or narrow views of activity with hands or of facial expressions during execution of an activity. Recording may also involve the use of two or more cameras capturing different angles and perspectives of an activity context (Fielding, Lee, & Blank, 2008). For example, Durkin (2006) used four small cameras mounted on a powered wheelchair to capture different perspectives of a child's learning to use it. Also, Xiao, Seagull and Mackenzie (2004) used two ceiling-mounted cameras, and a head-mounted camera worn by a team member, to get an overview of a trauma resuscitation team at work. The researcher's plan might therefore outline when, where and how to collect video data, which perspective(s) to use for recording the scene, who should operate the camera and how much data should be collected. The researcher should be prepared to modify or change the plan as analysis indicates new directions for collection of data.

Financial constraints for my research project initially forced me to position and operate the video camera during practice sessions. Later on in the project, as I became more experienced at videoing sessions, it proved beneficial to be in control of positioning the camera and choosing which perspective to use for the recording and from what distance. Currently, my preferred technique is to place a video camera equipped with a wide-angle lens on a tripod, on the periphery of the activity context. The aim is to get the best possible shot of the whole scene, without being overly preoccupied with the various features of video equipment or with the art of producing video films. The aim is thus to capture the ongoing interaction and communication between people and objects in the context of interest, with the least amount of intrusion. Sometimes the wide-angle lens is substituted with a zoom lens, and while this works it also needs comparatively more attention to use, although using a remote control device helps. Overall, this approach to recording helps me to avoid the risk of having too narrow a perspective of the activity in focus (Riley & Manias, 2004). I also find that it is important to let the camera run on even though nothing of interest appears to be happening, as later analysis can often reveal valuable and meaningful sequences.

Today digital video recordings, which can be stored on CD or DVD, allow for easy moment-by-moment access to marked key events and require relatively little physical storage space compared to taped recordings. However, a computer with special software might be needed for viewing and editing the footage. The fast pace of development of new equipment for recording, storing and editing means that there are more and more technological issues to take into consideration before deciding how to use video data in research. As the development of technology changes rapidly, my advice is to search the internet for information on the latest recommendations regarding equipment suitable for research purposes, and online guidelines, such as Derry (2007), can also be of great help.

Data analysis
The aim of the practice sessions is to facilitate participants' curiosity and initiative and to encourage them to explore the function and use of the powered wheelchair. The initial focus for analysis is therefore participants' reactions and behaviors when encountering the powered wheelchair, which is to them a new and unfamiliar piece of equipment. Early analysis involves intense viewing of the video recordings and requires one to be constantly alert and to stay open to all observable changes in the appearance of the participants as well as all changes in their reactions, behaviors and activities. People with a limited repertoire of communicative behaviors may use the same behavior in different situations with different meanings. Having linked one behavior with one particular meaning, it is therefore important not to assume that this meaning is always linked with this behavior. When the behavior is next recognized, one must put the first interpretation aside and interpret anew and then compare the new interpretation with previous interpretations. I have understood five different meanings for one example of communicative behavior and there could be more. Initially the descriptors of what I have observed are used as a base for interpretations of the participants' experiences and their reactions on interaction with the powered wheelchair. For example one participant was holding his arms close to his body when seated in the chair: his behavior was interpreted as a protective act due to the new and unknown situation. Another participant repeatedly hit the joystick, banged the tray or stamped on the footrests – his behavior was interpreted as explorative acts intended to find out what effects his actions could cause.

At first it was extremely hard to recognize any pattern between the different participants' behaviors and sometimes realization was delayed. For example, a certain act of mimicry was observed together with rocking of the upper body. "Rocking of body" was very obvious and quickly coded as "wants the chair to move". The incident "Mimic with mouth" (as when someone mouths the otherwise onomatopoeic "vroom-vroom" of a car) was less easily understood at first. But later, I did recognize what was going on and the observation was changed to "mimic showing knowledge of chair as moving thing". An incident might also come from a re-activated visual memory. Sometimes new incidents made me recall visual memories that I had previously not noted as incidents but which I recognized as such in the light of new incidents.

Possible interpretations of new indicators or changes in any category were also discussed with people closely related to the participants. Parents, staff and professionals were invited to watch short video sections capturing what I considered to be key events, and then offered their interpretations of what they had seen, based on their knowledge of the participant. We worked together, each re-viewing the recording and offering possible interpretations of the participant's recorded appearances and activities. Observations and interpretations of expressions of emotion, engagement and endurance, striving for mutual interaction, stimulation of own initiatives, adaptation of behavior and use of resources and environment were discussed individually or in supervision groups with the other facilitators. This help was invaluable, as it not only extended and endorsed or corrected my interpretations, but also helped me to become more acquainted with the participants who were not able to use language to express themselves.

At the beginning of the project a recording was made for almost every practice session, which meant around two hours of recording per week per participant. After a period of four months recordings were reduced to once every two weeks. Later on participants were recorded once a week the first month, twice a month the following six months and then four times a year. The sheer volume and complexity of the data can become overwhelming. Opinions diverge on if and how to computerize video recorded information in the field of qualitative research. Computer programs have been designed to assist with the analysis of qualitative multimedia by indexing, coding, annotating, structuring and editing video recordings; some are said to have been designed especially for use with grounded theory methodology (Parmeggiani, 2008). Computerizing and storing raw video recordings and edited video clips however, can take up large quantities of hard disc space. It also takes time to learn how to transfer the recordings to the computer and how to use the software. My advice is to use new technologies as long as using them will not be overly time-consuming and to make sure that the technologies are not allowed to dominate the procedure. It is also very important not to convert raw data to fragments. Later it may be necessary to view the data from a new perspective in order to view a particular incident in its context. If a sequence of events is fractured into many parts, rebuilding the sequence is likely to be difficult since manipulating fragments out of context quickly leads to confusion.

During analysis and at the point of "data overwhelm" the risk is that one withdraws into single-mindedness and just looks at the obvious. One way to overcome this is to focus on the detail that first catches the eye when viewing a recording – it might be mimicry or movements, or a certain way of keeping or avoiding eye contact. There is a debate in the video methods literature as to whether to transcribe or not and in what manner to do so (Goldman et al, 2007; Derry, 2007). In my experience it is important to translate the captured activity into words in the early stages of the research, and to make very thorough transcriptions of the observed activity, in order to get sensitized to seeing what sticks out in the mass collection of micro-events. This sensitizing helps one become aware of small details, changes and deviations in the action on the video recording.

Over time one becomes more sensitized to seeing incidents and the recognition of indicators becomes easier. The need to transcribe recordings dissipates as it becomes easier to make field notes from the recordings and the field notes become shorter and shorter. Human interaction and communication are performed at multiple levels, and each observer may see a slightly different world to everyone else, depending on what they have learned to see over their lifetime (Hall, 1984). I learned to overcome learned and preconceived ideas of what aspects to observe, and developed an ability to see in a multi-channeled way, meaning that my focus can be on more than one dimension simultaneously. One develops the talent to recognize communication and interaction at other levels than the obvious activity and verbal interplay which include nuanced expressions such as mimicry, gestures, prosody, pauses and emotional intonations. These recognitions are held in my memory as snapshots, or as short sequences, and bring color to the analysis. These abilities speed up analysis in two ways: firstly, fewer viewings of a particular recording are necessary in order to notice an incident and secondly, one becomes more adept at noting and comparing incidents.

Nevertheless, analyzing the practice context is very complex. Observing and field noting all the details in the different actors' interactions in order to make interpretations on how these interactions interplay and influence the outcome of a practice session is challenging. There are numerous details to attend to such as: posture, body language and facial expressions; signs of attention, focus and awareness; communicative behaviors, vocalization, speech and intonation; temporal pace and use of space; modes of touching, manipulating, handling and performing; and emotional reactions. When analyzing, focus on each participant's communicative expressions such as wording, prosody, facial mimic and body language. Also focus on the activity context and the way the participants approach, relate to, interact with or react to objects or those taking part in the practice session, for example the facilitator, the researcher, family and other staff or professionals. All these events are of interest when seeking to understand the participants' main concern and its resolution in the substantive field of interest.

In my study constant comparisons were made between field note and field note, and between field note or visual or audio memory to video recordings and developing codes. New observations from a video recording were also compared to older field notes or field notes from interviews. For example the incident of body-rocking together with the mouth-mimic was compared with an interview with acquainted persons who told me that the participant typically did this combination of activities, sometimes with sound added, when he was seated in a car. This strengthened the interpretation of the observations. The indicators and concepts abstracted by constant comparative analysis included posture; body movements; use of hands and feet; facial expressions; alertness; emotions; eye-contact and eye-fixation; tempo and pacing in interaction with objects and other people; understanding of cause-effect relationships; and own initiations of activities.

Since it is the emerging theory that influences where the researcher should sample next, any initial plan directing data collection should be modified as constant comparisons indicate new directions for data collection and analysis. This means being flexible about what should be recorded in familiar scenes, which new scenes to include and from which perspective to analyze the data. Theoretical sampling thus led to the inclusion of more participants with divergent characteristics related to age, level of cognitive deficit, degree of physical disability, additional motor and sensory disabilities as well as the observation of typically developing infants. The analysis of infant activities in a joystick-operated powered wheelchair provided a secure framework for comparison and organization of indicators and concepts extracted in previous analysis of participants with cognitive disabilities. The comparison with the typically developing infant activity provided a valuable stepping stone for the emergence of indicators of achievements regarding joystick use in participants with cognitive disabilities.

Guided by the analysis, I came to realize that facilitators impact upon a practice session and that I therefore needed to theoretically sample for facilitator impact. As the study progressed I had also made changes to my practice, substantially developing my facilitator methods and training other facilitators in the revised method. I realized that not only were the participants learning but also the researcher and the facilitators. I therefore revisited the video data to analyze interactions from six further perspectives: participant and researcher (participant's learning and researcher's learning); facilitator and participant (participant's learning and facilitator's learning); researcher and facilitator (facilitator's learning and researcher's learning). The theory explaining the eight phase learning process was already developed but analysis from these additional perspectives developed the theory and further explained that all participants go through a similar process of learning but at different levels of performance, abstraction and cognitive execution. The analysis of the interplay between the different actors took the conceptualization to a higher level, which leads me to claim that video data not only assists recognition of complicated indicators but also facilitates theoretical sampling in highly complex social situations.

Ethical considerations

Data capture techniques using audio-video are more active and intrusive than traditional methods of observation, and the recorded participant's lack of anonymity and privacy requires special considerations regarding how to use and edit video data (Riley & Manias, 2004). It is essential to provide participants with correct and accurate information about the purpose of the research and its possibilities and risks regarding confidentiality, conflicting interests and researcher safety (Morse, 2007). Also it is important to obtain the participant's or their spokesperson's informed consent of participation (Broyles, Tate & Happ, 2008). In the case of minors or vulnerable populations with decreased decision-making capacity, such as people with severe illness or cognitive disabilities, the ethical considerations are even more important (Stineman & Musick, 2001; Broyles et al, 2008). Derry (2007) also recommends the inclusion of statements of how the recordings will be safely stored, how long they will be kept and for which purposes they will be used. In my studies the participants signed a special document with two options. The first was the acceptance of use of the video recordings as data for the actual research only; the second was a voluntary agreement to utilizing parts of video recordings for demonstration, education and research in other communities. This approach to participant decision regarding the use of recordings is also recommended in recent guidelines on ethical concerns in video data collection (Derry, 2007).

Video data and grounded theory

Glaser (1998) contends that real-time interviews and observations of embodied behavior go together to unveil people's main concern in the field of interest (p.109). Video data is a multidimensional source providing a multichanneled flow of different dimensions or aspects that can be analyzed from multiple perspectives. Video data visualizes personal dimensions such as communication, emotion, motor performance, activity and physical and social interactions with the environment (Goodwin, 2003). Video data also visualizes spatial and temporal aspects that influence the personal dimensions and the inter-personal activities and interactions. Finally, more than one activity and/or interaction can take place simultaneously in a sequence of video. Hall (1969, 1973, 1984) agrees, arguing that film and video data provides multiple perspectives from which to analyze people's behavior and activity within the context in which they are active, in relation to how people use space (1969), time (1984) and how they relate to each other in different ways depending on their culture and when and where the activity takes place (1973). Since these perspectives are also important for understanding the concerns and behaviors of people in a substantive field of grounded theory research, it follows that video data can be extremely valuable in grounded theory research.

Each grounded theorist who considers using video data has to decide on what are likely to be the most fruitful ways of getting access to, or capturing recorded data in their specific field of interest. Too much or too little video data may hamper the emergence of patterns or theory. Use of computer software for the analyzes may speed up the actual analysis but at the risk of superficiality. The time used for learning the software may be better spent engaging with the data and the process of constant comparison. As Glaser (1998) warns: be aware of the technological traps (p.185). Video methods may be most useful and valuable if used in a considerate and thoughtful way, which respects both the participants and others captured on the recordings as well as the research aim. My view is that, by using technology such as video methods in their substantive field of interest, GT researchers can gain new insights into how people resolve their main concern in a specific context. Whether or not video data should be used depends on how the data are used for analysis and not on which methods are used for data collection and storage.

General considerations

Video data may be the primary source of data or may also complement other data collected in a research project. It may originate from different milieus such as physical events, video-conferences, networking over the internet, uploaded video clips from websites, or surveillance of environments such as city squares or airports. Or video data might be generated by the participants themselves, for example, one approach is to provide the participants with an open question and leave the choice of camera perspective and which events to record to them. These video diaries may unveil a participant's personal view of life over an extended period of time (Buchwald, Schantz-Laursen, & Delmar, 2009) or particular life experiences such as living with chronic illness (Rich, Lamola, Gordon, & Chalfen, 2000). Similarly, video recorded interviews with participants in their home environment have provided insights into their views of how the setting up of the private milieu influences the experience of having a life threatening illness (Downing, Jr., 2008 a), 2008 b)). Furthermore analysis of video data can lead to extracting important elements which can be used as the subject for further conversations. There are many ways in which video data may be collected and ultimately the choice of which method to use is dependent on the origin of the data, the goal inherent in the research methodology and the researcher's familiarity with digital technology. The relevance of transcribing or computerizing video data is similarly dependent on the research approach, differences in the participants' communication, language or culture and the aims of studying a specific context. Lastly, video data may be viewed, compared and analyzed not only by a lone researcher or a research team but also by a participant or a group of participants, separately or together. Who the viewer will be, depends on the context complexity, the activity in focus and the people acting in the field of interest.

Conclusion

Video data can be a marvelous source of information about people's activities and interactions in a particular field of interest. The permanence of the recordings allow them to be viewed by one or more persons, one or more times, in its entirety, repeatedly in short sections as clips, or collections of clips presenting meaningful portions of the video-stream, from single or multiple occasions of recording. The density, richness and multidimensionality of the raw recordings allow for analysis from many different perspectives of communication, activity, interaction with objects and others. Even recordings from surveillance of public spaces may be a data source to a grounded theorist. In my research project the opportunity to go back and compare a participant's new activities with earlier activities was invaluable, as was the opportunity to select specific parts of video recordings and have others view and interpret possible meanings of changes in activities. Without the audio-video recordings the subsequent assessments of each participant's achievements would have been impossible. With the recordings I have engaged in ground-breaking research and developed a theory that is internationally recognized (Nilsson, 2005, 2007, 2010). In doing so, I honor my debt to grounded theory and have demonstrated both the flexibility of the grounded theory method in its ability to absorb and adapt to new ideas regarding data collection, and its power in explaining complex social situations.

References

Broyles, L. M., Tate, J. A., & Happ, M. (2008). Videorecording in clinical research: Mapping the ethical terrain. *Nursing Research 57*(1), pp. 59-63.

Buchwald, D., Schantz-Laursen, B., & Delmar, C. (2009). Video diary data collection in research with children: An alternative method. *International Journal of Qualitative Methods, 8*(1), pp. 12-20.

Derry, S.J. (Ed.). (2007) *Guidelines for video research in education. Recommendations from an expert panel. Data research and development center, NOROC at the University of Chicago.* Retrieved from at http://drdc.uchicago.edu/what/video-research-guidelines.pdf

Downing Jr, M.J. (2008 a). Why video? How technology advances method. *The Qualitative Report, 13*(2), pp. 173-177.

Downing Jr. M.J. (2008 b). The role of home in HIV/AIDS: A visual approach to understanding human – environment interactions in the context of long-term illness. *Health & Place, 14*(2), pp. 313-322.

Durkin, J. (2006). *Developing powered mobility with children who have multiple and complex disabilities: Moving forward* (Doctoral dissertation). Department of Health Professions, Clinical Research Unit, University of Brighton, England.

Durkin, J. (2009). Discovering powered mobility skills with children: 'Responsive partners' in learning. *International Journal of Therapy and Rehabilitation, 16*(6).

Fielding, N., Lee, R.. M., Blank, G. (2008). *The Sage handbook of online research methods.* SAGE Publications Ltd.

Gallup Jr., G. G. (1994). Self-recognition: Research strategies and experimental design. In Parker, S.T., Mitchell, R. W., & Boccia, M. L. (Eds.), *Self-awareness in animals and humans.* U.S.A.: Cambridge University Press.

Glaser, B. (1998). *Doing grounded theory. Issues and discussions.* Mill Valley, CA, U.S.A.: Sociology Press.

Glaser, B. (2001). *The grounded theory perspective: Conceptualization contrasted with description.* Mill Valley, CA, U.S.A.: Sociology Press.

Glaser, B. (2003). *The grounded theory perspective II: Description's remodeling of grounded theory methodology.* Mill Valley, CA, U.S.A.: Sociology Press.

Goldman, R.., Pea, R.., Barron, B., & Derry, S.J. (Eds.). (2007). *Video research in the learning sciences.* Routledge.

Goodwin, C. (2003). The semiotic body in its environment. In J. Coupland & R. Gwyn (Eds.), *Discourses of the body.* New York: Palgrave/Macmillan, (pp. 19-42.)

Hall, E. T. (1969). *The hidden dimension.* New York: Anchor Books.

Hall, E. T. (1973). *The silent language.* New York: Anchor Books.

Hall, E. T. (1984). *The dance of life.* New York: Anchor Books.

Janhonen, S., & Sarja, A. (2000). Data analysis method for evaluating dialogic learning. *Nurse Education Today. 20*, pp. 106-115.

Morse, J. (2007). Ethics in action: Ethical principles for doing qualitative health research. *Qualitative Health Research, 17*, pp. 1003-1005.

Nilsson, L. (2005, July). The discovery of a theory of de-plateauing. How to use video-recordings to collect data during studies with people not able to speak or to use alternative communication. (Grounded theory methodology, Session B: Health care, Thursday July 7), *The 37th World Congress of the International Institute of Sociology*, Stockholm. Retrieved from www.lisbethnilsson.se/presentations.htm

Nilsson, L. (2007). *Driving to Learn. The process of growing consciousness of tool use – a grounded theory of de-plateauing.* (Doctoral dissertation). Lund University, Sweden.

Nilsson, L. (2010, May). Driving to Learn™ an intervention for people with profound cognitive disabilities. *The 15th Congress of the World Federation of Occupational Therapists,* Santiago, Chile. Retrieved from www.lisbethnilsson.se/presentations.htm

Nilsson, L., & Nyberg, P. (2003). Driving to Learn. A new concept for training children with profound cognitive disabilities in powered wheelchair. *American Journal of Occupational Therapy, 57*, pp. 229-233.

Parmeggiani, P. (2008). Teaching different research methods through the use of video analysis software for media students: A case study. *International Journal of Multiple Research Approaches, 2*(1), pp. 94-104.

Rich, M., Lamola, S., Gordon, J., & Chalfen, R. (2000). Video intervention/prevention assessment: A patient-centered methodology for understanding the adolescent illness experience. *Journal of Adolescent Health, 27*, pp. 155-165.

Riley, R..G., & Manias, E. (2004). The uses of photography in clinical nursing practice and research: a literature review. *Journal of Advanced Nursing, 48*(4), pp. 397-405.

Stineman, M.G., & Musick, D.W. (2001). Protection of human objects with disability: Guidelines for research. *Archives of Physical Medical Rehabilitation, 82*(2), pp. S9-S14.

Wang, R., Durkin, J., Nilsson, L., & Nisbet, P. (2010). Empowering children and adults with cognitive disabilities to learn skills for powered mobility: principles, evidence and recommendations. (Workshop, Best Practice guidelines BPGW10), *The 4th International Interdisciplinary Conference on Posture and Wheeled Mobility*, June 7-9 2010, Glasgow, Scotland. Retrieved from www.lisbethnilsson.se/presentations.htm

Xiao, Y., Seagull, F.J., & Mackenzie, C.F. (2004). Adaptive leadership in trauma resuscitation teams: a grounded theory approach to video analysis. *Cognition, Technology & Work*, 6, 158-164.

Yu, C., Zong, Y., Smith, T., Park, I., & Huang, W. (2009). Visual data mining of multimedia data for social and behavioral studies. *Information Visualization*, 8(1), 56-70.

7

Developing Grounded Theory Using Focus Groups

Cheri Hernandez
University of Windsor

One of the most important decisions considered by researchers when crafting a study design is the determination of the way(s) in which data will be collected to provide the greatest likelihood of meeting the study objectives or answering the research questions. Major types of data sources include: interviews of participants (internet-based or mailed surveys, individual interviews, telephone or 1:1 interview schedules, focus group method); observations (from observer only to full participation in the phenomenon being studied); as well as other written, oral or video resources (documents, chart reviews, library and web-based resources, and archival materials). Grounded theory is an inductive research method for the generation of substantive or formal theory, using qualitative or quantitative data generated from research interviews, observation, or written sources, or some combination thereof (Glaser & Strauss, 1967). Therefore, researchers can use any type of data, although most classic grounded theory researchers use interviews exclusively or in combination with one or more of the above data sources.

When interviews are used for data collection, the usual approach to grounded theory is to interview research participants, one at a time, with data analysis ongoing during and between interviews. This chapter is about an alternative approach to discovery of grounded theory using focus groups as the interview method, yet remaining true to grounded theory methodology. For the focus group method to be accepted as a useful technique for grounded theory methodology, three criteria for relevance/congruence must be considered – data source, method congruence, and grounded theory product. In other words, focus group interviews must be shown as relevant data sources; focus group technique should be congruent with grounded theory methodology, and a substantive grounded theory must be an achievable product when using the focus group interview method. All three of these criteria will be addressed in this chapter.

The focus group method is a generic research technique (Merton, Kendall, & Fiske, 1990, p. xxvii) used to collect qualitative data (Côté-Arsenault & Morrison-Beedy, 2005), not a research methodology. Therefore, grounded theory researchers need not fear that they are mixing research methodologies that may not be compatible. Glaser has repeatedly articulated the phrase "All is data" (Glaser, 1998), and has also specified it as a Glaser "dictum" (2003, p.

167). In addition Glaser (2001) has asserted that "interviews of various sorts" (p. 140) are among the options the grounded theory researcher has for collecting data. Therefore, the data from a data collection method of focus groups interviews is arguably a relevant type of data for use in grounded theory research. In addition, Glaser has recommended doing grounded theory by using datasets of earlier research, that is, secondary analysis (1978, p. 53; 2001, p. 55); consequently, previously collected focus group data have always been relevant sources of research data for grounded theory researchers. As early as 1967, Glaser and Strauss recognized that various researchers would encounter circumstances in their research environments that would lead them to use different strategies or techniques when using grounded theory methodology (p. 77) and this encouragement of new directions was later reiterated by Glaser (1978, p. 158). Therefore, in this chapter, the use of focus group interview data as original data source is proposed for use in generating a substantive grounded theory.

This chapter will describe how the focus group interview method can promote discovery of grounded theory, and the beginning explication of a substantive theory of reciprocal restricting in women with cardiac (heart) disease will be used to illustrate this process. First, a brief review of the history of the focus group method will be delineated to provide the necessary context to understand the usage, challenges, and benefits of the focus group method, in doing grounded theory research. The chapter will conclude with an overview of limitations and pitfalls that can be avoided, and potential benefits of using the focus groups to collect data in grounded theory research.

Historical development of focus groups

The focus group research method began as the focused interview for groups (Merton, Fiske, & Kendall, 1956) and was utilized by Merton and Kendall as early as 1943. The original intent of focused group interviews was "to assess an experience, not to tap a general opinion" (Merton et al., 1990, p. 25); however, the types of experiences to be appraised were participants' prior responses to a one-time stimuli, such as film, radio or print materials on topics like war. "The primary objective of the focused interview is to elicit as complete a report as possible of what was involved in the experience of a particular situation" (Merton et al., 1990, p. 17). For Merton's research team, "Qualitative focused group interviews were taken as sources of new ideas and new hypotheses, not as demonstrated findings with regard to the extent and distribution of the provisionally identified qualitative patterns of response" (p. xxii). Conversely, when focus group interviewing is conducted during grounded theory research, the researchers are looking for the problem and its resolution in ongoing situations such as experience of chronic illness, family issues, work life or organizational experiences, and so on. Some of the differences between classic grounded theory methodology and the focused group interview as first described by Merton et al. (1990) are pointed out in Table 1. These differences do not preclude the use of focus group interviews for grounded theory research but are important areas for researchers to consider and may warrant slight modifications in the conduct of these interviews.

Merton et al. (1990) outlined major tenets of their original approach. The following aspects of their original method continue to be utilized by researchers to this day: recruit groups of 6–12 participants (p. xxv); form homogenous groups for greater productivity (p. 137); adopt a non-directive approach to allow participants to respond regarding what is significant to them (p. 14)– that is, do not structure the situation or the response (p. 34); create an informal atmosphere (p. 38) much like a conversation among people involved in the same situation (p. 151); set up a circular seating pattern to promote group interaction and an informal atmosphere (p. 140); give introductory remarks to include the acknowledgement that there may be disagreement between participants due to diversity and there are no correct or incorrect answers (pp. 146, 147, 175); prepare an interview guide with general topic areas but not fixed questions to which one rigidly adheres (p. 48); remain on a topic area until participants have exhausted all relevant responses (p. 76); and have a skilled facilitator who recognizes typical problems and has the knowledge and skill to cope with each of these (p. 17). The above tenets are reasonably consistent with the type of approach that grounded theory researcher's use to create a comfortable and open atmosphere for participants they interview, and these tenets are also congruent with facilitating healthy group dynamics in a research situation. However, in grounded theory research methodology the approach is even more non-directive, with participants directing the interview through their responses to and ensuing discussion of, open-ended "spill questions", and the freedom they have to direct the interview in "the way that it should go" to explain their experiences.

Over the seven decades since the focus group interview technique was developed, various changes in the purpose and use of this method have been derived by researchers from many disciplines, as they have extended the uses of this technique to suit their research needs. No longer is the purpose of the focus group method limited to understanding the impact of one-time prior experience, but purposes vary from a one-time current experience (e.g., some types of market research) to evaluation of current experiences (e.g., for accreditation of educational institutions) to research that delineates, describes, or explains past and ongoing experiences of institutions, situations and chronic illnesses (e.g., the experience of cardiac disease in women). The research purpose will dictate changes in other aspects of focus group technique, such as number of participants per group, duration of the focus group, type and source of interview questions, focus on group consensus or multiple perspectives, and a variety of purposes, or desired end products of such interviews, including evaluation, description, explanation or development of a research instrument or intervention. The next section will delineate the way in

Table 1. Comparison of Focused Group Interview and Classic Grounded Theory Methodology

Pertinent Characteristics	**Focused Interview for Groups (Merton, Fiske & Kendall, 1990)**	**Classic Grounded Theory Methodology (Glaser & Strauss, 1967, Glaser, 1978, 1992, 1998, 2001, 2005)**
Purpose	Elicit as complete a report as possible about participants' prior responses to a one-time stimulus situation (p. 21).	Discovery of theory about a core category (Glaser, 2005) from data. Determine the problem and its resolution in a particular substantive (empirical) area (Glaser, 2001).
Type of Situation Explored	Recall of previous experiences to a stimulus event such as a film, radio broadcast or specific print material.	Begin with no preconceived problem (Glaser, 1992, 2001). Ongoing experience of illness, family, organizations, and other situations.
Review of the Literature	Pre-analysis done of the situation in which the subjects have been involved (p. 4). Researcher has already analyzed "the hypothetically significant elements, patterns, processes & total structure of this situation" (p.3).	Delayed until emergence of the core category (Glaser & Strauss, 1967, 1998). When literature review is mandated by academic contingencies, the researcher strives to be theoretically sensitive by holding all preconceptions in abeyance.
Interview Questions	Have an interview guide of pertinent topic areas and questions that reflect hypotheses derived from the pre-analysis of the situation (p. 41). Not a rigid interview schedule – allowance made for unanticipated responses (p. 48).	Spill questions used to get participant to begin, and then relevant questions will emerge as interviews proceed (Glaser, 1978).
Final Product	Full-bodied reporting of the range, specificity, depth, and personal context involved in the stimulus situation and responses to it, considered "essential to understanding of the nature & meaning of the responses" (p. 21)	Emergence of the problem and its resolution is what is pertinent (Glaser, 1978, 2005) Write-up of the relationships among the core category and its properties and the subcore categories (Glaser, 1978, 1992).

which the focus group technique was used to produce a substantive grounded theory in a study of women with cardiac disease.

Focus Groups in Grounded Theory

Focus group interviews have been recommended as appropriate for generating theory (Morgan & Krueger, 1993), a purpose that is congruent with that of classic grounded theory. In addition there are several potential benefits, described later, that have increased the usage of focus group interviews by grounded theory researchers. In this section, issues for consideration when doing classic grounded theory with focus group as data collection method will be delineated. These issues are discussed in the context of their use in a study of women living with cardiac disease. Two important aspects of grounded theory methodology are attention to theoretical sensitivity (Glaser, 1978) and use of the grounded theory's constant comparative method of data analysis (Glaser, 1965; Glaser & Strauss, 1967). These aspects remain unchanged when using the focus group method to derive grounded theory, therefore maintaining the integrity of grounded theory methodology.

Prior to choosing to use focus group interviews, grounded theory researchers must understand and utilize a variation of this technique that is most congruent with the development of a substantive grounded theory. Unlike most researchers, the grounded theory researcher begins with a general curiosity to know more about a particular area but does not have definitive research questions or stated (a priori) hypotheses (Hernandez, 2010). In a study of women who had experienced a heart attack or had angina, the objective was to understand the experience of living with cardiac disease. The original design of this study was to have a series of focus groups to collect data on the experience of heart disease in women, and a thematic analysis was planned. However, the researcher was an experienced grounded theory researcher who used grounded theory's constant comparative method of data analysis to analyze the focus group data.

Planning phase

Planning for focus group interviews takes more time than for individual interviews and recruitment is time-consuming (Krueger, 2006). The challenge is finding a date and time that will accommodate participants' work, family, and leisure schedules. Over-recruitment of participants is important due to the rate of "no shows," with anywhere from 10 -25% over-recruitment being recommended (Halcomb, Gholizadeh, DiGiacomo, Phillips, & Davidson, 2007; Kreuger, 1994). When more than one focus group is planned, varying the dates (weekend versus weekday) or times (morning, afternoon, or evening) may result in a more varied group of participants. In a focus group of women with cardiac disease, only seven of the hoped-for 10 women were able to attend at the specified date and time. When doing grounded theory, participants require sufficient time to discuss the aspects of their experience that they feel are important and relevant, so no more than 7 – 10 participants should be recruited for a three hour focus group, and less (4 – 6 participants) if there is only time for a two hour focus group. Researchers should carefully consider the number of participants they recruit for each focus group as there is a negative relationship between group size and subsequent participation of focus group members (Albrecht, Johnson, & Walther, 1993).

Focus groups can be run by one researcher but most require several researchers to be present during the focus group sessions. Research team members will have specific jobs such as the principal investigator, the focus group moderator (facilitator), and one or more note-takers (Côté-Arsenault & Morrison-Beedy, 2005, p. 175). Halcomb et al., (2007) recommended a minimum of two experienced researchers, the focus group facilitator and a field note-taker. Suggestions of what should be written up in the field notes include: main ideas being generated, significant quotes, important observations, and the notetaker's insights or interpretations (Krueger, 2006, p. 478), facial expressions and body language (Côté-Arsenault & Morrison-Beedy, 2005). The note-taker should also monitor the comfort level of participants and reflect this back to the facilitator as necessary (Halcomb et al.). Another key role is to monitor the status of the recording equipment and change audio or videotapes as needed. All three cardiac research team members attended the focus group interview: one was the group facilitator, another noted information on a flip chart, and the last member oversaw the dual audiotape equipment.

Glaser (1992) recommended the audiotaping of individual interviews (pp. 19 & 20) but more recently revoked this suggestion for solo researchers due to concerns such as delay in data analysis, volume of data, concern regarding delayed theoretical sampling, and negative impact on study timelines and budget (Glaser, 1998, pp. 107-109; Glaser 2001, pp. 54, 161, 166). However, Glaser (1998) acknowledged that audiotaping might be required with a team of researchers (p. 107), which is the usual case in focus group research. In addition, individual interviews are much slower paced than focus group interviews; the focus group facilitator has the additional role of monitoring the group dynamics to facilitate participation from all members, recognizing or 'reading' body language that reflects desire by certain participants who want to revisit or add to certain topics but may be reticent to interrupt, and so on. Therefore, audio or videotaping of focus group conversations is important to prevent the loss of essential data. In the cardiac study, a dual tape system was used to prevent data loss through audiotape defect or possible failure to detect audiotape stoppage. However, this dual system also allowed one set of audiotapes to be used for immediate data analysis and memoing while the other set was being transcribed.

When an expert transcriptionist is used, the audiotapes can be transcribed, validated, and analyzed long before recruitment is finished for the next focus group, and months before researchers could have conducted and analyzed individual interviews from a similar number of participants. As new recording technologies are emerging, and voice recognition software is becoming more sophisticated, transcription of data will become almost immediate. Other important reasons for audiotaping the focus group interview is the much faster pace of this data collection technique, fewer pauses for reflection as individuals have been waiting their turn to speak, and the multiple participants who are involved. Researchers cannot possibly take all this in and provide adequate field notes to capture what was said during two or three hours of such fast-paced conversation – failure to audiotape would result in 're-membering' information congruent with his/her preconceptions rather than what is relevant to the experience of the participants, therefore, resulting in decreased theoretical sensitivity.

Focus group method

Merton et al.'s (1990) original tenets of focused group interviews related to number and type of participants, and focus group facilitation method (atmosphere creation, introductory remarks, non-directive approach, and interview guide) continue to be relevant strategies for data collection. In addition, research approval and funding bodies mandate policies for research with human subjects, most notably the use of written, informed consent and guidelines regarding anonymity and confidentiality. Accordingly, subjects who are interviewed are required to be identified by pseudonym or number only, and this can be a code name chosen by the participant or assigned by the researcher. Alternatively, first names can be used and a number ascribed later and inserted into the focus group transcript in place of the participant's name. However the latter method is not recommended due to the loss of anonymity when first names are used during the focus group interview and the ensuing increased chance of identity recognition when real names are on the audiotapes, particularly in situations where those audiotapes will be used for secondary data analysis. The cardiac research received ethical clearance from the author's university and Merton's major tenets for focus groups were followed (See Table 2). The focus group was held with seven women between the ages of 35 and 75, four of whom had experienced a myocardial infarction and the remaining three had been diagnosed with angina. The 3-hour focus group was held in a meeting room of a university with closed door to promote privacy and discourage interruptions. Participants chose their own code name and these names were written on cardboard signs and placed on the table directly in front of them to facilitate the use of these pseudonyms throughout the focus group session.

For the duration of the focus group, participants were seated around a table with two audiotape recorders in the middle of the table. The group facilitator began the focus group by welcoming participants and requesting that they use their code names at all times for anonymity and to assist the data transcriptionist in attributing quotations to the correct code name. The

Table 2. Comparison of Focus Group Tenets and Cardiac Study Implementation

Pertinent Characteristics	**Focused Interview for Groups (Merton, Fiske & Kendall, 1990)**	**Women and Cardiac Disease Study**
Group size	6 – 12 individuals	7 women
Group type	Homogenous group for greater productivity (p. 137)	All participants were women living with heart disease (angina or heart attack).
Facilitator	Skilled facilitator who knows and can identify typical problems and is able to cope with these (p. 17).	Facilitator who is skilled in focus group facilitation as well as classic grounded theory.
Seating Arrangement & Atmosphere	Use of a circular seating arrangement to promote group interaction (p. 140) and to create an informal, conversational atmosphere (pp. 38, 140, 151).	Women were seated around a table in a comfortable, private room. Conversational rather than formal interactions.
Introductory Remarks	Acknowledge the potential for disagreeing opinions and that there are no right or wrong answers (p. 146-147, 175).	Welcomed both common and individual experiences and opinions, indicating that both were important to the study. Stressed the purpose was not to come to a common viewpoint.
Interview Guide and Approach	General topic areas, not fixed questions (p. 48). Non-directive approach to allow participants to respond with what is significant to them (p. 14).	Seven general 'topics' formatted into open-ended questions. Participants encouraged to respond to what is important to them because researchers did not have the experience of heart disease and, therefore, do not know what is relevant. Topics veered to other areas as participants relayed their experiences and reacted to those of others.

facilitator also stressed that the purpose of this meeting was not to get group consensus and that different experiences would be very important. To avoid having one or two participants monopolize the conversation, a strategy whereby each person was given the opportunity to answer each question was decided upon by participants. Participants would speak one by one, in order around the table, but different individuals could choose to begin the round of responses to a new question and then the order would continue from there. Participants were given paper and pen to jot down points to discuss later, in response to what other participants were saying or when they thought of something else they wanted to add to the conversation. When all participants had answered a particular question, those who had more to discuss were given as much time or as many opportunities to speak as they wanted. Although there was a list of general, open-ended study questions, participants were told during the introductory remarks that these were only a place to begin and may not really be relevant, in which case they could take the conversation in the direction that was relevant.

Data analysis

As previously stated, the original intent for data analysis of the cardiac study was to do a thematic analysis of the focus group data. This thematic analysis resulted in the reporting of the experience of women with cardiac disease as three major processes – delaying, denying, and restricting – with other subprocesses such as comparing (Hernandez, Williamson, & Kane, 2003, 2004). However, the principal investigator is an experienced grounded theorist who used grounded theory's line-by-line method of analysis, constantly comparing the data (Glaser, 1978, p. 57), a process which resulted in the beginning emergence of the substantive theory of *reciprocal restricting*.

When grounded theory is used with focus group data, the analysis method remains the same; the only difference might be in the potentially greater speed with which substantive codes are identified because focus group participants constantly compare their experiences with those of others during the focus group. In the cardiac study, substantive codes, both *in vivo* (such as going day-by-day, being careful, and pushing the envelope) and *sociological constructs* (such as denying, delaying, and minimizing), were identified during this open coding process. When critical ideas were 'sparked', the coding was interrupted to write memos about the codes that were emerging. Finally, the category which explains the behavior in the substantive area and enables the continual resolution of the main concern – the core category – emerged as reciprocal restricting. Women resolved their problem of living with cardiac disease (either previous myocardial infarction or angina) through a process of reciprocal restricting that involved placing restrictions on the impact of their cardiac disease while, at the same time, they acknowledged that the cardiac disease placed restrictions on their lives and lifestyle.

During the selective coding process, only those concepts that relate to the core category are coded. Through memoing the researcher continued to review the substantive codes, placing them in larger categories, and finally recognized that they all fit together through a theoretical code from the interactive coding family (Glaser, 1978, p. 76). This interactive theoretical code of mutual effects was one of several theoretical codes - process, strategy, and interactive families - found during data analysis. All three of these coding families were previously identified by Glaser in 1978, as theoretical codes that could be discovered in grounded theory research. There were six major strategies or tactics that women utilized to restrict the impact of their cardiac disease, and within each of these major strategies there were several processes that underpinned the strategies. The six strategies were delaying, denying, focusing, pushing the limits, minimizing, and forcing a comeback. Delaying and denying were strategies used by women who experienced symptoms from their cardiac disease while it remained undiagnosed, and denying continued to be used after the diagnosis of angina or heart attack. On the other hand, four 'strategies/tactics' could be attributed to the heart disease as it imposed restrictions on the lives of these seven women: vying for attention, limiting, curtailing, and forcing changes. Within each of these four categories there were sub-processes that reflected the strategy. Although these strategy and process theoretical codes were important, they were lesser codes that could not subsume or bring together all of the other substantive codes into an integrated theory. Only the interactive theoretical code of mutual effects was demonstrated to be the conceptual model of the relationship of the core category, reciprocal restriction, to its properties and other (non-core categories) and, therefore, was determined to be the overall theoretical code for this study.

The above section has demonstrated that the essential processes of grounded theory methodology can be followed, and that the grounded theory 'product' of a substantive grounded theory can emerge, when focus group interviews are used as data collection method. The next section will describe the potential limitations of using focus groups to collect data, so that researchers can prevent these from negatively impacting their grounded theory studies.

Limitations of Focus Groups for Data Collection

There are some limitations to the technique of focus group interviewing which can exist irrespective of whether or not grounded theory is the research methodology being used. These potential limitations are addressed in this section.

Culture/Group dynamics.

Focus groups may not be acceptable to some cultures (Holcomb et al., 2007). The development of negative group dynamics can limit participation or freedom of expression and thereby negatively impact on the credibility of the final substantive theory. However, this pitfall can be overcome through the use of a skilled facilitator and providing introductory remarks that give sufficient orientation as to the research purpose and participant 'role'. Another strategy is to avoid having participants attend more than one focus group: participants may be more willing to discuss topics when they anticipate never seeing fellow-participants again. The use of serial focus groups is also discouraged to prevent the occurrence of negative group processes such as power dynamics or "group think" that could possibly jeopardize the integrity of the research (Côté-Arsenault & Morrison-Beedy, 2005).

Theoretical sampling.

The most important problem that could occur in the use of focus groups, when used in conjunction with grounded theory methodology, is the potential problem of not being able to theoretically sample. This might occur if researchers are unable to recruit participants whose contributions can saturate the emergent codes/categories. Unlike grounded theory with individual interviews, participants have to be chosen 'up front' so that all can participate in the focus group; this means that participants are chosen before any substantive codes/categories have been found, and therefore, participants cannot be chosen on the basis of theoretical sampling (sampling based on the emerging theory), as is usual when doing individual interviews. One strategy that can promote theoretical sampling is the use of the flip-chart as one of the researchers' note-taking methods, as well as the use of paper for note-taking by participants. One researcher should be assigned to jot down words or phrases, used by individual participants, on a flip chart that is readily visible to all focus group participants. Having key points on the flipchart allows subsequent speakers to add to the key points to reflect their experiences. Also, providing note cards or writing paper for participants to capture 1) important information they forgot to mention in a previous response, or 2) ideas that were 'sparked' by other speakers' comments, allows them to voice this information when the opportunity permits. In addition, theoretical sampling can be promoted by analyzing each focus group interview prior to holding the next one and utilizing pertinent topic areas that emerge from prior focus groups. Finally, some researchers have mitigated this potential problem by combining focus group interviewing and individual interviewing in the same study.

Deviation from grounded theory methodology.

The combination of grounded theory methodology and focus group interviews has become increasingly popular. Glaser (1998) encouraged researchers to take grounded theory in their own directions but to remain within the parameters of grounded theory methodology. In this chapter, the author argues that the use of focus group interviews as one type of data collection method is one such direction and is one that does not jeopardize the integrity of grounded theory methodology. The example of a focus group conducted with seven women with heart disease was used to demonstrate the adequacy of the type of data that can be collected using the focus group technique, the ability to follow the constant comparative method of analysis, and the emergence of a substantive theory of reciprocal restricting. However, if the focus group interviewing technique is not operationalized, or the data collected are not analyzed, in a manner congruent with grounded theory methodology, then the research product may not be a well-integrated substantive grounded theory. For example, according to Webb and Kevern (2001), nursing researchers who combined focus group interviews with grounded theory methodology and published their work between 1990 and 1999 neglected to follow major canons of grounded theory methodology: failing to simultaneously collect and analyze data (most frequent problem), and coding using thematic analysis, a preferred strategy of many focus group researchers, rather than the constant comparison that produces a core category, its properties, and related non-core categories. These problems continue to show up in more recent literature. For example, in research on risk perception in South Asians, with Type 2 diabetes, living in England no mention was made of a core category (Macaden & Clarke, 2006). Therefore, grounded theory researchers who wish to use the focus group technique as a data collection method should be aware of these two major pitfalls and remain true to the grounded theory process.

Benefits of Focus Group Interviews in Grounded Theory Research

In recent years, focus groups have gained popularity as an interview and data collection method in disciplines such as nursing, health, and social sciences (Holcomb et al., 2007; Webb & Kevern, 2001). There are several benefits of focus group that have promoted this increased usage.

Time and budget resources.

One major advantage of the focus group interview is the assertion that it is less time-consuming and more cost-effective (Holcomb et al., 2007) than individual interviews. For example, a focus group of six to ten participants will take approximately three hours, whereas the same group of people interviewed separately would likely require at least triple that time and would be spread out over the period of several weeks or months, depending on researcher time constraints and participant availability. These benefits alone are noteworthy and may reflect the attraction of this method for graduate students with major time constraints, or researchers with minimal or no research funding.

Participants.

Another advantage of focus groups is the possible inclusion of a more varied group of individuals: often participants who are reticent to speak one-on-one with a researcher may be willing to participate in a group of their peers, as "inhibitions often are relaxed in group situations" (Kreuger, 1994, p. 34). Others may simply welcome the opportunity to compare notes with people in similar situations; the most frequent comment that the author has heard from former participants is how much they learned and/or enjoyed learning from the other participants in the group. Attending one focus group may also be an attractive option for those who believe that individual interviews may be more frequent and thus more time-consuming or when participants are reluctant to have interviews conducted in their homes.

Some participants may prefer to attend a focus group rather than an individual interview because they believe that their responses will be more anonymous in the context of a group. Also, some individuals may be more willing to express feelings and experiences that they would be too timid to bring up or discuss in an individual interview (Morgan & Kreuger, 1993).

Grounded theory emergence.

A major benefit of data collection by focus group, for grounded theory researchers, is the potential for the core category to emerge more quickly during focus group interviews. Occasionally, a 2- or 3-hour focus group with a skilled facilitator and grounded theory researcher is enough to evoke the emergence of major categories and many of their properties or even of the core category itself. This potential benefit was discovered in the research with the seven women who had cardiac disease - the core category emerged, along with 10 major categories and their properties after only one 3-hour focus group interview. The various reasons for the rapidity of core category emergence are unknown but a few are posited here: 1) when individual interviews are used in grounded theory, the core category frequently emerges well before 10 participants have been interviewed; therefore, when researchers listen to 6-10 participants constantly comparing their experiences during a focus group session, the emergence of major categories and the core category can happen quickly; 2) use of a flip chart and participant notes to capture important information during the focus group can promote theoretical sampling; and 3) the preconscious processing that results from the preceding circumstances.

Conclusion

The focus group has become increasingly popular in virtually every discipline because of its many strengths as a method of data collection, and the variety of research circumstances in which this technique can be used. In this chapter, the author has demonstrated the use of focus group interview to develop a substantive grounded theory, the theory of reciprocal restricting, in research with woman living with cardiac disease. The fit of focus group interviews with grounded theory methodology has been shown as it pertains to the relevance of 1) the type of data that can be obtained, 2) the ability to follow grounded theory's constant comparative method of data analysis, and 3) the emergence of the grounded theory product of a substantive theory. Grounded theorists need not fear that the use of the focus group method would be a violation of the tenets of grounded theory, because it is simply one more method of obtaining data, and the grounded theory dictum is that "All is data" (Glaser, 1998). Potential limitations and benefits of focus group interview method have been described so that grounded theory researchers can avoid possible pitfalls and take advantage of focus group benefits. Using focus group interviews to generate grounded theory can be very effective when research team members carefully plan and organize the interview situation; the focus group facilitator has well-developed facilitation skills and constantly strives to create an informal and comfortable environment in which all participants feel free to speak, interact and be heard; and grounded theory methodology is carefully followed.

References

Albrecht, T. L., Johnson, G. M., & Walthier, J. B. (1993). *Understanding communication processes in focus groups.* In D. L. Morgan (Ed.), *Successful focus groups: Advancing the state of the art* (pp. 51-64). Newbury Park, CA: Sage.

Côté-Arsenault, D., & Morrison-Beedy, D. (2005). Maintaining your focus in focus groups: Avoiding common mistakes. *Research in Nursing & Health, 28,* 172-179.

Glaser, B. G. (1965). The constant comparative method of qualitative analysis. *Social Problems, 12,* 436-445.

Glaser, B. G. (1978). *Theoretical sensitivity.* Mill Valley, CA: Sociology Press.

Glaser, B. G. (1992). *Emerging vs. forcing: Basics of Grounded Theory analysis.* Mill Valley, CA: Sociology Press.

Glaser, B. G. (1998). *Doing Grounded Theory: Issues and discussions.* Mill Valley, CA: Sociology Press.

Glaser, B. G. (2001). *The Grounded Theory perspective: Conceptualization contrasted with description.* Mill Valley, CA: Sociology Press.

Glaser, B. G. (2003). *The Grounded Theory perspective II: Description's remodeling of grounded theory methodology.* Mill Valley, CA: Sociology Press.

Glaser, B. G. (2005). *The Grounded Theory perspective III: Theoretical coding.* Mill Valley, CA: Sociology Press.

Glaser, B. G., & Strauss, A. L. (1967). *The discovery of Grounded Theory.* New York, NY: Aldine.

Halcomb, E. J., Gholizadeh, L., DiGiacomo, M., Phillips, J., & Davidson, P. M. (2007) Literature review. Considerations in undertaking focus group research with culturally and linguistically diverse groups. *Journal of Clinical Nursing, 16,* 1000-1011.

Hernandez, C. A. (2010). Getting grounded: Using Glaserian grounded theory to conduct nursing research. *Canadian Journal of Nursing Research. 42,* 150-163.

Hernandez, C. A., Williamson, K., & Kane, D. (2003, May). *Women's experience of symptomatic heart disease.* Poster session presented at the 4th Advances in Qualitative Methods conference, Banff, AB.

Hernandez, C. A., Williamson, K., & Kane, D. (2004, May). *Women's experience of symptomatic heart disease.* Paper presented at the annual conference and meeting of the Midwest Nursing Society, St. Louis, MO.

Krueger, R. A. (1994). *Focus groups: A practical guide for applied research* (2nd ed). Thousand Oaks, CA: Sage.

Krueger, R. A. (2006). Analyzing focus group interviews. *Journal of Wound, Ostomy & Continence Nursing, 33,* 478-481.

Macaden, L., & Clarke, C. L. (2006). Risk perception among older South Asian people in the UK with Type 2 diabetes. *International Journal of Older People Nursing, 1,* 177-181.

Merton, R. K., Fiske, M., & Kendall, P. L. (1990). *The focused interview: A manual of problems and procedures* (2nd ed.). New York, NY: Free Press.

Merton, R. K., Kendall, P., & Fiske, P. (1956). *The focused interview.* Glencoe, IL: Free Press.

Morgan, D. L., & Kreuger, R. A. (1993). When to use focus groups and how. In D. L. Morgan (Ed.), *Successful focus groups: Advancing the state of the art* (pp. 3-19). Newbury Park, CA: Sage.

Webb, C., & Kevern, J. (2001). Focus groups as a research method: A critique of some aspects of their use in nursing research. *Journal of Advanced Nursing, 33,* 798-805.

8

The Utility and Efficacy of Qualitative Research Software in Grounded Theory Research

Michael K. Thomas
University of Wisconsin-Madison

An interesting property of technology is that we seem to be locked in a love-hate relationship with its seductive ironies. While it empowers it also threatens. While it excites and inspires, it also intimidates and threatens. It realigns existing power structures and extinguishes established norms, cultures, and ways of being in the world. And while technology always comes with promises, these promises are largely messianic and evangelical (Nye, 2006; Feenberg, 2002). Technology enthusiasts normally reference benefits of the pillars of modernism: efficiency, predictability, an emphasis on quantity over quality, and surrendering to non-human control in pursuit of these (Ritzer, 1993). Technology promises to make our lives better and even has been thought of as the chief driver of progress itself with the imagined future being ever better than the present.

The central problem with the application of technology is its nuanced alignment. The tool must be nuanced in its application and must be appropriate for its contextual exercise. Here the appropriate use of technology is the use of computer software and other technological tools in the service of classic grounded theory research. At first blush, pursuing classic grounded theory (Glaser & Strauss, 1967; Glaser, 1978, 1998, 2009) with technology-rich tools may seem somewhat contradictory. I contend that, when technology is used thoughtfully and with reflection, it can strongly support the use of classic grounded theory as a methodological tool for inductively developing data driven theory.

Technology

Before launching into a discussion of grounded theory and issues surrounding its appropriateness for technology-rich environments, it is necessary to explain what I mean by "technology-rich." The word "technology" derives from the Greek term *technos* meaning *technique* or *way of doing something*. Dusek (2006) combines different definitions of technology as, "the application of scientific or other knowledge to practical tasks by ordered systems that involve people and organizations, productive skills, living things, and machines" (pg. 35). Technology, however, has come to mean in the contemporary period a way of doing something using tools that are recently invented, complex in their creation, and remain somewhat unreliable and tentative in their implementation and utilization. Here, a refrigerator is defined as an appliance, something that springs from technology but is so reliable and useful that it becomes transparent, fading into the background of our lives. Computers, on the other hand, are very new, complex and are, indeed, life changing but they retain a certain lack of transparency in our lives. They are NOT fully integrated into our lives, and they remain somewhat mysterious for people. An important note here is that technologies are not tools unless they are used for work or, in other words, are used to solve problems. They themselves must be solutions to work problems. They should not be thought of as solutions in search of problems.

Theory

The term theory derives from the Greek term *theoria*, which means a proposition. In a more scientific way a theory is an integrated set of hypotheses that has collective predictive and explanatory power. Theories aim for abstractability, which means that the propositions offered by the theory should be generalizable. The best theories are universal. They always "work." The term *theory* is used in other ways. It is used in a scientific sense like the theory of gravity or the theory of evolution. It is also used (erroneously) to mean a guess. It is also used in an ideological sense. A theory may be a lens for foregrounding certain elements in a data set and backgrounding others. The development of critical theories uses theory as a magic scepter that can bring out some things and not others in data. Adherence to a theory can be a badge of allegiance to particular causes. Theory is a term for knowing. We may arrive at knowing in certain ways. We may produce truth propositions logically or by way of experience. Logical propositions are deduced from other established or agreed upon propositions. Experiential propositions are induced by way of recognizing patters in our environment. "A process consists of unfolding temporal sequences that may have identifiable markers with clear beginnings and endings and benchmarks in between" (Charmaz, 2006, pg. 10). A pattern is a recognized recurring event with common characteristics of previous events. In grounded theory and in qualitative research generally, the goal of the research is not to test hypotheses but to generate them inductively rather than to deduce them logically. Latent processes and patterns in data should be identified and named so that they can be interrogated and integrated into inductive theory. To the extent possible, any preconceived notions should be left aside. Of course this is not really possible in the strictest sense but we *try* to remain as open as possible in the initial phases. "Great white father" theories should not be brought in (Glaser, 1998, pg. 71).

Affordances and appropriateness

In determining whether or how a tool could or should be used, we must explore its affordances and its appropriateness and interrogate why we wish to use it. Is it a solution in search of a problem? Who are the winners and losers in its application? Computers and software and all tools have affordances. These are characteristics and characteristics come with implications or consequences. Grounded theory itself is a methodology that has certain affordances that have implications or consequences. There are some things that can be done with it and some things that cannot. Like the patsy and the salesman, we must consider the technophile and the Luddite as we think about the affordances of technology rich tools. The technophile is given to technological utopianism. Technology is good and there is no problem that technology can not be applied to. On the other extreme is the Luddite who abhors technology and catastrophizes dystopian eventualities. In making informed decisions about the use of technology for grounded theory we simply must think in terms of appropriateness and about affordances. We must not be Luddites and we must not succumb to unbridled technophilia. I neither wish to advance a sense of utopianism nor distopianism with the use of qualitative research software. We should simply ask, 'Does the tool afford appropriate uses?' The caution we must heed is that a tool's characteristics drive its use. The man who buys a new sledgehammer will search for things to smash with it. There is a reciprocal relationship between the tool and its characteristics and the user of the tool. This is why the affordances of the tools must be interrogated as we evaluate their appropriateness for particular purposes. Now we must turn to what is appropriate in grounded theory and then align these principles with the affordances of the software tools we explore.

The methods of grounded theory

For the sake of brevity, I will conflate the methods of grounded theory into sampling procedures, coding procedures, memoing procedures, constant comparison, and writing. Sampling is the selection of means for the collection of data. Usually in grounded theory studies this involves choosing research participants to interview or choosing contexts in which the researcher will immerse herself. In statistical research, the assumption of randomness is essential in sampling because statistical claims are made with respect to the representativeness of the sample of a larger population. Claims made about the sample may be reasonably extrapolated to the population. In qualitative research, claims of representation by way of sample are shifted to the 'consumer' of the research. Though grounded theory is a qualitative research tradition that does claim generalizability of resultant theoretical claims, the claims are not statistical.

Coding is the procedure during which a researcher assigns a parsimonious but meaningful name to a chunk of data. When the data is fractured in this way, the recurrences of incidents may be noticed and hence latent patterns in the data emerge. Once they are named, they may be compared, contrasted, and further interrogated. Open coding is assigning such names to the data without any preconceived set of codes. The researcher asks herself while coding, "what pattern does this incident indicate?" or "what is this an example of?" or states, "Ah ha! Here it is again it is..." (Glaser, 1978). Once patterns have emerged, the researcher is able to identify concepts to be further interrogated. This allows for theoretical sampling or returning to the field to collect data this time guided by what the data that was already analyzed yielded. It also allows for selective coding which means using the name of a pattern that has already emerged to find additional incidents of this category in the data. This tests the usefulness of the code. Glaser suggests "code for it" to see if it "codes out." Codes may be collapsed into one another and thus categories become larger and larger referring to more and more data. This process of constantly comparing codes to each other and back to data is called constant comparison (Glaser & Strauss, 1967). It is a sifting and winnowing process that continually refines the emergent concepts. Eventually a single concept emerges as most important and this becomes the core category or core variable.

Theoretical codes may be used to interrogate a concept and help to tease out the relationships between codes. There are many theoretical codes or coding families. The most discussed are the 6 Cs (causes, conditions, contexts, contingencies, consequences, and covariances) (Glaser, 1978, pg. 74). Some that I have found to be particularly useful as well are stages, typologies, limits and cycles.

An essential procedure in grounded theory that somehow gets missed in many articles is memoing. Memoing is taking notes about how coding is proceeding and what concepts may be emerging. It is basically a place for the researcher to think. This takes place by way of writing. Memos should be written in an informal way and memos should be sortable. Later when writing of the theory occurs, memos are sorted into categories and are written up. Glaser has said that one does not write a book but rather "writes up" a book on a theory.

To summarize, a tool that would help with grounded theory must assist in induction. It must support conceptualization over description. It must assist in open coding, selective coding, and theoretical coding. It must allow users to experiment with relationships between concepts. It must support different types of coding and allow for linkages between codes, memos, and data so that it would support constant comparison. It must support sorting of codes of different kinds and support memo writing and memo sorting. It should also support writing and the creation of visualizations of concepts, hypotheses, and models. It would be useful if it can use different types of data (not just textual). It would be useful if mathematical correlations could be tested with it. It should allow users to connect the emergent theory with existing theories that are related in the literature. It should support multiple users and it must be reliable, secure, and work seamlessly so that it does not get in the way. The best tools are those that are appliances rather than technologies. They should be so easy to use that the user is able to focus on the task at hand and not have to struggle with the tool.

Advantages and issues

The affordances of qualitative research software tools provide many advantages for the grounded theory researchers. These will be explored here along with issues that should be considered. Again, the underlying assumption here is that tools carry with them affordances that influence the user and therefore the analysis itself.

Quantification

Qualitative research software (QRS) makes the quantification of textual data easy. It is easy to say how much text has been coded in a certain way. It is easy to make claims about how many codes were used and how many times they were used. An issue here however, is that quantification may be irresistible to the researcher and the grounded theory researcher should understand why this quantification is being done and how it relates conceptually to the emerging theory. Is it being done simply so that it can be reported as such? Is there an attempt to sound more scientific? Are covariances being discovered that integrate into the theory or may be further interrogated? It is common for qualitative researchers to bring a preexisting theoretical framework to the research that comes with a coding scheme. These are baskets that are then filled with data. The implication of such an exercise is that some categories will be more filled than others and it may be that some data suffer from the platypus problem in that they don't fit into the categories easily. This is not grounded theory. In grounded theory, there are no preexisting baskets. Categories must be built inductively. While quantification can be an immensely useful tool for the grounded theory researcher, it is not an end in itself and must earn its way into the theory. The researcher using QRS must have data driven conceptual reasons for quantifying and should not be muscled by the software.

Recall of coded text

QRS connect data to codes, memos, and documents very easily. This is useful in that the researcher can quickly and painlessly see all of the text that has been coded in a certain way and this makes quoting from the data very straightforward. Quoting research participants is a common practice in qualitative research reports. Quoting lends the research report a sense of authenticity and can help to convince readers that the categories were fair and appropriate. It also adds color to the data. Computer programs help the researcher to locate specific passages of text, images, and moments in audio or video recordings. Codes may be assigned and linked to specific images or passages and may then be retrieved later as part of a global analysis. Computer programs allow for the sortability of memos and allow them to be linked to codes and other memos and moments in data and these connections can easily be seen. Computer programs create an audit trail which aids in making claims about rigor and validity.

One problem with such linkages is that they may fossilize. Tentative linkages may begin to feel true even though they may have been tentatively created. Faith in a database may easily be misplaced. The grounded theory researcher must remain flexible and such linkages using software that appear on the screen may seem more real, tangible or even factual than those made on paper. Linkages must be well-memoed so that such digitally driven fossilizations of connections are tempered.

Grounded theory also emphasizes that a study is a study of a concept or a small set of concepts. It should focus on the conceptual rather than on description as is the case in other traditions such as ethnography. In this way, grounded theory papers tend to avoid very much quoting from data and instead focus on the inner workings of the proposed inductively developed theory.

A sense of rigor

QRS imbue research with a sense of rigor. It may be that even reporting that QRS was used is enough for research reviewers to be convinced of the rigorous systematic nature of this research enterprise. While this sounds like I'm treating it as a socially vested fiction, it is not necessarily so. All research must be convincing and convincingness is part of validity (see Lincoln & Guba, 1985 on convincingness as a kind of validity in qualitative research). But the grounded theory researcher is concerned with discovering latent patterns in data and interrogating them for the purpose of generating concepts and eventually producing inductive theory. The constant comparison method itself is used to validate the theory (Glaser, 1998). The question that the grounded theory researcher must ask herself is "Am I simply using QRS to blind my readers with a sense of rigor that is unfair?" A problem for the whole field of qualitative research is that if QRS becomes so ubiquitous and uniform, it may set up inappropriate expectations of the research. Forcing qualitative research into a single approach is a serious potential problem. If everyone uses the same software, some traditions could be inappropriately marginalized.

Imbuing with scientism

QRS may be used not only to imbue qualitative research with a sense of rigor but also with a sense of being scientific. In this way, research looked at from such an objectivist point of view could be seen as better simply because QRS was employed in the study. This is dangerous. The very things that many in the qualitative research community feared about scientific research to begin with are brought back by way of the software. The criticism of modernistic emphasis on efficiency, objectivity, the emphasis on the quantitative over the qualitative, predictability, and surrender to non-human control is that it is dehumanizing. Postmodern research begins with suspicion of science as acting as though beyond reproach and free from human problems such as bias. Many have interrogated this in science. The very thing that qualitative researchers run from is brought back in with the software. The grounded theory researcher must ask herself, "What am I losing by using this QRS?" or "Am I trying to be more scientific in a way that is not appropriate?"

Connecting memos, codes, and data

One great aspect of QRS that has already been mentioned is the ability of these tools to create connections between data, codes of various types, and memos. QRS may, in this way, greatly aid the researcher in constant comparison. While this is useful, we have already seen that this may fossilize these connections and codes and memos may lose spontaneity and their free form nature thus interrupting the theoretical sensitivity of the researcher.

Making the research about the tool

I conducted a workshop once on the QRS NVivo. The participants all had brought their data and had great expectations about what the tool could do for their research. This faith is misplaced because the software does not do the analysis. Further, all of the efforts then revolved around technical issues rather than the workshop participants' data. "How do I insert a link?" "How can I save my project so that other people can see it?" It is essential for the grounded theory researcher that all of these functionalities become transparent. QRS tools can have steep learning curves and this itself may make it inappropriate for some researchers. At the same time, I caution qualitative researchers not to give up too quickly on using software tools because of this. Any tool requires a little practice to gain the requisite facility.

The emphasis on coding rather than memoing

On many occasions I have been asked by researchers who have used QRS to take a look at their project and help them make sense of it. This is a strange request. They have usually coded many documents and have produced many codes in the process. Now they are left starring at a database of codes and documents and don't know what to do next. Almost always this is a twofold problem. One is that they simply have too many codes and have not integrated them or they have too many codes because they have one incident codes and have named an incident rather than a pattern. But also this can be a problem of coding without memoing. In grounded theory, memoing is where the action is. Coding must be interrupted to memo. I have noticed that when I use QRS, I somehow get stuck coding and I am less likely to interrupt my coding to memo. I believe that the affordances of the software promote coding over memoing. While this is not a reason to avoid QRS altogether, the grounded theory researcher must be vigilant against coding without memoing.

Auto coding

QRS allows users to auto code data. Specific words or phrases can be set to be tagged or coded in an exacting way. Particular utterances, speakers, parts of images, or even segments of recordings can be automatically coded so that they become part of the database. These can then be queried to note relationships between categories. While this can be immensely useful it is also risky because the temptation to auto-code wherever possible can be very great particularly when faced with mountains of data. The problem is that the very act of auto-coding decontextualizes the category. In this way a pattern is not discovered; it is preconceived and then forced onto the data. This is contrary to the inductive and theoretically sensitive nature of grounded theory. The very rules that make auto-coding possible should earn their way in to the theory and contextuality should never be lost.

Creating visualizations

Because software is particularly good at counting, it can be used to create wonderfully compelling visualizations of quantitative data. These can be so compelling that they have even been called "info porn" (see the periodic table of visualizations). Qualitative concept mapping with imbedded links to codes, memos, data, and documents are made easy with QRS. Many researchers insist that they "naturally" are visual people or find such visualizations and their making to be very useful. Visualizations may not only be used to display data or report on the research but themselves may be created as memos to explore the data. Visual memos can be very powerful as tools of analysis in grounded theory. While such visualizations are indeed powerful, they may obviate as much as they elucidate. They are also not a substitution for writing the concepts and explaining them with well written prose that explain their nuance and subtly. Concept mapping can also privilege path analysis style hypothesizing of conceptual relationships between categories. While this may be very appropriate it could lead to the forcing of the notion of covariance and the question of statistically significant covariance. This may or may not be appropriate. Again, the tool must not muscle the grounded theory researcher. The tool must be used critically and thoughtfully and its affordances acknowledged and its role in the research acknowledged.

Coding digital media

Because many QRS packages now include the ability to code digital media such as images, films, and audio recordings, QRS is likely to be increasingly used for grounded theory research. Images for example may be coded in different ways with sections of images being coded in one way and other sections in another. This is not just for interview transcription anymore. Again, these may be connected to multiple codes, memos, documents, and other data. A technical problem with this is that I have found that the cognitive load this places on even new computers can be considerable and I have had problems with crashes. While this may be solved with increasing computing power it brings up a couple cautions worth mentioning here. One is that software packages tend to test the cutting edge of computers' abilities. This means that as they do more, they need more power and newer versions of software tend to require ever increasing power. Also, new versions of software packages can be prone to bugs that have not been fully vetted by software designers. This perpetual cycle of upgrading can make software packages prone to permanently being beset with some problems. Learning the software can be difficult and new versions can have surprisingly steep learning curves. The grounded theory researcher must be on guard against being drawn into this without asking the simple question "Is using this QRS more trouble than it's worth?"

Sharing analysis with multiple researchers

This is not as easy as it sounds but really is a great advantage of using QRS. Today, researchers may be in different countries and need to have access to data and analysis. This is only possible electronically. A problem with this is that it privileges settling on preexisting coding schemes. Another is that there have to be multiple instantiations of the software being used and data has to be kept in a place where this is reliable and secure. Tools such as Google Documents are making this easy and free. One trend that results from the affordance of sharability of data sets and analysis artifacts is that some online journals are beginning to ask that a research article include links to its data set. This raises issues of ethics and privacy of both research participants and researchers. Should all memos written in a study be made public? There is also a question then of who owns the data. While academic journals typically take the copyright of an article rather than the author, it is not clear if this should be the case for datasets. As it is easy to create digital data, it is easy to copy it and share it and there are many less than obvious implications for this.

Dealing with literature

QRS can help coding and memoing about the literature. In this way, the literature becomes the data. If this is done after the emergence of the conceptual categories, this is useful. An important aspect of grounded theory is that literature reviews are not conducted before the research is done. The traditional literature review is done before the research to set up hypotheses to be tested in the research. Since grounded theory insists on induction and on leaving aside preexisting theoretical frameworks, in the early part of the research the literature review is inappropriate. This is NOT however to say that grounded theory researchers do not attend to the literature. This attention comes after rather than before the research. QRS can make this integration easy as elements from the literature can be tagged or coded and later these connections can be drawn upon during the writing phase of the research.

Remaining Issues

Issues remain with the use of software for grounded theory analysis. Software is notorious for creating a specific location of the research. Are you suddenly tied to your office so that you can use a particular machine for the work? The expense of the software must also be considered. Many people hire consultants or graduate students to help them go through the data while trying to simultaneously do grounded theory. In grounded theory, someone who helps code shares in the creation of the theory and should be credited with this.

When using QRS, create frequent back-ups and use cloud computing as a strategy to save data. There are many online and even free services that can be used to store data online. While this should be balanced with privacy concerns and should not be taken as foolproof, they are useful when the researcher may reasonably expect to be able to get online easily and have access to the data.

Advantages	Issues
Quantification	Quantification may be irresistible
Recall of coded text	Faith in the database may be misplaced
Imbuing with a sense of rigor	GT is not findings or verification
Imbuing with scientism (objectivism)	The dehumanizing feel that many researchers complain about with quantitative analysis is brought back in
Connecting memos, codes, and data	Memos lose spontaneity and their free form nature and connections may fossilize
Concept mapping with links	Concept maps may obviate as much as they elucidate. This also privileges path analysis style hypothesizing of conceptual relationships.
Sharing analysis with multiple researchers	This is not so easy and privileges settling on preexisting coding schemes.
Auto coding	Auto coding decontextualizes. Coding is done differently with a machine than without.
Coding multiple digital media (it's not just interview transcripts anymore)	This can easily be done without software raising the question whether software use is more trouble than it is worth.
Importing literature to code and link	Not all literature is digitized making another step necessary for QRS to be useful. This can easily lead to data overload as we saw with the creation of transcriptions.

Summary of some specific tools

While it is inherently risky to write about specific software titles in manuscripts such as this one owing to the ephemeral nature of computer technologies, these are some software titles that merit mention. Also mentioned are some titles that represent new trends in software designed for the purpose of supporting qualitative analysis. I emphasize again that the affordances of the software are what the researcher must focus on and that the affordances of the tool have implications for any output from the tool. I also emphasize that it is not useful to have either a technophilious or a technophobic view of these tools. Grounded theory is a methodology that is used for the inductive development of theory. Whatever aids the researcher in developing hypotheses that are grounded in the data is fair game. Just as Glaser would say, "All is data" (Glaser, 1998), all tools should be used in the service of the analysis of the data, data management, and writing. It is not my intention here to review these tools to recommend some over others but simply to provide a brief overview of the tools for what they can do rather than to explore what one should do with them. Software guides tend to focus on what the program can do and tend to emphasize ease of use, the learning curve, and its power for efficient and effective qualitative data analysis. They do not have much to say about particular traditions or approaches in qualitative data analysis and certainly not grounded theory. These tools are not substitutes for learning methodology. Rather, the researcher must have a solid grasp of grounded theory before being able to make sophisticated and appropriate decisions about using qualitative research software.

Analysis Tools.

The following are tools specifically designed to support qualitative research. All of them support coding text and building associations between texts and codes. NVivo is one of the more popular qualitative research package and was originally designed with grounded theory in mind. Developed by the Australia based company QSR, it evolved from NUD*IST. With NVivo it is now possible to not only analyze text but media files and import bibliographic data for references. XSight, also developed by QSR is simpler than NVivo and is designed for shorter duration, less involved, research. ATLAS.ti is another robust tool for qualitative research used around the world that supports multimedia analysis and an array of visualization tools. Dedoose is ironically titled and is really designed for mixed methods approaches. It allows users to weight codes. Weight does not have to be how much text was coded in a particular way but rather this feature allows the researcher to tag how heavy or important a code is. Visualizations are connected to the data and documents can be tagged with "face sheet" data or what is called in Dedoose "descriptor sets." AnSWR is also designed for integrating quantitative and qualitative approaches and allows for coding text and coding graphics. ALCESTE (Analyse Lexicale par Contexte d'un Ensemble de Segment de Texte) combines textual analysis with statistical modeling. EZ-TEXT allows for multiple users to import code books so that different coders can use the software at the same time. Qualrus has an inter-coder agreement function built into the top level functionality of the program. Hyper Research, The Ethnograph, and MAXqda are other analysis tools worth exploring.

Mapping Tools.

In addition to analysis tools, there are also software packages designed to allow users to explore or demonstrate relationships between categories in a manner akin to the concept map. Free Mind, written in Java, is concept mapping software. It is simple but stable and free. UCINET is another concept mapping software that functions with NetMap. This allows users to create visualizations of relationships between Nodes. ArcGIS is used for geographic data interaction. SoNIA (Social Network Image Animator) is a Java based program that allows for the creation of animations or video that can show relationships between networked nodes. This can be very powerful for displaying relationships over time. This can be used by the grounded theory researcher to develop ideas about how data fit together in a temporal sense. Tools for visualizations and modeling are becoming more powerful and more accessible for researchers everywhere.

Media Utility.

There are some tools that grounded theory researchers may find useful for dealing with media. iMovie or MovieMaker are useful for editing simple videos. Audacity is useful for editing or copying audio files and changing their file types and Virtual Dub is useful for editing video and adding subtitles. Transana is a software tool made at the University of Wisconsin that associates text with video. It allows users to transcribe and synchronize transcriptions with media. Transcriber and Praat are software tools designed to align transcription text with specific moments in audio files. Praat includes tools for phonological speech analysis. There are many other such tools that provide visualizations of speech and vocalization patterns.

Bibliographic Software.

This is not part of grounded analysis but bibliographic tools have become ubiquitous. EndNote, ProCite, and RefWorks are examples. These are programs that allow users to create a database of items in literature references. A user can create an entry for a journal article, and then automatically create a table of references and decide on multiple citation styles so that the program can output lists of references in different styles. The beauty of this is that libraries can be shared. Also, online libraries allow for these to be downloaded when users find items online. Zotero is a browser plug-in that creates database entries right from a browser.

Writing tools.
Currently there are several pieces of software that assist users in writing tasks. This is accomplished by allowing users to break up text into smaller chunks and link them together by way of embedded links. In this way, a sortable collection of memos may be linked, arranged and rearranged. Templates are available to aid in this. These tools map quite well on to the grounded theory practice of sorting memos. Scrivner, for example gives users a screen that is a cork board and index cards that may be titled, arranged on a cork board. It also has a nice feature of giving the writer word goals. If, for example I need to write a daily 1,000 words, as I type, it will display with a progress icon how close I am to achieving the goal. Though originally for MAC users only, Scrivener recently became available for Windows users.

Another such tool is called WritersCafé. This program also allows for memo sorting to support the development of longer manuscripts but it also has the feature of offering many genres such as the three act play, the murder mystery and so on. It also allows for sorting of words on the screen which is a powerful creative inducer. Again these may be connected by links. It is hard to emphasize the power of this tool. It calls itself "playground for the imagination." Story Space, like WritersCafé is designed for the development of creative writing texts. It is an environment that allows users to "play" with unstructured writings.

Final Thoughts

Glaser (1998) endorses the use of grounded theory methodology in part by insisting that it "works." This test might also be applied to the use of technology-rich tools in service of grounded theory analysis. Does it "work" or does it get in the way? Again I offer a heuristic set of questions for the grounded theory researcher to use when making decisions about using QRS. Is the tool worth the trouble? Does the tool muscle the analysis? Why am I really using it? By remaining reflective, sensitive, and open, the classic grounded theory researcher may make thoughtful and informed decisions about the appropriate use of qualitative research software.

References

Charmaz, K. (2006). *Constructing grounded theory: A practical guide through qualitative analysis.* London: Sage.

Dusek, V. (2006). *Philosophy of technology: An introduction.* Malden, MA: Blackwell Publishing.

Feenberg, A. (2002). *Transforming technology: A critical theory revisited.* Oxford: Oxford University Press.

Glaser, B. (1978). *Theoretical sensitivity: Advances in the methodology of grounded theory.* Mill Valley, CA: Sociology Press.

Glaser, B. (1998). *Doing grounded theory: Issues and discussions.* Mill Valley, CA: Sociology Press.

Glaser, B. (2009). *Jargonizing: The use of the grounded theory vocabulary.* Mill Valley, CA: Sociology Press.

Glaser, B. & Strauss, A. (1967). *The discovery of grounded theory: Strategies for qualitative research.* Chicago: Aldine.

Lincoln, Y. S., & Guba, E. G. (1985). *Naturalistic inquiry.* Beverly Hills, CA: Sage.

Nye, D. E. (2006). *Technology matters: Questions to live with.* Cambridge, MA.: The MIT Press.

Ritzer, G. (1993). *The McDonaldization of Society.* Thousand Oaks, CA: Pine Forge.

9

De-Tabooing Dying in Western Society: From Awareness to Control in the Dying Situation

Hans Thulesius
Lund University

In this chapter I will present an emerging grounded theory of *De-tabooing dying*, which suggests a structural and attitudinal change toward death and dying in Western society. This work has its roots in the grounded theory monograph *Awareness of Dying* (1965), which introduced GT to the world. I propose that open awareness of dying is contingent upon, and simultaneously, the first step of de-tabooing, which is followed by the control of dying. I will consider the tension between the attitudes of the legislative and the public in some Western societies toward the taboo, and the taboo-transcending changes made to some legal structures. I will also present a taboo of challenging the autonomy and the self-ruling of people in general and the dying patient in particular. A taboo positioning reaction is finally presented in which control of dying is confronted. At the same time the taboo of challenging autonomy and self-ruling affects the positioning, which often results in a de-paternalization of the role of physicians. The chapter will end with a short section on the limitations of the study and a brief conclusion.

In 20th century Western society a change in attitude toward death and dying developed into a taboo on the subject of dying (Walter, 1991), where a taboo, which results from social custom or emotional aversion, is an inhibition or ban on *saying* or *doing* something. The taboo is being confronted in a process of de-tabooing the subject of dying. This chapter discusses de-tabooing and presents taboo positioning as a property of de-tabooing. Glaser and Strauss's seminal work (1965) with dying patients in hospitals revealed that each dying situation occurred within a context of awareness defining who among the health professionals, chaplaincy, family, friends and the patient knew what about the nature of the patient's death and the likely timing of the death. Awareness contexts were explained as *closed awareness*, *open awareness* and *suspicion* (of dying) *awareness*, and different people engaged in different behaviors according to the awareness context. Glaser and Strauss's study marked the beginning of a new norm of open awareness contexts prevailing in the dying situation in hospitals in the US and in Western society more widely. With open dying awareness contexts now the norm, the main driver in the dying situation has become one of *controlling the dying process*. The strategies available to control the nature and the timing of death include withholding life support, withdrawing life support, euthanasia, assisted suicide and palliative care. Since withholding or withdrawing life support, euthanasia and assisted suicide all involve a third party enabling the death of the patient, legal challenges are made against these third parties. In response, patients seek to gain the legal right to achieve their own death with the help of another. Thus, the drive to challenge society's taboo on dying comes from those in the dying situations. Society is now being forced to confront its taboo and amend the prohibition on enabling the death of another.

Controlling the dying process

Reversible and irreversible strategies

In open awareness contexts dying patients learn about their diagnosis and prognosis, which potentially increases their autonomy in the dying process. Open awareness contexts can thus be justified by personal autonomy arguments: people have the right to know even the worst of news about their own body and life to be able to plan their remaining days and to work toward controlling the ordeal of dying. Therefore, de-tabooing of the discourse on dying involves an increased interest in strategies for gaining control of the dying process, strategies that have either a reversible (e.g. palliative sedation) or irreversible (e.g. euthanasia, physician assisted suicide) outcome, (see below).

In an open awareness context, a creative communication between caregivers, patients and significant others is made possible, and so it becomes easier for nurses and physicians to work to improve the quality of life of the dying person by controlling pain and other suffering and to comfort the person dying. The decision to use morphine and sedatives to control severe pain and anxiety is less fraught since drug addiction is moot for a person with a limited life expectancy; indeed, if a dying person is suffering unbearably despite sustained efforts to control the symptoms, it is possible to induce sedation deep enough to cause the dying person to enter into a temporary sleep. This is often called terminal or palliative sedation and is a reversible control of dying as compared to the irreversible outcome of euthanasia and assisted suicide (Rietjens et al., 2006). Palliative care is characterized by an effort to control the sufferings of dying without the intention to either shorten or increase the life span of the dying person but rather to augment the dying person's quality of life at the end of it (Sepúlveda, Marlin, Yoshida, & Ullrich, 2002). Palliative care developed from the hospice care philosophy of the 1960s and is an advancement of what Glaser & Strauss (1965) called comfort care.

Irreversible strategies for controlling the dying process cause the death of the patient and include withdrawing or withholding life support, euthanasia and assisted suicide. All of these strategies involve a third person. Society's responses to one person causing or enabling the death of another are codified in legal structures. These structures are being challenged by the issues caused by the gradual de-tabooing of dying; the concern to control the dying process and the strategies used.

Structural signs of de-tabooing

A common argument in the 1960s against open awareness contexts in the dying situation was that if a patient were to know about his or her bad prognosis, he or she might commit suicide. Recent data lend some support to this argument since the relative suicide rate is higher in cancer patients (Misono, Weiss, Fann, Redman & Yueh, 2005) and among ALS (Lou Gehrig 's disease) patients (Fang, et al, 2008) as compared to the general population. Evidently, open awareness contexts also raise the awareness of the option of ending one's life in a controlled manner.

In most Western countries it is legal to control dying using the irreversible strategies of withholding or withdrawing life support treatment, but illegal to control dying by assisted suicide, or by euthanasia. Euthanasia, however, became common in the Netherlands in the 1970s after the case of a physician ending the life of her own dying mother following her mother's repeated requests. This is an important event in the evolution of Dutch euthanasia practices (DeWachter, 1989) and Dr Geertruida Postma was sentenced to a short imprisonment and one year probation. The judge, however, suggested a decriminalization of euthanasia and in 2002 this practice was regulated in Dutch law under certain conditions. That year a similar law was introduced in Belgium and in 2008 euthanasia and assisted suicide was de-criminalized in Luxembourg. The three countries collaborate in the Benelux union and share legislation by the Benelux Court of Justice since 1975. In the state of Oregon physician-assisted suicide was legalized in 1998; similar laws followed in Washington in 2009 and in Montana in 2010. In Switzerland assisted suicide (not by physicians) was legalized as far back as 1918. Structural signs of detabooing the control of dying therefore include both the growth of palliative care and the development of liberal legislation on euthanasia and assisted suicide in the Benelux countries, Switzerland, and in the US states of Oregon, Washington, and Montana.

Tension between structures and attitudes

In a recent survey of attitudes to end-of-life decisions in six European countries and Australia, 84 percent of Swedish physicians would never be willing to "administer, prescribe or supply drugs with the explicit intention of hastening the end of life on the explicit request of a patient" (p. 2) as compared to 15 percent in the Netherlands where euthanasia is decriminalized (Löfmark et al., 2008). In Italy the question was expected to be so abhorrent to public opinion, the taboo on dying so strong, that it was not included in the survey (Löfmark et al., 2008). The Dutch position on euthanasia and assisted suicide has a history that is intertwined with the Dutch history of tolerance (De Wachter, 1989; Emanuel, 1994). A culture of pragmatism and a lack of historical mishaps of "mercy killing" explain that the word "euthanasia" is not contaminated in the Netherlands as opposed to neighboring Germany. The Nazis performed *gnadentod*, a typical euphemism for the killing of at least 200,000 German and Austrian physically or mentally handicapped citizens by medication, starvation, or in the gas chambers between 1939 and 1945 (Proctor, 1988). As a consequence, negative attitudes to euthanasia and assisted suicide dominate the German opinion (Cohen, Marcoux, Bilsen, Deboosere, van der Wal & Deliens 2006). Moreover, in a European comparative study of public attitudes toward euthanasia from 1981-1999 (Cohen et al., 2006), Germany was the only country with a negative trend in the development of attitudes toward euthanasia. Most physicians in countries where euthanasia and assisted suicide are illegal are reluctant to perform euthanasia or supply means for assisted suicide.

In Sweden, as well as in the UK and parts of the USA, a majority of the public seems to be pro-euthanasia and assisted suicide (Cohen et al., 2006) yet the political majorities and the medical profession in these countries are against. The physicians align with the legislation that prohibits euthanasia and assisted suicide, which position could be viewed as a paternalistic stance affected by what is politically correct.

A growing taboo of challenging autonomy

Medical paternalism is defined by Oxford's *Concise Medical Dictionary* (2010) as "an attitude or policy that overrides a person's own wishes (autonomy) in pursuit of his or her best interests." Yet, in acknowledging a patient's right to know of his or her impending death, a new taboo is developing where challenging the self-ruling aspects of autonomy becomes less acceptable. Personal autonomy is a strong concept in the bioethical discourse of today. Self-decision-making is seen as a human right in accord with societal democratic ideals and this was evident in our data with hundreds of persons claiming "autonomy" as the main argument for control of dying in general and irreversible control of dying in particular. The following quotes from the survey "Every person is the master of his own life" and "If a person should decide anything it is about her own life and death" are examples of this attitude. "Self-ruling should be a natural part of health care also when you are very ill" is another quote from our survey data with the same attitude occurring in the interviews. Not surprisingly, to challenge the dying patient's right to self-rule and control the dying process was an anathema to many respondents in our study. Thus, a competitive interaction, or a wrestle, between the taboo against the control of dying and the taboo against challenging self-ruling aspects of autonomy results in de-paternalism, where the traditional medical paternalism of control of dying is losing ground to the self-ruling challenging taboo. This is evident in the attitudes of physicians in the Benelux countries with a large majority being in favor of euthanasia and assisted suicide (Löfmark et al., 2008).

De-paternalism is an important concept as it is both an indicator of de-tabooing the subject of dying and a consequence of the taboo wrestle conducted across many dying situations. As such it links the general attitudes of western society to the specific 'dying situation'; the control of dying and the attitudes of the protagonists. It was from a study concerned with exploring attitudes toward the control of dying and to euthanasia and assisted suicide in particular that taboo positioning, a taboo confrontation reaction, was discovered.

Taboo positioning

The data for the study comprised interview data from more than 50 formal and informal interviews with laymen, nurses and physicians from Sweden, Denmark, Norway, United Kingdom, Ireland and Canada, and the United States; open written survey comments from 470 Swedish physicians and lay people; internet forum postings from 40 North American contributors to internet forums discussing euthanasia and assisted suicide; and samples from the literature on euthanasia and assisted suicide. Two similar postal surveys were designed, each consisting of 15 multiple choice statements and questions based on an end-of-life scenario. Space for open-ended comments followed every survey item. A total of 1200 physicians from psychiatry, surgery, general practice, geriatrics, and internal medicine specialties were randomly selected for the first postal questionnaire, whilst the second questionnaire was distributed to 1200 randomly selected individuals living in the County of Stockholm, Sweden. The response rate was 74 per cent for physicians and 58 per cent in the general population. All data was analyzed using GT procedures according to Glaser & Strauss (1967) and Glaser (1978, 1992, 1998, 2001, 2003, 2005, 2007) and as reported in recent studies (Thulesius, Håkansson & Petersson, 2003, 2007; Thulesius & Grahn, 2007; Thulesius, Sallin, Lynöe & Löfmark, 2007).

The taboo struggle between control of dying and challenging autonomy was observed in both our interview and survey data. The taboo positioning process starts with emotional positioning followed by reflected positioning, a labeling wrestle, stipulated positioning and eventually defending the new normal.

We identified a positioning resulting from the interaction between the taboo connected with euthanasia and assisted suicide, and a taboo under development prohibiting a challenge to patient autonomy and self-determination. Between the two tendencies there are power and values at stake dealing with the question "who should ultimately control dying?" Should it be the healthcare staff including physicians or should it be the patients themselves? Thus a positioning between the old and the new taboo often starts with an immediate emotional positioning for or against euthanasia and assisted suicide that also involves conceptual confusion - many respondents, both the public and physicians, did not make a distinction between euthanasia, assisted suicide, palliative sedation, withdrawing or withholding life-sustaining treatment. Our analysis thus suggested that when people are confronted with control of dying issues their first attitude reaction is cognitively bypassed and reflecting comes later. Thereafter follows reflected positioning where the emotional reaction is reflected on and then rationalized. Also, there is a labeling wrestle by the use of euphemisms e.g. "blessed relief" and dysphemisms, for example "murder."

The words used to describe a taboo are often euphemisms, but also dysphemisms can be used. According to Webster's Dictionary a euphemism is "a substitution of an expression that may offend or suggest something unpleasant to the receiver with an agreeable or less offensive expression." The opposite of a euphemism is a dysphemism "a derogatory, offensive or vulgar word or phrase to replace a (more) neutral original." If one side in the interaction struggle gains dominance, there will be stipulated positioning, including ethical reasoning and "soft laws." This is sometimes formalized into "hard laws," followed by defending the new norm for the control of dying such as euthanasia and physician assisted suicide. In this defense struggle it is almost an anathema to challenge patient autonomy and self-determination.

Emotional positioning is an emotionally triggered non-cognitive reaction resulting in a positive or a negative attitude to the control of dying taboo issue. When a taboo is confronted the reactions are usually an emotional and immediate gut-feeling with no time for reflection. Emotional positioning takes place before the taboo issue is fully defined and therefore it involves conceptual confusion about the type of control of dying being discussed and thus conceptual separation capacities seem decreased or even disabled. Euthanasia, physician assisted suicide, and withholding life-sustaining treatment are not seen as separate entities and figuring out what these concepts mean is often difficult. Therefore people confuse what they are positioning themselves toward. One physician said: "conceptual confusion is always present when it comes to issues surrounding death and dying. There is never clarity when these areas are discussed." Although we never asked about withdrawal of life sustaining treatment many survey and interview respondents, physicians and the public alike, answered as if this was synonymous to euthanasia and assisted suicide. This concept mix-up was an early part of the taboo positioning. This pattern of immediate emotional positioning, seemingly without previous cognitive reflection, was seen repeatedly in the survey data of both the public and the physicians (Helgesson, Lindblad, Thulesius & Lynoe, 2009) and in the interviews. A reflective taboo positioning as being for or against or being hesitant to a suggested taboo transcendence follows the emotional reaction; involving reasoning for and against by exampling, comparing, and rationalizing. Eventually an attitude positioning to a pro, con or hesitant stance occurred.

Reflected positioning involves a cognitive juggling of the control of dying taboo by reasoning for and against it. A similar wrestling with the taboo toward challenging patient autonomy takes place. In this way the position taken is rationalized and explained. Or, as one Swedish nurse put it: "You take on an attitude based on emotions and after that you construct a theory that defends that position." Eventually a taboo positioning to a pro, con or hesitant attitude stance occurs. Some people change their attitude from being pro to being hesitant or from being against to being slightly positive toward the different types of control of dying. Eventually the respondent may do a repositioning if the arguments are strong enough for or against the taboo. A respondent from the public illustrates this: "When I come to think of it [euthanasia], it is probably not good to legalize it even if I would like it for myself." Another respondent, a physician, was listening to a case story of real life euthanasia and at first presented a positive immediate emotional reaction toward euthanasia. Then she started to reflect. She discussed time framing of euthanasia and compared it with abortion's time frame. She also argued pro euthanasia using utilitarian arguments: if euthanasia was not made legal then it would be performed illegally (a similar argument to that used to argue for legal abortion of which she approved); euthanasia could then be legal within certain time frames and other conditions. Finally she took a stance based on arguments against legalization and concluded: "since there is an alternative way [palliative care] of resolving the issue of insurmountable suffering then this [euthanasia] is not necessary"

Reasoning for or against a taboo transcendence with many different types of arguments is a central part of reflected positioning. Euthanasia and assisted suicide proponents might use "hyper-autonomy" or body ownership arguments: "I own my body and therefore I rule over my death as well as over my life." This is an extrapolation of self-ruling life into self-ruling death. Euthanasia and assisted suicide opponents may defend socionomy and societal control, (Bell, 1969), against autonomy, self control, with arguments from philosophy, theology or humanism. A physician interviewee quoted the philosopher Kant, who said, "If suicide is allowed then everything is allowed," and the physician thought this was a wise argument against physician-assisted suicide.

Religion plays a role in the reasoning since all irreversible dying control strategies are heavily tabooed, and either unaccepted or illegal in Muslim, Jewish and Catholic cultures. The participants also reasoned by defending or redefining the professional role and tasks of physicians. One physician claimed, "it's not part of the professional role [of physicians] to take lives," while another said "The physician profession is a service profession. The self-ruling of patients should apply." One public respondent said, "It is not the task of physicians. Insecurity is always present despite the criteria [for physician-assisted suicide]," while another suggested "It should not be just any dr, but special physicians [performing physician assisted suicide]."

Exampling by telling personal stories of control of dying was common among our respondents. Many argued for euthanasia and assisted suicide as a way of avoiding symptoms by referring to painful and undignified experiences of their own family members. They did not want anyone else to "suffer as grandma," therefore irreversible control of dying was justified as a way of relieving lingering ordeals. In an American Internet discussion forum one participant wrote, "My father who had ALS chose Remington [rifle] assisted suicide. In a note he wrote, I knew where I was going, I just decided to take a faster train." Examples of euthanasia and assisted suicide from the Netherlands and Switzerland were mentioned by physician respondents: "I have relatives who paid in advance to a Swiss clinic, so I like the idea of discussing the issue"; "I know about EXIT in Switzerland and I don't like it"; "A patient can always go to Switzerland"; "It shouldn't be done by palliative physicians but by certain "execution" institutes like in Switzerland/Netherlands"; as well as by the public: "I have a work mate who has good experiences from Holland [with euthanasia]";"Think about how good it is in Holland, I hope the government open their eyes [for euthanasia and assisted suicide]."

Several respondents mentioned pet euthanasia and it was argued that animals are treated better than humans when it comes to suffering control; a participant in an American Internet discussion forum wrote "I watched my grandpa die of cancer. He went downhill fast and was in so much pain leading up to his death. It's no way to go. When animals are suffering and going to die, we put them out of their misery, so why not for a person???" A physician respondent said, "Euthanasia is veterinary medicine and therefore veterinarians should do it."

Some respondents could accept euthanasia and assisted suicide with arguments from examples and particular situations. A health worker said: "I am opposed to [legalizing] euthanasia and assisted suicide but in certain cases I would consider it as in the case of a middle-aged patient with brain metastases who could not walk anymore. Also, my mother suffered end stage cancer with bad pain control. She was given an infusion with sedatives and morphine with increased doses until she died. That was good after all."

Counter arguments to the irreversible control of dying offered by euthanasia and assisted suicide were risks of abuse, the slippery slope, procedural failures, and professional role erosion. Also it was argued that euthanasia and assisted suicide is needless if palliative care and palliative sedation therapy is offered. A limited sleep is better than a permanent sleep. Swedish physicians expressed these opinions: "I think that good palliative care makes this debate redundant"; "Palliative care, possibly [palliative] sedation, should be continued [instead of physician assisted suicide]."

Labeling wrestle is the battle over discourse. The nature of taboos is that they are often labeled by dysphemisms, words that are experienced as derogatory, offensive and politically incorrect. There is a call for less offensive words, to replace the more offensive words in an attempt to achieve political correctness and in support of de-tabooing. When different strategies for controlling dying were labeled in offensive terms, these were countered by using euphemisms that were more politically correct and vice versa. Since a euphemism is often used to 'politically correct' something which is banned (Burridge, 1996) and since a synonym to political correctness is cultural sensitivity (Andrews, 1996), a property of de-tabooing could thus be called 'politically correcting'. Arguing with euphemisms or dysphemisms that label euthanasia and assisted suicide emerges during emotional and reflected positioning. The naming of the different strategies for the control of dying is closely related to the definition of these issues. If "murder" is a dysphemistic labeling of euthanasia then "dignified death" is a euphemistic labeled reflection of the same incident. One physician respondent noted that: "[It is] wrong to use the negatively loaded words murder and suicide" and a respondent from the public stated "I would prefer a nicer word than suicide." Labeling wrestle may also be part of the defending new normal, as discussed below.

Stipulated positioning means that the transcendence of the taboo is conditionalized, in other words different conditions have to be met in order to transcend the taboo of dying control. This stipulating is commonly observed as the de-tabooing proceeds even with those being in favor of euthanasia and assisted suicide and eventually turns into a formalized legal positioning, as in the Benelux, Oregon and Washington. To stipulate when euthanasia and assisted suicide is acceptable requires safety and legal criteria. In order to protect against abuse of the controlled dying many conditions have to be met. Autonomy and self-ruling is essential. Controlled dying must be something the patient wants for himself or herself. The decisions should not be caused by others' influence. A physician respondent emphasized that "It is OK [with physician-assisted suicide] if the law requires that it is the definite wish of the patient." Likewise, the patient must suffer from an incurable deadly disease and have a short life expectancy. Ultimately, the control of dying has to be approved by skilled physicians and legal authorities. Thus, one of the ultimate de-tabooing exponents of control of dying in a society is the legal stipulation of the taboo transcendence. A Dutch professor of law thus explained why the Dutch euthanasia law was more than 90 pages: "This is what happens when we transcend a taboo. Then we need to regulate it with much precaution."

Defending the new normal occurs in societies with liberal laws toward euthanasia and assisted suicide where the self-ruling challenging taboo has won the wrestle with the control of dying taboo. When the taboo of euthanasia and assisted suicide has been legally transcended, as in Benelux, Oregon, Washington, and Montana, many people are defending the new normal because to challenge it has become anathema. A Dutch physician said "People now see it [euthanasia and assisted suicide] as a human right," and to question human rights is taboo in a democratic society. This is obvious in the Dutch public debate where criticism against euthanasia and assisted suicide is viewed as almost antidemocratic. Euthanasia and assisted suicide criticism is thus a re-tabooing attempt while defending the new normal is a de-tabooing activity.

The growing interest in the control of dying thus seems to indicate that a taboo is being transcended and losing its power in the western world. The comment from a participant in the Swedish survey on physician-assisted suicide (Helgesson, Lindblad, Thulesius, & Lynöe, 2009) illustrates the taboo toward irreversible control of dying: "It feels weird and scary to answer these questions." Yet, the taboo is about to loosen up: in the same survey many respondents, both physicians and the public, spontaneously appreciated the survey that brought up such a difficult issue (Helgesson et al., 2009): "Finally, an interesting questionnaire"; "Complicated but very important survey"; "Thank you for letting me participate"; "Great that you research this issue" were some statements illustrating gratitude toward the frankness of the investigation. Yet, death and dying is still not something that we discuss freely, given the taboo nature of the topic; when a Swedish physician and member of parliament was asked why palliative care still was under prioritized she answered "So many do not want to talk about death."

De-tabooing in the literature

De-tabooing dying started with *Awareness of Dying* (1965) and converges with *Balancing End-of-Life Cancer Care*, another grounded theory (Thulesius, Håkansson, & Petersson, 2003, 2007). Balancing explains problem-solving strategies of health care staff and physicians and offers a comprehensive perspective on end-of-life care and how dying is controlled between cure and comfort care. The Balancing outcome is characterized by Compromising, at best an optimized situation, at worst a deceit. De-tabooing control of dying explains attitudinal and legal changes regarding end-of-life and involves different types of Positioning between the taboos of control of dying and paternalism. Compromising is a property of Positioning, which in turn is an important part of Balancing between the many different care options at the end-of-life.

After I had introduced the de-tabooing theory to a Swedish psychiatrist and suicide researcher, he suggested that a de-tabooing of the suicide discourse had begun after the start of a government zero-vision program re suicide (Beskow, 2009). He then encouraged a de-tabooing of suicide with increased resources for education and research in the field (Beskow, 2010). This influence is an example of the contagious power of a GT concept thus affecting the literature in an area outside of where it was discovered. Support for the importance of de-tabooing control of dying was given by Seales (2000). He draws similar conclusions as those presented in this chapter: such that open awareness of dying has paved the road for a control of dying that can be met by either euthanasia and assisted suicide or palliative care (Bernheim et al., 2008).

When searching for de-tabooing in the literature outside of the medical context, it was found mentioned in an article about Spanish film after the Franco dictatorship referring to de-tabooing sexuality. Directors such as Pedro Almodovar and Bigas Luna are given as examples of de-tabooers (Pappova, 2009). Also, when analyzing extreme right wing voting in Western Europe the authors suggested a de-tabooing process caused by a more restrictive immigration policy which makes it less taboo to criticize immigrants and thus vote for a right wing party (Lubbers, Gijsberts, & Scheepers, 2002). De-tabooing could thus explain part of the phenomenon of the growth of right wing voting in Europe this last decade. Tabooing was further mentioned in reference to Game Conservation in the Amazons. Tabooing the hunting of a particular type of animal would help to preserve that species from being depleted (Hames, 2007). Yet, no de-tabooing /tabooing subcores or conceptual properties are presented in any of these texts, indicating a low theoretical level of the conceptual use of tabooing/de-tabooing. It can also be argued that this study, by asking participants to consider the issues, is an act of de-tabooing and has become part of the emerging de-tabooing process, and on publication will form part of the discourse on de-tabooing.

Limitations

One may argue that the subject of dying is not in the process of being de-tabooed or that dying and the control of dying has never been a real taboo topic. The taboo reaction of emotional positioning and the following cognitive reflecting and stipulating theorized here was yet seen in a large number of people from our interviews, survey, and Internet forum data. The immediate emotional reaction to a taboo (we call it 'emotional positioning') is part of the normal psychological attitude reaction; the attitude process starts with an emotion – the right brain hemisphere decides "likes or dislikes," and impressionism rules. This observation is supported by popular neuroscience findings and theories emphasizing how important emotions are for decision-making (Damasio, 1994). For example, a theory of moral judgment according to which both cognitive and emotional processes play crucial and sometimes mutually competitive roles resonates well with the emotional positioning presented in the present study (Greene, Nystrom, Engell, Darley & Cohen, 2004).

The data relating to the attitudes of the public in this study came from the survey comments collected from people living in Sweden's largest urban area. Had the sample also collected data from those living in rural areas, the views may have been less positive to euthanasia and assisted suicide since euthanasia is twice as common in Dutch urban areas as in Dutch rural areas (Marquet, Bartelds, Visser, Spreeuwenberg, & Peters, 2003).

Conclusion

Today a de-tabooing of dying, and particularly the control of dying, is suggested to be the core explanation to most current issues in the discourse on dying in western society. De-tabooing started with open awareness contexts in the dying situation. Thereafter follow concerns to control the dying process by strategies that are either reversible (palliative care) or irreversible (euthanasia and assisted suicide, withholding or withdrawing life support treatment). Personal autonomy is high up on the list of arguments for euthanasia and assisted suicide that can be expressed as "Life ownership rights." De-tabooing control of dying therefore involves a tabooing of challenging self-ruling aspects of autonomy. Where once control had been in the hands of the health professionals there is a transfer of power to the patient resulting in the de-paternalization of physicians.

De-tabooing also includes taboo positioning - a confrontational reaction to the control of dying. The first step, 'emotional positioning,' involves confusion between the different strategies available to control dying. Positioning then becomes cognitive by reflecting and reasoning by giving examples of cases of different types of control of dying, a labeling wrestle using euphemisms and dysphemisms, and stipulating when irreversible control of dying is acceptable, eventually in the laws of societies where euthanasia and assisted suicide is permitted under certain conditions. These laws are then defended and to criticize them is politically incorrect and thus part of the new taboo of challenging autonomy.

Palliative care aims to minimize suffering at the end-of-life by controlling symptoms while focusing on living. Euthanasia and assisted suicide relieves suffering by delimiting a life of hopelessness and pain or anticipated pain in patients with incurable disease in a controlled and timely, but irreversible way. Both these control of dying paths emphasize the importance to involve relatives and significant others in the care and the decision-making surrounding the care. Both also have relief of suffering and patient autonomy and self-ruling as goals but when death takes place self-ruling is extinguished. A reversible control of dying with the option of palliative sedation thus protects self-ruling and personal autonomy better than the irreversible control of dying by euthanasia and assisted suicide.

References

Andrews, E. (1996). Cultural Sensitivity and Political Correctness: The Linguistic Problem of Naming. *American Speech 71*, 389-404.

Bernheim, J.L., Deschepper, R., Distelmans, W., Mullie, A., Bilsen, J. & Deliens, L. (2008). Development of palliative care and legalisation of euthanasia: Antagonism or synergy? *British Medical Journal 336*, 864-867.

Beskow, J. (2009). Deeper knowledge of suicidality necessary. Cognitive perspectives on the debate about the zero vision. *Läkartidningen 106*, 1917 1918. Swedish.

Beskow J. (2010). Let us break the suicide taboo! *Läkartidningen 107*, 960-9611. Swedish.

Burridge, K. (1996). Political correctness: euphemism with attitude. *English Today 12*, 42-43

Cohen, J., Marcoux, I., Bilsen, J., Deboosere, P., van der Wal, G. & Deliens, L.(2006). Trends in acceptance of euthanasia among the general public in 12 European countries (1981-1999). *European Journal of Public Health 16*,663-669.

Damasio, A. (1994). *Descartes' error: Emotion, reason and the human brain.* New York, NY: G.P. Putnam's Sons.

De Wachter, M. A. M. (1989). Active euthanasia in the Netherlands. *Journal of the American Medical Association 262*, 3316-3319.

Emanuel, E.J. (1994). Euthanasia. Historical, Ethical, and Empiric Perspectives. *Archives of Internal Medicine 154*, 1890-1901.

Fang, F., Valdimarsdóttir, U., Fürst, C.J., Hultman, C., Fall, K., Sparén, P., Ye, W. (2008). Suicide among patients with amyotrophic lateral sclerosis. *Brain 131*, 2729-2733.

Glaser, B.G. (1978). Theoretical sensitivity: Advances in the methodology of grounded theory. Mill Valley, CA: Sociology Press.

Glaser, B.G. (1992). Basics of grounded theory analysis: Emergence vs. forcing. Mill Valley, CA: Sociology Press.

Glaser, B. G. (1998). *Doing grounded theory: Issues and discussions.* Mill Valley, CA: Sociology Press.

Glaser, B. G. (2001). The grounded theory perspective: Conceptualization contrasted with description. Mill Valley, CA: Sociology Press.

Glaser, B. G. (2003). The grounded theory perspective II: Description's remodeling of grounded theory methodology. Mill Valley, CA: Sociology Press.

Glaser, B. G. (2005). *The grounded theory perspective III: Theoretical coding.* Mill Valley, CA: Sociology Press.

Glaser, B. G. (2007). *Doing formal grounded theory.* Mill Valley, CA: Sociology Press.

Glaser, B. G., & Strauss, A. L. (1965). *Awareness of dying.* Chicago, IL: Aldine.

Glaser, B. G., & Strauss, A. L. (1967). The discovery of grounded theory: Strategies for qualitative research. Chicago: Aldine.

Glaser, B. G., & Strauss, A. L. (1971): *Status passage: A formal theory.* Mill Valley, CA Sociology Press.

Greene, J.D., Nystrom, L.E., Engell, A.D., Darley, J.M. & Cohen, J.D. (2004). The neural bases of cognitive conflict and control in moral judgment. *Neuron 44*, 389-400.

Hames, R. (2007). Game conservation or Efficient Hunting, in *Evolutionary perspectives on Environmental problems.* In Penn, D.J. & Mysterud, I. (Eds.) New Brunswick, NJ: Transaction Publishers.

Helgesson, G., Lindblad, A. B., Thulesius, H. & Lynoe, N. (2009). Reasoning about physician-assisted suicide: analysis of comments by physicians and the Swedish general public. *Clinical Ethics 4*, 19-25.

Löfmark, R., Nilstun, T., Cartwright, C., Fischer, S., van der Heide, A., Mortier, F., Norup, M., Simonato, L. & Onwuteaka-Philipsen, B.D. (2008). EURELD Consortium. Physicians' experiences with end-of-life decision-making: survey in 6 European countries and Australia. *Bio Med Central Medicine 6*, 4.

Lubbers, M., Gijsberts & M., Scheepers, P. (2002). Extreme right-wing voting in Western Europe. European *Journal of Political Research 41*, 345-378.

Marquet, R. L., Bartelds, A., Visser, G.J., Spreeuwenberg, P. & Peters, L. (2003). Twenty five years of requests for euthanasia and physician assisted suicide in Dutch general practice: trend analysis. *British Medical Journal 327*, 201-202.

Misono S., Weiss N.S., Fann, J.R., Redman, M. & Yueh, B. (2008). Incidence of suicide in persons with cancer. *Journal of Clinical Oncology 26*,4731-4738.

Pappova, P. (2009). Multiculturalism and a Search for Identity in Spanish Film Production after the Fall of Francoism. *Ars Aeterna 1*, 63-71.
Proctor, R. N. (1988). *Racial hygiene: Medicine under the Nazis.* Cambridge, MA: Harvard University Press
Rietjens, J. A., van Delden, J. J., van der Heide, A., Vrakking, A. M., Onwuteaka-Philipsen, B.D., van der Maas, P. J. & van der Wal, G. (2006). Terminal sedation and euthanasia: a comparison of clinical practices. *Archives of Internal Medicine 166*, 749-753.
Oxford's Concise Medical Dictionary (2010). Oxford UK: Oxford University Press.
Seale, C. (2000). Changing patterns of death and dying. *Social Science and Medicine 51*, 917-30.
Sepúlveda, C., Marlin, A., Yoshida, T., Ullrich, A. (2002). Palliative Care: the World Health Organization's global perspective. *Journal of Pain and Symptom Management 24*, 91-96.
Taboo. (2011). *Visual Thesaurus.* Retrieved from http://www.visualthesaurus.com/
Thulesius, H., Håkansson, A. & Petersson, K. (2003). Balancing: a basic process in end-of-life cancer care. *Qualitative Health Research 10*, 1353-1377.
Thulesius, H., Grahn, B. (2007). Reincentivizing – a new theory of work and work absence. *Bio Med Central Health Services Research 7*, 100
Thulesius, H., Håkansson, A. & Petersson, K. (2007). Between comfort and cure - basic balancing strategies in end-of-life cancer care. In Holton, J., & Glaser, B., (Eds.), *The grounded theory seminar reader.* Mill Valley CA: Sociology Press.
Thulesius, H., Sallin, K., Lynöe, N. & Löfmark, R. (2007). Proximity morality in medical school--medical students forming physician morality "on the job": grounded theory analysis of a student survey. *Bio Med Central Medical Education 7*, 27.
Walter, T. (1991). Modern death: Taboo or not taboo? *Sociology 25*, 293-310.

10

On Translating Grounded Theory When Translating is Doing

Massimiliano Tarozzi
University of Trento

I am a social researcher in education teaching qualitative methods, particularly grounded theory, at the university. Although I am neither a translation studies expert nor a linguist, I have been dealing with theoretical issues related to translation in grounded theory (GT) for a number of years. The following thoughts stem from the fact that I am the Italian translator of *The Discovery of Grounded Theory*[1] and have had the opportunity to discuss some of my questions about the book with one of its authors, Barney Glaser. This seminal text required two years of scrupulous labor to translate, as is often the case with a well-known book about a complex subject.

Although it is not my native language, the majority of the methodological literature I read is in English, as is the case of all the non-English native speakers operating in an academic world. This literature is also the main reference for the methodology courses I teach, which are conducted in Italian. Finally, I often prepare papers and presentations about my Italian investigations by transforming my interviews, codes, categories and memos into English, the lingua franca of international research. Working as a member of international research teams has prompted me to deal with translation theory as well. When I wish to share my research data with non-Italian-speaking colleagues or, more specifically, when I ask them for an external audit of analyses originally written in Italian, I have to translate data and codes from Italian to English (or from English to Italian when I present international results to an Italian audience).

Such consistent linguistic exercise, very common among non English native speakers, prevents me from taking research-related translation issues for granted. As a bilingual researcher, I have come to consider the use of language a non-neutral research tool. Moreover, in this chapter, I argue that dealing with translation while doing GT should be viewed as a resource rather than a difficulty, since it offers the analyst an additional tool for analysis. Ironically, in the constructivist climate of the moment, where language's key role is somewhat overestimated, it surprises me that translation issues with few exceptions (Shklarow, 2009) are taken so much for granted or ignored. In this chapter, I want to make the reader aware that GT translation can be integral to research.

Summary and rationale

I intend to put the microscope under the microscope. In other words, I wish to take into account the cultural and linguistic implications of translation in research, particularly when using GT. If I posit the translation process as a non-neutral tool, according to GT it can be regarded as "data" and not as a means. This chapter aims to discuss two main topics: 1) some methodological suggestions ensuing from my experience of translating the founding text of GT, especially in the light of recent literature regarding translation studies and; 2) some methodological implications of a bilingual focus in doing GT.

I think it is worth mentioning here that I am not a radical constructivist, either at the epistemological level or in terms of translation studies. Although it is hard to ignore the key role of language in constructing, rather than simply conveying, meanings, I do not believe that problems raised by the translation process should force one to resort to a radical relativism and deem it impossible to say the same thing in two languages. Derrida's (1995) deconstructionism in translation denies the possibility of the equivalent meaning of a word in one language and its correspondent in another, leaving the translator as the suspect author of an autonomous work, far from the original.

Translators, according to Umberto Eco, are inclined to the generic possibility of equivalence (fidelity), but limited as to "propositional content" (Eco, 2003a). Eco claims that literal translation is impossible, as the failure of automatic translators demonstrates. However, in translating, he believes one can say almost the same thing. This "almost" is a propositional content and the result of linguistic and, above all, cultural negotiation (Eco, 2003b): "To translate means to understand the internal system of a language and the structure of a given text in that language, and to build up a double of the textual system which, under a certain description, can produce analogous effects on the reader on several levels: semantic and syntactic, stylistic, metrical, phonosymbolic and also emotional, where textually intended" (Eco, 2003a, p.16, my translation).

To illustrate the impossibility of a translation devoid of linguistic and cultural negotiation processes, Italian readers can witness the pratfalls of automatic translation into Italian by applying either translate.google.com or Yahoo's Babelfish to the first paragraph of *The Discovery of GT*. Readers of English may also enjoy the retranslation into English of Babelfish's robotic Italian (Table 1).

In this chapter, I would like to show what kind of negotiation processes need analysis and interpretation in translation, according to the recent literature of translation studies. What is more, I argue that dealing with such issues is an exercise that can lead to a better understanding of the nature of the method and how to use it.

Table 1: Translation and re-translation of the *Discovery's* initial paragraph:

Original	Automatic Translation	Human-interpretive translation	Automatic Re-translation
The discovery of grounded theory Most writing on sociological method has been concerned with how accurate facts can be obtained and how theory can thereby be more rigorously tested	la scoperta della teoria a terra La maggior parte di scrittura sul metodo sociologico è stato interessato con fatti precisi come può essere ottenuta come teoria e può quindi essere più rigorosamente testati	La scoperta della grounded theory Quasi tutti i testi di metodologia della ricerca sociale si sono occupati prima o poi di come ottenere risultati fattuali precisi e quindi di come una teoria possa essere rigorosamente testata	the discovery of the theory to earth The majority of writing on the sociological method has been interested with precise facts like can be obtained like theory and can therefore be more rigorously heads

Translating is doing GT

In one of his *Essays*, Italo Calvino wrote that "Translating is the true way to read a text" (Calvino, 1995). He recommends the exercise of translation to all good readers because he believes that one truly reads an author only by translating his/her work into a different language and comparing it with the original or by comparing different versions of the same text. The operations related to interlinguistic translation allow one to interpret and, thus, fully understand a text. When this is not only a non-fictional text, but a sociological classic showing a revolutionary research method, such understanding can become a unique way of discovering the significance of that method's key concepts.

That is why a process of understanding is implicit in every translation. Or, as George Steiner put it, "understanding is translation" (Steiner, 1975). The acts of analysis and interpretation are not only embedded in translation between two texts, but also between the two cultural and linguistic encyclopedias in which they are rooted.

The idea that every translation is both a transfer of words from one language to another and an intercultural connection is not only a product of the post-modern cultural turn in translation studies. In the first century B.C., Cicero wrote in *De optimo genere oratorum* that it is not appropriate to translate word for word (verbum pro verbo). What is necessary is to keep alive the power and the efficacy of the terms, even when this requires the translator to move away from the original word.

In order to translate a text like *Discovery*, familiarity with the original language is not enough, it is essential to know the topic. When Cicero translated Demosthenes and other Attical orators from Greek to Latin, he claimed it was necessary to do so *ut orator*, as an orator himself. Similarly, I rendered Glaser's and Strauss's book into Italian first as a social researcher and then as a translator. Although his command of English was better than mine, the professional translator who began *Discovery* became quite often enmeshed in misunderstandings and misconceptions typical of those not acquainted with the sociological topic and the cultural and scientific setting which brought the book to light.

I believe it is not enough to know the cultural milieu of the text in order to adequately translate an abstract and not always fluent methodological book. What is essential is that the translator have direct knowledge of and specific experience with the research so that he or she can clarify terms and concepts that would otherwise remain ambiguous or de-contextualized. That is the value of having lived experience, as opposed to being widely read, in this or other methods. One has to have undergone the same process in order to deeply understand and accurately translate the meaning of endogenous expressions, textual examples and nuances.

For instance, to translate the adjective "theoretical" in Italian was very tough, because of its ambiguity. But this is a key adjective in the book that qualifies both the sampling and the attitude of the researcher (theoretical sensitivity). In Italian, it can be rendered either as "*teorico*" or as "*teoretico*." The latter is a typically philosophical adjective, which is equivalent to "speculative" as opposed to "practical" on one side and "empirical" on the other; the former is a commonly used form indicating what is relative to the theory, almost equivalent to "conceptual." I used both: for "sampling" *teorico* and for sensitivity *teoretica*. Because I wanted to connote sampling as closely related to the processes of conceptualizing and data analysis. Therefore *teorico* worked better as a word less abstract, speculative and philosophical but still associated with the emergence of theory. On the other hand, sensitivity is a philosophical virtue, rooted in the Greek word *theorein* and *teoretica* emphasized this aspect.

Table 2: Comparing two different translations of "Theoretical"

<table>
<tr><td>Theoretical sampling</td><td rowspan="2">Theoretical</td><td>Teorico</td><td>common use, equivalent to "conceptual".</td><td>Campionamento teorico</td></tr>
<tr><td>Theoretical sensitivity</td><td>Teoretico</td><td>philosophical equivalent to "speculative"</td><td>Sensibilità teoretica</td></tr>
</table>

Contemporary translation studies concur in considering translation an act of intercultural mediation. The language-culture link has several implications:

1. If ignored, it may cause misunderstandings. Meanings always reflect cultural models: the Italian proverb "*Chi dorme non piglia pesci*" (He who sleeps doesn't catch fish), which is similar to "The early bird catches the worm." But the Italian idiom can be extrapolated to include laziness and presupposes the understanding that laziness is a socially and culturally negative attitude. Otherwise it cannot be understood. Similarly the English expression "It's raining cats and dogs" cannot be understood by someone who, even knowing the Queen's English, does not share a common cultural background.
2. More frequently, the semantic power of a translation that ignores cultural settings is impoverished and it looses dramatically its semantic power. A good example is the translation of a key term like "grounded," which doesn't have an exact equivalent in Italian. "Grounded" is untranslatable mechanically (Babelfish translates it as "*teoria al suolo*," literally, theory to the ground/soil). As it is, grounded has countless shades of meanings. The past participle of the verb to ground can mean "rooted or based," but also "a ship or boat that touches the bottom of the sea and is unable to move off," or "an aircraft not allowed to take off: to teach first rudiments, to prepare the background of a drawing." So grounding a theory in data has radical as well as material reverberations; it is the vital, occasionally even violent, rooting of creative epistemology in the fertile humus of experience. At the same time, it needs to be a rooting so precise and punctual that it can serve as the basis for further constructions, the ground on which to build complex formal theories. A theory like this is not only based on facts or empirically derived from data. To grow organically, it requires the mulch and compost of lived experience.
3. I have analyzed these shades of meaning, without taking them for granted, only because I had to translate the text. However, interrogating and searching the semantic area of the term "grounded" allowed me to better understand its relevance, meaning and workability as well as the way to apply this key notion which qualifies the specific nature of the method inaugurated by Glaser and Strauss. In this sense, translating was the "true" way to read *Discovery* and to grasp the meaning of some of the method's key concepts: What is a theory? What does it mean to create a theory? How can a researcher work theoretically? What is theoretical sensitivity? All these issues were raised by the need to develop all the meanings intertwined in the word "grounded," which before I took for granted and used naively, without any special attention.

4. Finally, I wish to stress that translating is a way to keep in contact with both the culture-source and the culture-target. Translating *Discovery* was a way to introduce GT in Italy. This purpose was very clear to those who promoted this cultural enterprise. In this sense, Strati, the book's editor, and I debated whether the translation should pay philological respect to the original, now considered a classic, or would be better regarded as a living work with practical value for contemporary Italian readers.

This is a typical translator's dilemma of which there are distinguished historical examples like Martin Luther's translation of the Bible into German. Umberto Eco observes that in describing his work, Luther used *übersetzen* (to translate) and *verdeutschen* (to Germanize) interchangeably. In doing so, evidently, Luther saw translation as cultural assimilation.

When translating a text from another culture, there is always a choice between "domesticating" and "foreignizing" (Venuti, 1998). When domesticating a translation, translator and editor decide to play down cultural differences as much as possible by bringing the original text within the philological parameters of the target culture, eliminating every roughness and vanishing the translator. It is a kind of cultural assimilation work, ethnocentric to some extent, in which the dominant culture prevails; on the contrary, foreignizing means purposely maintaining some estrangening elements of the parlance of the culture of origin, which, though they may undermine the overall fluency of the text, serve to remind the reader of its difference and distance from the host culture.

Such a distinction is important per se. It could be very helpful for a GT user to think about the cultural implications of the translation process. Moreover, the aforementioned alternative has been adopted and applauded by postcolonial scholars as a way of underlining the importance of avoiding cultural assimilation to a dominant Western model. In our case, the risk was the opposite. As Italians, we were in jeopardy of being encompassed by the double hegemony of the Anglo-Saxon culture of social science and the current dominance of the English language in the scientific community. We were translating *Discovery* from American English, the dialect of the sociological mainstream. Texts coming from a powerful culture tend to be translated with key words of the original language intact, even if this makes them less intelligible for a target-culture audience.

In fact, (in contrast with some other European languages[2]) we decided not to translate the expression "grounded theory" into Italian but to leave it in English, in the title and in the text. This expression is now so widespread among academics and laymen that to translate it would be to create a kind of pseudo- foreignness or an estrangening or even comic effect. There is no need to artificially introduce a new expression after more than forty years of saying it in English and not only for historical reasons. While searching for terminological correspondence, the translator must renounce some word's properties, saving only those relevant to the context. As we have seen, "grounded" is a term too rich in semantic variations to render into Italian with a single word. Since this is the key-notion and the core concept of the whole book, we preferred to use the original English, without choosing only one meaning in Italian words. A translator forgoes his duty, in this case, honorably, without surrender, since this is part of the processes of intercultural negotiations that every translation requires.

Apart from postcolonial claims, the domesticating vs. foreignizing distinction reminds us of Eco's dictum: "A good translation is always a critical contribution to the understanding of the translated work" (Eco, 2003a, p. 247). It is also why there is never a unique possible translation or a universal lexicon for translators. The meaning of a word or proposition is not only a linguistic construction; it is also pragmatic, historical, semiotic and, in a broader sense, cultural. Enlarging the practice of translation to the semiotic sphere (before it was narrowed to mere linguistic practice) is due mainly to Roman Jakobson (1959). However, in the case of the translation of a text about research methodology (and particularly this book), the translation process does not limit itself to invading the semiotic sphere. It has just as much to do with the understanding and the use of a method of social research. To some extent, translating *Discovery* was like doing GT, as an extension of Glaser's claim that writing the book was ipso facto GT. In our conversation (Tarozzi & Glaser, 2007), speaking about the sense of the Italian translation and referring to the genesis of *Discovery*, he said: "The book itself is a grounded theory. It wasn't thought up. It was based on doing *Awareness* and *Time for Dying*. So it was grounded in research. That has tremendous grab".

By the same token, translating is doing GT because every translation is a form of interpretation, an investigation of meaning, a rigorous inquiry aimed at understanding a text (rather than a phenomenon of the real world). Moreover, according to the hermeneutic perspective, it extends the nature of a text to the whole reality. From Heidegger to Gadamer, interpreting is translating.

As I said, I do not belong to the hermeneutic tradition. As far as I am concerned, there is no perfect coincidence between interpreting and translating, but one cannot deny that translation is a form of interpretation that uses language as a medium, and that interpretative acts always precede it (Eco, 2003a). Even though translation does not overlap the practice of research, the two are so closely related that they can be a mutual source of useful methodological suggestions, especially within a GT framework.

The following table outlines correspondences between these two parallel processes:

Table 3: Similarities between translation and GT analytic processes

TRANSLATION PROCESS	GT PROCESS
Reading (the source text)	Data collection and open coding
(Semiotic) analysis of the source text	Focused/selective coding
Interpretation	Theoretical coding
Elaboration in the target text	Integrating theory
Writing in the target text	Writing the report

Translation is a process of understanding meanings that requires the translator to exercise interpretive and analytical acts in the source language as well as elaboration and writing in the target language. I believe it is not an exaggeration to compare the translator's preliminary reading to data collection and simultaneous open coding and the subsequent semiotic analysis as well as the interpretation phase to advanced and theoretical analysis.

Moreover, translation requires negotiation skills to constantly mediate the inevitable gap in equivalence between cultural and linguistic systems. These negotiation skills, which are not systematized by guidelines, procedures or structural educational courses, remind me of some of the characteristics of theoretical sensitivity which have to do with momentum, insight and seeing possibilities.

Coding in another language as analytic resource

A second point I said I would deal with in this chapter is the fact that doing research in another language is a powerful analytic resource, when the researcher is using an inductive and comparative method aimed at generating theory. In 1959, Roman Jakobson wrote an essay "On linguistic aspects of translation," making an important contribution by comparing translation to other disciplines like semiotics, cultural anthropology, narrative studies, etc.

In this essay, Jakobson distinguished three types of translation:

1. Endolinguistic (or intralinguistic). The interpretation of signs throughout other signs of the same language. It is translation within the same linguistic system, by means of reformulation; for example bachelor=not married; or transcription of an oral message in its written form. This is called rewording.
2. Interlinguistic. The interpretation of signs throughout the signs of another language. It is the transposition of signs by means of different linguistic systems. This is translation proper.

3. Intersemiotic. The interpretation of verbal signs throughout non-verbal sign systems, like the transposition of a novel into a movie. Within this type of translation lies also the Ekphrasis: the exercise of the ancient rhetoric which consisted of a written translation of a visual work of art.[3]

These three types share the same characteristic: every full equivalence between the cultural and linguistic systems of the starting and arrival text is impossible, as is clearly demonstrated by the huge difference that exists when a translated text is retranslated into its original language. As if, for example, someone would re-translate in Italian the classical French version of Dante's *Inferno*, edited by Emile Littré. The final effect would probably be far from the original (Steiner, 1975).

However, Jakobson's three types of translation interest us because they reinforce the correspondence between the translation process and the GT analytic process and reveal it as a powerful analytic resource. For example, the transformation-into-text of data taken from facts, events and phenomena can be understood as an operation of intersemiotic translation. When I use observation as an instrument of data collection, I am transmuting the acts of some subjects within a context into a text that can be elaborated and analyzed. It is an exercise similar to the Ekphrasis. The transcription of an interview is always a translation act, whether recorded in field notes (as Glaser suggests) or tape-recorded. Some claim that the transposition of an oral message in its written form is a transcription or a notation and only a translation in the metaphorical sense (Mounin, 1965). However, regarding the transcription of interviews aimed at an analysis within a GT, I believe that every transcription is also a translation. There is nothing automatic about transcription. It is the first analytical level, since it is an interpretive job that reduces complex verbal and non-verbal communication to a unique textual dimension (Tarozzi, 2008, p. 86).

Moreover, every time we use a different language as a research instrument, by interviewing, coding and writing in another language, there are evident difficulties in transposing these data and thoughts into another linguistic-cultural context, as well as remarkable benefits and extra resources, that need to come with some warnings.

In sum, at the narrowest methodological level, we cannot take for granted the question of translation from a non mother-tongue in doing GT. In particular, coding in another language requires continuous acts of interlingual translation which increase our facility to comprehend, add sophisticated interpretive instruments, refine analysis. Every interlinguistic translation is the result of elaborate acts of de-coding, in the source language, and re-coding in the target language, which occur at several levels: semantic, syntactic and pragmatic. Therefore these continuous processes of de-coding and re-coding support and make more effective the various coding phases that GT requires for data analysis.

For example, a few years ago I took a troubleshooting seminar with Barney Glaser to which I brought my research data to discuss with him and the group. Having to translate my first chaotic code map (see Figure 1, below) into English put into effect the subtle operation of interlinguistic translation to which I refer above, allowing me to clarify it, to let relevant categories emerge, to individuate principal links among them, to define ambiguous concepts by identifying some of their properties and even to recognize, although still in a rough form, the conceptual area in which the core category was embedded.

Figure 1: First code map of a GT research

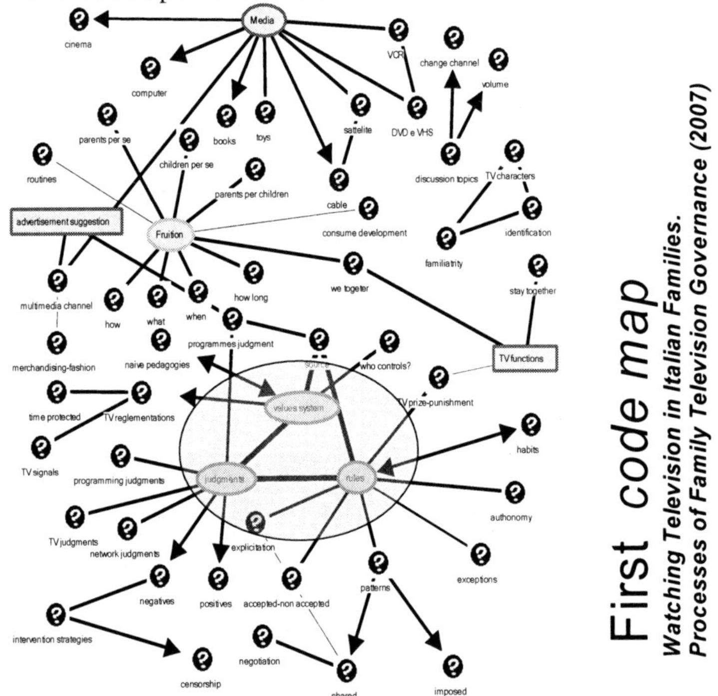

In this case, the usually boring and time-consuming work of translating the data and open coding became an added and welcome instrument of analysis. Because translation always presupposes a process of understanding-interpretation-analysis, it can represent a precious new instrument in the researcher's hand to deal with data. It is another field of constant comparison that represents and strengthens the heuristic foundation of GT. We can add the comparison among different sign systems which express data and categories to the ordinary comparison among data, data and categories and categories among them. This will produce new conceptualizations, promote the emergence of insight, and train the theoretical sensitivity.

Coding in Italian, coding in English

While doing GT in a language different than the mother tongue is generally a resource for the researcher, we have to take the characteristics of the language seriously into account. Italian is particularly suitable for supplying careful, rich and refined descriptions. For that reason, I think Italian is more suitable than English for the first phases of research and memos. That may be why English has been called an "isolating (or analytic) language," while Italian is more "inflectional" (Comrie, 1989). Italian has several declensions to express grammatical relations, relational categories such as gender and number for nouns and lots of variation in verb conjugations that enlarge the possibility of precise description.

On the other hand, English is a more conceptualizing language than Italian, and has greater propositional power. Therefore, it seems more suitable for making propositional statements, binding concepts, expressing complex and tricky categories with synthetic nomenclature. Because of this, I prefer English for more advanced coding, where it is necessary to label concepts. In the early stages of analysis (open and initial coding) Italian is particularly suitable because it corresponds more closely to the original data. The more the analysis proceeds into selective and theoretical coding, the more English becomes appropriate for sorting and conceptualization.

Table 4: Isolating vs. Inflectional languages and their use in GT analysis

English	Isolating	Analytic	Suitable for advanced coding and memoing
Italian	Inflectional	Descriptive	Suitable for early coding and memoing

One could enter into prolonged discussions about the comparative, cultural and anthropological, implications of using a language that tends to organize, systematize and code the world within propositions which trap concepts into true-false assertions, rather than employing a more descriptive, narrative language which is both creative and versatile. Here I will limit myself to warning researchers about the consequences of the use of a language which tends to organize the world in one way or another, in terms of the construction of an interpretive theory of a certain phenomenon. In either case, I warrant it worthy of further research that two investigations of the same object, with the same method but carried out in different languages, will produce slightly different theories.

With regard to the implications of typical features of language in research practice, the use of the gerundive "-ing form" is emblematic of GT. First Glaser (1978), and then Charmaz (2006), have invited its use as a form to express categories in coding and in writing memos. According to Charmaz, it "fosters theoretical sensitivity because these words nudge us out of static topics and into enacted processes" (Charmaz, 2006, p. 136).

A gerund is the present participle of a verb used as a noun implying action. But this form does not exist in Italian or in many other languages. In its place, we use the infinitive form which freezes our conceptual labels into rather icy and fix descriptions. While coding in Italian, we must make do with these more static grammatical forms, which neither connote the dynamic movement of a concept, nor disclose the action's power.

Very often, while we are working in Italian, it occurs to me to code in English, exactly because of the flexibility of the language in expressing synthetically dense concepts and in order to better emphasize processes. Otherwise, in Italian, we have to resort to nouns which are more flexible or to use expressions which lack grammatical accuracy but allow us to preserve the intensity and evocative power of a code. As English is more synthetic than Italian and condenses meaning with fewer words, it is perfect to create titles and slogans, as well as categories. On the other hand, these syntheses are less precise than in Italian, which gives the researcher more linguistic pixels to denote a concept or a category with greater clarity and expressive power.

Conclusion

In the past forty years, the GT methodology has proposed and propagated a specific research language. Currently, expressions like "theoretical sampling," "core category," "saturation," and "constant comparison" are in worldwide circulation and have become part of the technical language of social science. They have also contributed in delimiting the originality and uniqueness of this methodology.

Nevertheless, Barney Glaser sees the worldwide circulation of the GT "jargon" as a trivialization of his method (Glaser, 2009). Ironically, jargon legitimates and credentializes as it trivializes and narrows the complexity of the world. Glaser sees the considerable success of GT as, thus far, tied to the widespread use of its legitimizing expressions: a nomenclature which otherwise would have survived a more substantial use of the method itself. After four decades, he believes these expressions are worn out and have lost their "grab" or original conceptual power. I am not sure whether or not this is true but I believe that, by translating these conceptually dense expressions in a different cultural and linguistic system, we can revitalize that power by simultaneously preserving and renewing their semantic meaningfulness.

If, when translating these key notions for contemporary social research, we cannot find the word that most faithfully corresponds to the original, we can say almost the same thing. This "almost" includes not just the aforementioned negotiation processes but also the possibility of renovating the method itself by critically rethinking it from a perspective 10,000 kilometers (6,300 miles) and 44 years away but, experientially, very close to the place where it was generated.

References

Arduini, S. & Stecconi, U. (2007), *Manuale di traduzione. Teorie e figure professionali.* Roma: Carocci.

Bassnett-Mcguire, S. & Trivedi, H. (Eds.) (1999). *Post-colonial translation. Theory and practice.* London: Routledge.

Calvino, I (1995). *Saggi 1948-85.* 2 vols. Edited by M. Barenghi. Milano: Mondadori, Tradurre è il vero modo di leggere un testo. Tomo II. pp.1825-1831.

Charmaz, K. (2006). *Constructing grounded theory: a practical guide through qualitative analysis.* London: Sage Publications.

Comrie, B. (1989). *Language Universals and Linguistic Typology, 2nd ed.* Blackwell: Oxford. Translation from the Italian *Universali del linguaggio e tipologia linguistica.* Bologna: Il Mulino, 1983.

Derrida, J. (1995), Des tours de Babel, in Graham (ed.), *Differences in translation.* Ithaca: Cornell University Press, pp. 209-248 (trad. It. In Nergaard (ed.), *teorie contemporenee della traduzione.* Milano: Bompiani. 1995, pp. 367-418).

Eco, U., & McEwen, A. (2001). *Experiences in translation. Toronto Italian studies.* Toronto: University of Toronto Press.

Eco, U. (2003a). *Dire quasi la stessa cosa.* Milano: Bompiani.

Eco, U. (2003b). *Mouse or rat?: translation as negotiation.* London: Weidenfeld & Nicolson.

Glaser, B. G. (1978). *Theoretical sensitivity: advances in the methodology of grounded theory.* Mill Valley, CA.: Sociology Press.

Glaser, B. & Tarozzi, M. (2007). Forty years after Discovery. Grounded Theory worldwide. Barney Glaser in conversation with Massimiliano Tarozzi, In *The Grounded Theory Review*, Special issue, November 2007, pp. 21-41

Glaser, B. (2009). *Jargonizing. Using the Grounded Theory Vocabulary.* Mill Valley, CA: Sociology Press.

Jakobson R (1959), On linguistic aspects of translation. In *Saggi di linguistica generale*, Feltrinelli, 2002 (ed. or 1963)

Mounin, G. (1965). *Teoria e storia della traduzione.* Torino: Einaudi.

Shklarow, S. (2009). Grounding the Translation: Intertwining analysis and translation in cross-language grounded theory research. *The Grounded Theory Review*, vol.8, no.1, pp.53-74.

Steiner, G. (1975). *After Babel: aspects of language and translation.* New York: Oxford University Press.(trad. it. Dopo Babele. Milano: Garzanti. 2004)

Tarozzi, M. (2008). *Che cos'è la grounded theory.* Roma: Carocci.

Venuti, L. (1995). *The translator's invisibility.* London: Routledge. Trad it. L'invisibilità del traduttore. Roma: Armando, 1999

Endnotes

[1] Edited by A. Strati, translation by M. Tarozzi, *La scoperta della Grounded theory*, Roma: Armando, 2009.
[2] In German, 'grounded theory' remains in its original English form but can also be translated as "gegenstandsverankerte Theoriebildung." In French, it is rendered as "l'analyse par théorisation ancrée." In Spanish, it is "teorìa fundamentada," in Polish, "teorii ugruntowanej" and, in Swedish, "Grundad teori."
[3] See, for example, the description of the famous picture Las meniñas by Velasquez, in the introduction of Foucault's book, *Les mots et les choses.*

Part III

Historical and Philosophical Grounding

11

Lessons for a Lifetime: Learning Grounded Theory from Barney Glaser

Kathy Charmaz
Sonoma State University

This chapter looks back at Barney Glaser's early teachings of grounded theory from the viewpoints of doctoral students who studied with him at the University of California, San Francisco (UCSF) in the 1960s and '70s. Barney's sequential grounded theory courses formed the core of the innovative sociology program at UCSF. The small size of his classes (around ten or fewer), workshop format, collective responsibility to advance each student's analytic work—and particularly Barney's incisive analytic mind—made the classes exciting, inspiring, and productive. At that time—as now—devoting almost two years to formal courses in qualitative *analysis* was astonishing. Some doctoral programs in sociology offered courses in field research but typically concentrated on entry into the setting, membership roles, and data collection, not on analysis. In his writing and teaching, Barney stood at the forefront of making analytic strategies explicit and of developing useful conceptual analyses that addressed events and issues in the field setting.

What did studying with Barney for a sustained period mean for the early students' lives and careers? As their careers are ending or have ended, this chapter focuses on their reflections about what they gained from Barney and what they took with them throughout their careers. The chapter highlights the voices of these students, their recollections, and experiences. I draw on their responses to an open-ended questionnaire that I sent to as many former students as I could locate and also on several published statements.

"Getting It"

In the day-to-day lexicon of Barney's grounded theory seminars, getting it meant that students not only understood grounded theory but could use it in the moment. Understanding the logic and form of grounded theory, however, challenged numerous early students to rethink and discard their earlier preconceptions of research and theory. Grounded theory not only undercut dominant 1960s assumptions of how to conduct social research, it also undermined conventional sociological notions of society and social relationships of that day. At a time when logico-deductive inquiry prevailed, at least as prescribed in textbooks, grounded theory began with inductive logic and viewed social reality as consisting of processes. Getting it meant that numerous students had to grapple with research and theory in new ways.[1] Tom Lonner captures the difficulties students had with getting it:

> Many in my class came prepared with quite structured views of social, community, and institutional process. Indeed, most felt that order was fixed and change was problematic, as opposed to the reverse. Most had some difficulty in seeing the world as being constantly constructed according to interesting patterns and under specific conditions. As a result, few in my class graduated, if memory serves, due to an inability to see reality through alternative lenses. I must admit that I shared this limitation early in my first class with Barney. I remember presenting my first analytic foray in class and Barney said, "Tom, you are not getting this. If you continue to not get it, you will not pass this class." That scared me to death as I thought I had been getting it. There was more than a bit of Schadenfreude among my classmates about this brief encounter. But I did pass the class and incorporate everything that Barney, Anselm [Strauss], Lenny [Schatzman], Fred [Davis], and Carroll [Estes] drilled into me.

I recall Barney and Anselm expending substantial effort and time to help students and new scholars to "get it" and bring them into the grounded theory fold. At that time, the classes sometimes included a doctoral or postdoctoral student from Education or Public Health at other universities and occasional visitors. Their presence seemed to kindle Barney and Anselm's zeal to reveal the limitations of other forms of research and to show them the advantageous grounded theory way. They found a ready audience in students who had already discovered the limitations of traditional research. These students got it. David Hayes-Bautista shows how his interests led him to seek qualitative methods and new forms of theorizing.

> As my original training had been in engineering, I had a pretty good grasp of quantitative methods (I wrote my first computer program in 1963, on an analog computer). I specifically wanted to learn this thing called qualitative methods, and Barney's class was an excellent preparation. I was very unhappy with the theoretical models used to analyze Latino health and behavior at the time (and I still am---see my recent book *La Nueva California: Latinos in the Golden State* [University of California Press] for a recounting of my intellectual journey) and wanted to gain the skills to create new theoretical models. This course was just the thing!! I was able to do what I wanted, in a theoretical development sense. And I've been doing so ever since.

Getting it—and keeping it—was not always smooth. Students came to grounded theory from different starting points and got it at different rates.

Some gifted students drifted away; some never got any of it, and a few questioned whether grounded theory would be accepted in their respective disciplines. But others persevered. Phyllis Stern (2009; Stern & Covan, 2001) remembers that during the first session of their grounded theory seminar, Barney asked the nursing students to drop the course to reduce the class size. Stern turned out to be the lone nurse who persisted in staying in the class. The strength of her commitment became apparent years later when she disclosed that she not only felt unwelcome in Barney's class, but also felt like an imposter in higher education (Stern, 2009). In her response to my query, she states:

> At first I was completely confused by all the sociology jargon, but I have long believed that propensity leads to understanding, and little by little, I got it. The sessions were by and large exciting, as we analyzed one another's data.

Eleanor Covan began her sojourn into grounded theory as a critic and questioner of the method. Stern and Covan (2001) wrote about Covan's resistance to adopting grounded theory after having gained strong quantitative training:

> Glaser's "trust me" approach did not convince her that qualitative findings based on grounded theory would be well-received. This attitude is common in students enrolled in process learning courses....Now that we have learned it, it is clear to us that process learning is the only way to teach grounded theory, but we find students become as confused as we were initially. (p. 22)

Those students who sooner or later resonated with Barney's approach embarked on an intellectual journey that changed their lives. Barney's way of teaching took us much further in our work than what we had gleaned from *The Discovery of Grounded Theory*. Odis Simmons' experience of Barney's analysis seminars captures the recollections of many:

> I recall them as being the most exciting intellectual experience of my life. In these seminars Barney went beyond *The Discovery of Grounded Theory*, introducing and developing many of the ideas that ended up in *Theoretical Sensitivity* and his subsequent books.

Making It

Making it in the grounded theory classes meant more than demonstrating one's worth as a graduate student. It meant acceptance in one's discipline. But how to get there? Publications. Learning grounded theory from Barney paved a path to publishing. Barney offered new views of sociology, of theorizing, and of success and in the process imparted lessons for "making it" as budding researchers. He eschewed many academic expectations and conventions, ranging from serving on university committees and preparing lectures to avoiding the influence of earlier theories and research. Barney warned us against the dangers of copious scholarship and of becoming the kind of academic who gave meticulous attention to disciplinary literatures but published little. Barney's advice about service and teaching did not fit the academic situations in which some of us found ourselves, but his view of theorizing did. It was remarkably democratic. In the 1960s, a sharp divide existed between sociological theory and research, and theory remained the purview of anointed elites. Barney punctured the awe accorded to theorists as "great men [sic]" He saw grounded theory as offering the tools for us to become theorists and, by extension, great men—and women. Phyllis Stern put it succinctly, "He told us we could become famous," as indeed she did.

Barney demanded but demystified publication. We should and could publish. From the beginning, Barney and Anselm told us that they expected us to write books, not dissertations. My cohort, the first in the Sociology Graduate Program, took this expectation seriously but none of us who wrote grounded theory dissertations published them soon after finishing. Barney's emphasis on publishing may have shifted to include articles by the time Carolyn Wiener entered the program. She states:

> He said the first day that no one would complete the GT class unless they published at the end, and I, in my naiveté, believed him. Consequently, I had an extensive back and forth with the editor of *Social Science & Medicine* one summer (when neither Anselm nor Barney were available) over the fine points of the article, eventually winning with its publication. Needless to say, no one else in the class took him seriously about the necessity to publish.

Wiener's first article became widely respected and continues to be cited as an important contribution to the sociology of illness. In the 1960s and '70s, only a small fraction of sociologists published anything. Graduate students seldom presented conference papers then, much less published them. Yet early UCSF sociology and nursing students began to publish articles (see, for example, Beeson, 1975; Bigus, 1972; Calkins, 1970, 1972; Hayes-Bautista, 1976; Suczek, 1972; Wiener, 1975a; Wilson, 1977) and then later, books. Jeanne Benoliel [Quint] (1967), Robert Broadhead (1983), Marsha Rosenbaum (1981), Carolyn Wiener (1981) and Holly Wilson (1982) all published their grounded theory dissertations as books. Several graduates of the Sociology Program have died but left the imprint of both Barney and Anselm's teaching in their books (Biernacki, 1986; Cauhape, 1983; Star, 1989).

Barney's teaching reached beyond the borders of qualitative research. Jobs for nurses with doctorates opened in the 1970s as universities developed new doctoral programs in nursing, although qualitative methods remained contested. Few academic jobs existed in sociology but numerous UCSF graduates constructed innovative careers in research, policy analysis, non-profit organizations, and government. Barney's conviction that quantitative researchers could adopt grounded theory methods resonated with students who later became quantitative researchers. Some taught themselves how to do survey research; others like David Hayes-Bautista returned to using large quantitative data sets:

> Barney starting talking, towards the end of the seminar, about using quantitative data with a grounded theory approach. Basically, that is what I have been doing for nearly 30 years. In fact, most people think I am just a quant nerd, but I use quantitative data very differently from other folks. Thus, I have been working up the Latino Epidemiological Paradox model by simply repeating Barney's old questions: so, what do I see in the data? The difference is that now my data sets consist of mortality data, birth data, hospital discharge, census and other large, population-based data sets. I guess I am considered somewhat of a maverick in medical educational circles, but it all comes back to Barney's training.

Appreciating Barney's Teaching

Students appreciated what Barney taught and how he taught it. Learning grounded theory kindled revelations in students unfamiliar with qualitative research. It also gave those of us who arrived with qualitative research experience explicit strategies for developing, refining, and checking *our* ideas. The format of the classes contributed to how we learned the method. Nursing courses may have relied on learning by doing in classroom settings, but 1960s sociology doctoral programs did not. Moreover, Barney's innovative method of engaging students in theory construction in class sessions turned the conventional sociology graduate seminar inside out and, simultaneously, encouraged students' analytic thinking. Typical graduate seminars across the nation focused on discussing and critiquing existing literatures rather than eliciting students' fresh ideas about their research.

Each week, Barney's class attended to one student's work and we worked on coding, comparing, and categorizing data that he or she had distributed a week before. Barney appointed another student to be the note-taker, so that the presenter could concentrate on the discussion of the emerging analysis. Our discussions moved quickly as we raised and answered analytic questions. Barney solved the 1960s graduate school game of "one-upping" seminar classmates through brutal personal "put-downs": Show how smart you are by raising your classmate's theoretical analysis to a more abstract level. One student summarized the value of Barney's seminar format: "The weekly seminar model is brilliant."

Barney's teaching style provided analytic direction while fostering students' emergent discoveries of methodological strategies as well as with data analysis. I recall Barney and Anselm each saying that students would adopt and adapt grounded theory to fit their research and practical problems. Odis Simmons remembers an occasion when Barney induced students to do just that:

> Before he explained what a memo is in grounded theory Barney instructed us to "Go home and write some memos." Students looked at one another quizzically until someone asked, "Barney, what's a memo?" Barney replied, "It doesn't matter, just go home and write some." I understand now that he wanted us to discover how to memo in the manner that fit our individual minds. My grounded theory students today often want clear, detailed instructions, models and/or examples before assignments. Having learned from Barney's teaching the value of allowing students to struggle a bit so they can reach their own "aha" realization, often to their consternation I purposefully give minimal advanced "how to do it" instructions. I find that this promotes deeper learning and provides something to build from, regardless of their level of understanding.

Students who "got" grounded theory from Barney gained methodological tools that served them well—when they used them and honed their skills. Barney had long asserted that grounded theory had much greater versatility than only a method of doing qualitative research. Tom Lonner's inventive career exemplifies the power of using grounded theory for diverse purposes and projects:

> Barney gave me tools that fit me like a glove, tools that enabled me to rapidly build a career in applied policy and practice research in real communities (e.g., social impact analysis and remediation in Alaska Native villages, community control of social service programs in Washington State), on real issues (e.g., applied subsistence policy in Alaska, alcohol control in Native Alaska, Indian child welfare compliance by Washington State), in real institutions (e.g., hospitals and managed health care plans in California), and the administration and interpretation of cultural presentation (I was director of the Alaska State Museums). I ended up, of course, believing my own findings and [being] fairly effective in expressing them in policy arenas where they became law, regulation, policy, and practice, a generally positive result (from my perspective).

Several 1970s graduates worked with Anselm for many years as fulltime qualitative researchers in medical sociology. Like others who had the opportunity to work closely with both Anselm and Barney, these students found that lessons from Barney reemerged in their work long years after the classes. Carolyn Wiener recalls:

> He was a good teacher of the method as he saw it, which was more disciplined than Anselm's approach, and I didn't really understand how he was influenced by his quantitative training until years later when I started to work with students. I feel I benefited from the amalgam of the two of them. Barney's emphasis on substantive theory leading to formal theory and strict insistence on a basic social process are reflected in my article that came out of his method class, "The Burden of Rheumatoid Arthritis" and my thesis/book, *The Politics of Alcoholism.* Anselm encouraged more free-wheeling flights of imagination, "blue skying" he called it (for example, comparing medical machinery to home appliances in order to tease out the properties of the former). It's also important to me that Barney is responsible for my obtaining a pre-doc in Berkeley's School of Public Health. This provided me with a grueling year in a quantitative setting, where I was constantly faced with explaining GT and defending qualitative research. As hard as it was, the experience refined my understanding of GT, and I was gratified to learn years later that the people who were hardest on me ended up recommending my book to students!

Barney's ability to read students' data and develop an analytic frame about a social process became legendary. Some students relied on his genius to see them through. Barney could devour data and develop the outlines of an analysis within the short span of a seminar session. Both Barney and Anselm were interested, accessible, and immediately involved in their students' work. The first cohort of doctoral students might have received even more attention than later cohorts, although students from many cohorts noted their responsiveness. If I gave something to Anselm and Barney during the day, one or both of them would call to discuss it that evening. By the time I was writing my dissertation, I wanted to develop my own analysis, so I did a considerable amount of analytic work and writing before showing anything to Barney or Anselm. Typical of their response to our work, when I did give them chapters they both responded with excitement and encouragement.

During the five years between when I entered the doctoral program in 1968 and finished in 1973, frictions had developed among the faculty. Barney figured prominently in them. A couple of faculty members had come to question the value of grounded theory and used my dissertation defense to air their reservations. My grounded theory dissertation, *Time and Identity: The Shaping of Selves of the Chronically Ill,* pleased Barney, Anselm, and my external reviewer but rather irritated some other faculty. The methodology was not explicit. The theoretical frame drove its structure, not an engaging description of the data. The lengthy, abstract analysis contained a sad story and made it a slow read to the consternation of one examiner who preferred humor and good writing. But the most critical question almost stopped me: "How could you develop this abstract, theoretical, *Teutonic* analysis from only 55 interviews?" Before I could respond, Barney replied, "Because it's not just about 55 interviews; it's about her whole life." And so it was. I had grown up in a family in which my father and uncle's heart disease cast long shadows over our lives. My short career as an occupational therapist who worked in physical rehabilitation had etched memories into my consciousness and two subsequent ethnographic studies in rehabilitation had left lasting impressions. I remain grateful to Barney for his insight and for recognizing that my experience supported constructing my theoretical analysis instead of preconceiving it.

Like many of Barney's recent students, Phyllis Stern maintained her colleagueship with Barney over the years. She says, "So why are Glaser's protégés so loyal? Because he's so *there*. If you think you're stumbling, you can call him, or e-mail him, and get his advice" (Stern, 2009, p. 62). For many of us from the early years, Barney's advice formed a cornerstone of the foundation of our professional lives. Thank you, Barney.

References

Beeson, D. (1975). Women in aging studies: A critique and suggestions. *Social Problems,* 23(1): 52-59.
Benoliel [Quint], J. (1967). *The nurse and the dying patient.* Chicago: Aldine.
Biernacki, P. L. (1986) *Pathways from heroin addiction: Recovery without treatment.* Philadelphia: Temple University Press.
Bigus, O. E. (1972). The milkman and his customer: A cultivated relationship. *Urban life and Culture,* 1(2): 131-165.
Broadhead, R. (1983). *The private lives and professional identity of medical students.* New Brunswick, NJ: Transaction Books.
Calkins, K. (1970). Time: Perspectives, marking and styles of usage. *Social Problems,* 17(4): 487-501.
Calkins, K. (1972). Shouldering a burden. *Omega,* 3(1): 16-32.
Cauhapé, E. (1983). *Fresh starts: Men and women after divorce.* New York: Basic Books.
Charmaz, K. (1973). *Time and identity: The shaping of selves of the chronically ill.* Unpublished PhD Dissertation. University of California, San Francisco.
Glaser, B. G. & Strauss, A. L. (1967). *The Discovery of Grounded Theory.* Chicago: Aldine.
Hayes-Bautista, D. (1976). Modifying the treatment: Patient compliance, patient control and medicine. *Social Science & Medicine* 10: 233-238.
Hayes-Bautista, D. (2004). *La Nueva California: Latinos in the Golden State.* Berkeley, CA: University of California Press.
Hesse-Biber, S. N. (2007). Teaching grounded theory. In A. Bryant & K. Charmaz (Eds.), *Handbook of grounded theory* (pp. 311-338). London: Sage.
Rosenbaum, M. (1981). *Women on heroin.* New Brunswick, NJ: Rutgers University Press.
Star, S. L. (1989). *Regions of the mind: Brain research and the quest for scientific certainty.* Palo Alto, CA: Stanford University Press.

Stern, P. N. (2009). In the beginning Glaser and Strauss created grounded theory. In J. M. Morse, P. N. Stern, J. Corbin, B. Bowers, K. Charmaz, & A. E. Clarke. *Developing grounded theory: The second generation* (pp. 23-29). Walnut Creek, CA: Left Coast Press.

Stern, P. N. & Covan, E. K. (2001). Early grounded theory: Its processes and products. In R. S. Schreiber & P. N. Stern (Eds.), *Using grounded theory in nursing* (pp. 17-34). New York: Springer.

Suczek, B. (1972). The curious case of the "death" of Paul McCartney. *Urban life and Culture*, 1(2): 61-76.

Wiener, C. (1975). The burden of rheumatoid arthritis. *Social Science & Medicine*, 9: 508-516.

Wiener, C. (1981). *The politics of alcoholism: Building an arena around a social problem* New Brunswick: Transaction Books.

Wilson, H. S. (1977). Limited intrusion: Social control of outsiders in a healing community. *Nursing Research*, 26:105-111.

Wilson, H. S. (1982). *Deinstitutionalized residential care for the mentally disordered: The Soteria House approach*. New York: Grune & Stratton.

Endnote

Now over forty years later, the logic of grounded theory is much more widely known although teachers may still struggle with students' entrenched views of deductive inquiry (see Hesse-Biber, 2007).

12

An Integrated Philosophical Framework that Fits Grounded Theory

Alvita K. Nathaniel
West Virginia University, School of Nursing

Glaser and Strauss published *The Discovery of Grounded Theory* nearly 50 years ago—instantly challenging many dogmatic beliefs about research methodology. Grounded theory was an innovative new inductive method that leads to the discovery of theoretically complete explanations about phenomena—a dramatic departure from customary deductive research of the mid-20th century. Countering popular beliefs about qualitative research, the grounded theory method demonstrated rigorous scientific standards and produced systematic, non-biased emergence of new truths. Furthermore because of its conceptual nature, grounded theory defied the arbitrary separation between theory and research (Charmaz, 2000). Unfortunately, neither Glaser nor Strauss articulated a philosophical foundation for the method. So, through the years various authors have proposed piecemeal explanations of the method's ontological, epistemological, and methodological underpinnings, thus promoting erosion and remodeling of the grounded theory method and creating a variety of notions about the method's philosophical foundation. The purpose of this paper is to propose an extant, integrated, philosophical framework that fits the classic grounded theory method and undergirds its rigorous scientific processes.

Why is it important to identify the philosophical foundations of a research methodology? If carefully attended, the first principles, assumptions and beliefs of a given philosophy contribute the ontology and epistemology to a *methodology* and hold it together. This provides structure, logic, and cohesion. Methodology carries through to the *method*, which includes practical steps or procedures such as data gathering, coding, and analysis and also language, images, relationships, and meanings. Thus, the philosophy's assumptions and beliefs imbue the day-to-day practical application of the method and its eventual product. This engenders research that is ethical, logical, truthful, and cohesive—earmarks of good scholarship. Consider a simple metaphor in which a type of map (climate, road, topographic, political, or geologic) represents the philosophical foundation of a method, the cartographer represents the researcher, and the completed map represents the product of research. Each map, though depicting important features of the same physical area, has a unique purpose, language, meaning, appearance, audience, and practical use. When an observer looks at a map, there is a certain trust that the map represents what it is supposed to represent and is factually correct. The same should be true of research. When left adrift without a philosophical foundation, PhD candidates and experienced grounded theorists alike are free to develop non-cohesive, illogical theories and to remodel the method to fit their own ideas of what it should be. Therefore, this paper articulates a comprehensive philosophical perspective, which will provide grounded theorists with a clear and systematic basis with which to develop logical, cogent, congruent, and fully integrated theories.

This paper will demonstrate that classic grounded theory is highly consistent with Charles Sanders Peirce's (pronounced "purse") philosophy of pragmatism, his epistemological and ontological assumptions, and the correlative principles of scientific method. Further, the paper proposes that Peirce's philosophy may have indirectly influenced both Glaser and Strauss. In fact, similarities between Peirce's philosophy and grounded theory are so strong that one can speculate that Peirce's philosophy provides the provenance for grounded theory. This paper does not seek to modify classic grounded theory in any way. Indeed, classic grounded theory as originally described by Glaser and Strauss and later by Glaser should not be modified. It is internally consistent and sound and ultimately important because good grounded theories consistently illuminate "new truths."

Pragmatism and the emergence of classic grounded theory

Pragmatism

Charles Sanders Peirce (1839-1914), America's most prolific philosopher, wrote 12,000 published and another 80,000 unpublished pages. His writings have been scattered and difficult to collect and categorize (Burch, 2008). The topics on which Peirce wrote have an immense range. He wrote about mathematics, physical sciences, economics, and the social sciences. He is most widely known as the father of pragmatism (which he later re-named pragmatacism). Peirce's philosophy and logic defy simple description. Although he was influenced by Aristotle, Kant, Hegel (Burch, 2008), and Scotistic realism (Peirce, 1955), Peirce developed a unique and highly integrated philosophy.

Peirce's writings may be difficult to understand because of his complex writing style and use of invented words. For example, when discussing metaphysics, he uses terms like *quale, representamen*, and *interpretant* (Peirce, 1868). In a letter to one of his patrons, Peirce writes about *tone, token,* and *type; dynamoid* object, *immediate* object, *abstractive* object, and the *necessitant dynamoid* object of *collective*—none of which have the common language meaning in the context of his writings (Peirce, 1906/1992). So, even though Peirce is highly respected by contemporary philosophers, his work has been essentially inaccessible to many people outside the discipline of philosophy.

An ambitions man, Peirce attempted to write an overarching philosophy with an integrated systemization that enfolded an assemblage of his theories.

His goal was to construct a system describing universal totality (Houser & Kloesel, 1992b). Peirce was considered an evolutional philosopher who was always open to the revelations of experience and was prepared to change his theories accordingly. Over time, Peirce's systematic philosophy coalesced from a number of distinct theories and doctrines that he wove together into a rational whole (Houser & Kloesel, 1992b).

Grounded theory

When Glaser and Strauss originated the grounded theory method, they were already highly accomplished scholars. From the beginning, both men gave much credit to the teachers who influenced them, yet neither of them fully discussed grounded theory's philosophical foundations. In publications they did not mention Charles Sanders Peirce or William James, but they often wrote that the method was influenced by Glaser's study of quantitative and qualitative math at Columbia University under Lazarsfeld, Glaser's study of explication de text at the University of Paris, his study of theory construction under Merton, and Strauss's study of symbolic interactionism under Blumer at the University of Chicago (Glaser, 1998; Glaser & Strauss, 1967). These early influences can be linked to many of the key processes of the method, but cannot serve as its overriding philosophic basis. Even though Glaser claims that classic ground theory is a-philosophical (Glaser, 1998), close scrutiny reveals that the classic method as initially described by Glaser and Strauss (1967) and further by Glaser (1978, 1992, 1993, 1998, 1999, 2001, 2002, rev. 2007, 2005a, 2005b; Glaser & Tarozai, 2007) is amazingly consistent with Peirce's writings. As will be shown in the following sections, it is not surprising that classic grounded theory and Peirce's philosophy are congruent, since it is likely that Peirce was a direct or indirect influence on several of Glaser and Strauss's mentors.

Contrary to popular belief, symbolic interactionism is not the philosophical foundation of classic grounded theory. However, Glaser accepts this ontological view as one possible sensitizing framework among many. Interestingly, there is strong evidence that the symbolic interactionist perspective was influenced by a small piece of Peirce's philosophy. Much of Peirce's philosophy was carried forth by his friend and patron, William James. To make his work more accessible, James attempted to interpret Peirce's meanings so more people could comprehend his work (Redding, 2003). This is important to the purpose of this paper because both James and Peirce influenced John Dewy and George Herbert Meade, who in turn influenced Herbert Blumer of the Chicago School—who is credited with symbolic interactionism (Blumer, 1969; Redding, 2003). Blumer strongly influenced Strauss.

Paul Lazarsfeld and his work on qualitative analysis was an early influence on Glaser (Glaser, 1998, 2005b; Glaser & Strauss, 1967). Lazarsfeld's research strategies are very similar to those used in the grounded theory process. Although to a lesser degree than Blumer, Lazarsfeld was also influenced by Peirce. Lazarsfeld credited James and Dewy (who were both strongly influenced by Peirce) with pragmatic ideas about behavior and the scientific method. He claimed that the pragmatists tended to condense two arguments: "that habits are expressed and in turn affected by actual behavior, and that habits, as inferential concepts, have somehow to be defined and 'measured' by behavioral terms" (Lazarsfeld, 1972, p. 24). This leads Lazarsfeld to questions about indicators and their probabilistic relation to underlying characteristics. He says that, "by using a number of them [indicators] we hope that our classification will be correct...." (Lazarsfeld, 1972, p. 25). Lazarsfeld termed this type of referential classification of concepts *the diagnostic procedure*, a concept highly related to scientific method and completely integral to grounded theory.

There is also some indirect evidence that Peirce may have influenced Robert Merton, another of Glaser's mentors. Peirce wrote about statistics as a study of chance. Merton became captivated with the use of statistics as a foundation of social research. He considered statistics a decisive factor by which to judge the value of a social research. Merton explicitly included Peirce's protégé, William James's, ideas in some of his subsequent writings (Merton, 1949/1968, 1994). For example, Merton's concept of self-fulfilling prophesy (Merton, 1949/1968) is similar to that discussed by William James in *The Will to Believe* (James, 1898). Peirce's indirect influence on grounded theory seems to be far reaching.

Ontology

The starting point of any scientific method must be its ontological perspective. Ontology is defined as a branch of metaphysics that embraces such issues as "the nature of existence and the categorical structure of reality" (Honderich, 1995, p. 634). Ontology is particularly important to a discussion about grounded theory and will be examined in light of three pivotal ideas that Glaser and Peirce have in common: objective reality, latent patterns, and the human perspective.

Peirce's Ontological View

Reality. Similar to Glaser, Peirce defined real as "that which is not whatever we happen to think it, but it is unaffected by what we may think of it" (Peirce, 1871, p. 88). Peirce wrote that there is a true answer or final conclusion to every question, toward which every person is constantly gravitating. He also proposed that the final opinion is independent of all arbitrary and individual thought, therefore everything which is thought to exist in the "final opinion" is real (Peirce, 1871). In other words, Peirce believed that every piece of scientific evidence adds to what was previously known and moves toward a complete picture of truth and reality. As the process unfolds, what is *real* in a practical sense consists of both the object and the investigator's ability to understand and communicate it.

Latent Patterns. Like Glaser, Peirce recognized a tendency of phenomenon to behave in a certain way. Peirce hypothesized that the primordial universe was chaos, and has been progressing to a state of organization. This gave rise to his concept of habit-taking. He proposed that "there is an original elemental, tendency of things to acquire determinate properties. . . ." (Peirce, 1886/1992, p. 243). With this evolutionary view, Peirce described his system as follows: 1) first are original events; 2) second are laws which produce sequences; and 3) third are mediating elements between chance, which brings forth original events, and law, which produces sequences. Additionally, he believed that the tendency of habit taking must have gradually evolved and would therefore tend to strengthen itself. He identifies thirdness (or representation) as that which is operationalized in nature. To this concept he advanced the theory that thirdness allows predictability in nature, and thus scientific method.

Human Perspective. According to Peirce, understanding the natural world and communication between people can only occur through the use of signs or symbols. His comprehensive philosophy of symbols is termed semiotics. Peirce identified three types of signs. The first kind of signs are likenesses, or what he terms *icons*. The function of icons is to convey ideas of the things they represent by simply imitating them. The second kind of sign is indications, or *indices* which show something *about* the object in terms of their physical connection with them. The third kind of signs are general signs or *symbols* which are associated with meaning through their common language usage (Peirce, 1894/1992). It is this third kind of sign, termed *symbols*, that takes into account the human perspective. As described by Peirce, symbols are understood both by the utterer and the listener. Each understands the symbol via his or her own interpretation. Peirce delineates three levels of interpretation (or interpretant). In the language of Peirce, the *intentional interpretant* is a determination in the mind of the utterer; the *effectual interpretant* is the translation in the mind of the interpreter; and the *communicational interpretant* (or *cominterpretant*) is a determination fused in the minds of the utterer and interpreter—which allows for true communication (Peirce, 1906/1992). As discussed in a following section, classic grounded theory is consistent with Peirce's view of symbols in that it recognizes the import of each person's interpretations (perspectives). Further, it is this small, but important portion of Peirce's philosophy that sets the stage for subsequently developed symbolic interactionist theories.

Peirce conceived of concepts as the organizing structure by which one understands indicators and other small bits of information. Peirce wrote that the "function of conceptions is to reduce the manifold of sensuous impressions to unity, and that the validity of a conception consists in the impossibility of reducing the content of consciousness to unity without the introduction of it" (Peirce, 1868, p. 1). Thus the function of concepts, according to Peirce, is to reduce numerous empirical data into one meaningful whole. Concepts are known to be valid only because one cannot reduce the content of consciousness to a meaningful whole without the introduction of them. Peirce suggests a gradation among concepts, stipulating that one concept may unite the manifold of sense and another might be required to unite the concept and the manifold to which it is applied, and so on.

The ontology of classic grounded theory

In an attempt to clarify the grounded theory method, many have published opinions about its ontological position. Ontology is important because judgments about the nature of truth and reality determine how data are gathered, analyzed, and presented. Since Glaser and Strauss rarely discussed ontology, published speculations about the ontological position of grounded theory abound. These speculations may be responsible in some measure for erratic attempts to remodel the method by others. Many people have published a broad spectrum of conflicting ideas about grounded theory (Annells, 1996; Benoliel, 1996; Boychuk-Duchscher & Morgan, 2004; Cutcliffe, 2000; Greckhamer & Koro-Ljungberg, 2005; Haig, 1995; Kinach, 1995; Lomborg & Kirkevold, 2003; McCann & Clark, 2003; Mills, Bonner, & Francis, 2006; Reed & Runquist, 2007; Seaman, 2008). These attempts to clarify the method have led to a confusion of conflicting labels including realist, constructivist, critical realist, objectivist, relativist, interactionist, positivist, post-positivist, and others. As scholars work to classify the method, it becomes clear that a vacuum was created by an initial absence in seminal classic grounded theory literature of the underlying ontological assumptions that underpin the method. This absence of ontological positioning has led to hidden scientific practices (Greckhamer & Koro-Ljungberg, 2005).

Inferences from Glaser's words help to clarify the method's ontological position. Glaser recognizes that 1) there is an objective reality that can be observed 2) inasmuch as it is possible, the researcher gathers data from the perspective of the research participant and 3) grounded theory sheds light on latent patterns.

Objective Observations. Like Peirce, Glaser recognized that there are events which can be objectively observed. In 1966 before the method was fully developed, Glaser and Strauss described the collection, coding and analysis of data as follows: "Whether the fieldworker starts out in the confused state of noting everything he sees, because everything may be significant, or whether he starts out with a more defined purpose, observation is quickly accompanied by hypothesizing" (Glaser & Strauss, 1966, p. 56). In *Theoretical Sensitivity* Glaser writes, "Generating good ideas also requires the analyst to be a non-citizen for the moment so he can come closer to objectivity and to letting the data speak for itself, and further from issue orientations implicit in the data which can dictate a biased view of it" (Glaser, 1978, p. 8). He goes on to write, "At first the analyst may feel that his non-preconceived field work yields only scattered observation. But as soon as he starts to comparatively analyze the data—preferably the first day—codes emerge yielding theoretical leads and theoretical sampling is off to a start" (Glaser, 1978, p. 46). These passages show clearly that Glaser recognizes an objective reality, separate and apart from the researcher.

Latent patterns. Development of hypotheses and theories assumes that there are predictable patterns that can be observed. Glaser writes, "In GT, a concept is the naming of an emergent social pattern grounded in research data. For GT, a concept (category) denotes a pattern that is carefully discovered by constant comparing theoretically sampled data until conceptual saturation of interchangeable indices. It is discovered by comparing many incidents, and incidents to generated concept, which shows the pattern named by the category and the sub-patterns which are the properties of the category....GT is a form of latent structure analysis, which reveals the fundamental patterns in a substantive area or formal area" (Glaser, 2002, p. 4). The grounded theory method corrects for error or bias through constant comparison and abstraction, which further clarifies the underlying latent patterns (Glaser, 2002, rev. 2007). Therefore, all grounded theories depict discovered latent social patterns.

Participant Perspective. While Glaser denies that grounded theory deals with co-constructed reality, he clearly recognizes the importance of grounded data from the perspective of the research participant. He calls it a "perspective-based" methodology (Glaser, 2002, rev. 2007). Grounded theory seeks to understand the main concern and its resolution from participants' perspectives—from their words and behavior. Glaser says that the researcher tries to objectively figure out "what's going on" and then to conceptualize it. As noted above, this conceptualization brings into focus previously unnamed latent patterns. Conceptualization is a pivotal point for both Peirce and Glaser.

Epistemology

Epistemology is the branch of philosophy that is concerned with the "nature of knowledge, its possibility, scope, and general bias" (Honderich, 1995, p. 242). If an objective reality exists, as suggested by both pragmatism and grounded theory, how does one comprehend it? This is an important question because assumptions embedded in epistemic positions have implications pivotal to each step in the research process. The epistemology of pragmatism and classic grounded theory are very similar: Both rely on classification and clustering of symbols (indicators) to comprehend concepts; both recognize that each person understands and interprets symbols from their own unique perspective; both propose that reality can be known ideally, through use of a self-correcting scientific process; and, both utilize deduction, induction, and abduction as a means of discovering new knowledge.

Peirce's Semiosis. Semiosis is connected with Peirce's epistemology in that the objective world can only be understood through the interpretation of symbols. The *symbol* connects the person with the object through what Peirce describes as "the symbol using mind" (Peirce, 1894/1992). Symbols have meanings, significances, and interpretations. Signs are symbols that denote qualities, relations, features, items, events, states, regularities, habits, and laws.

Peirce identifies different types of signs, which are in relationship to each other (Peirce, 1894/1992). Indicators, or *indices*, are one kind of sign. These show something *about* the object in terms of their connection with them. Indices make the process of understanding possible by placing the object in dynamic relation (Houser & Kloesel, 1992b; Peirce, 1894/1992). Indices are smaller bits related to the object or experience. This is the initial interface of the world with the mind–the unstudied, unexplained, immediate snapshot. Indices place the object into the context of one's experience.

Semiosis and grounded theory. Peirce's semiosis is congruent with grounded theory. In grounded theory, indices are the empiric bits that investigators work with as they organize, categorize, conceptualize, and hypothesize. When clustered meaningfully, indices define and describe each concept and distinguish it from others. Through them one understands the concept. According to Glaser, once a concept or category is fully understood, the indices become interchangeable. For example, how does one recognize a friend? There are many indicators, including physical appearance, voice, fragrance, predictable behavior, and so forth. To recognize the friend, one need not identify every possible indicator. The sound of the friend's voice is enough to identify him or her. This indicator is interchangeable with his all other indicators. Each is a sign of the friend. Any will indicate the person's identity and all of them together will give a more clear and precise picture of the friend.

Pragmatism and grounded theory. A few extant contemporary publications recognize the underlying pragmatist assumptions of grounded theory (Annells, 1996; Kushner & Morrow, 2003; Lomborg & Kirkevold, 2003; Mills et al., 2006), but none closely examine the method in light of Peirce's original philosophy. Like grounded theory, Peirce's philosophy has undergone many transformations in the hands of his successors. However, Peirce's original theory of pragmatism is congruent with the epistemology and ontology of grounded theory. In laying the groundwork for pragmatism, Peirce described the highest type of apprehension as follows: "Consider what effects, which might conceivably have practical bearings, we conceive the object of our conception to have. Then our conception of these effects is the whole of our conception of the object" (Peirce, 1878, p. 132). He further posits that the "whole function of thought is to produce habits of action" and "to develop its meaning, we have. . . to determine what habits it produces, for what a thing means is simply what habits it involves (Peirce, 1878, p. 131). "The entire intellectual purport of any symbol consists in the total of all general modes of rational conduct which, conditionally upon all the possible different circumstances and desires, would ensue upon the acceptance of the symbol" (Peirce, 1905, p. 346). In an uncommon attempt to simplify his writing, Peirce quoted William James's definition of pragmatism as, "the doctrine that the whole 'meaning' of a concept expresses itself either in the shape of conduct to be recommended or of experience to be expected" (Peirce, nd/1992, p. 400). How does this relate to grounded theory? Making the same point as Peirce, Glaser simply asks, "What is going on?"

Although known as the father of American pragmatism, Peirce was actually a physical scientist throughout his life. When Peirce wrote that the meaning of a concept consists of the entire set of its practical consequences, he meant that a meaningful concept must have some sort of experiential "cash value." The concept must lend itself to being recognized by a collection of empirical observations. Peirce believed that the entire meaning of a concept consisted in the totality of possible observations (Burch, 2008). Thus, Peirce created a philosophy of science that set the stage for Glaser's conceptualizations of observable patterns, interchangeable indices, tentative hypotheses, and modifiable theories.

Methodology

Methodology is the philosophical study of the scientific method, sometimes called the "science of science" (Honderich, 1995). The validity of a particular method of inquiry rests upon its close adherence to the underlying philosophical principles—its ontology and epistemology. Peirce developed a philosophy that fully and cogently integrates ontology, epistemology, and methodology. Further, Peirce's philosophy rationally undergirds Glaser's classic grounded theory. Both Peirce and Glaser understood the value of deduction, induction, and abduction to the scientific method and both believed that everything is explicable in a general way.

Congruent with Glaser's concept of process, Peirce proposed that every event has a cause. He reasoned that 1) if we admit that every event has a cause, we are bound to grant that every fact has an explanation; 2) explanation is the adoption of a simpler supposition to account for a complex state of things; 3) explicability has no determinate and absolute limit; 4) as one looks back, one sees that those things that are now heterogeneous, were once homogeneous; 5) *chance* has a part in creating heterogeneity from homogeneity–i.e. in establishing order from chaos; and 6) in infinite regress, the further we go back, the more indefinite the laws, the closer we come toward a complete explanation (Peirce, 1883/1992).

Peirce claimed that neither probability nor belief are valid for determining science, rather scientific inquiry completes itself in the reiteration of the three steps of abduction, deduction and induction. For Peirce, deduction "proves that something *must* be, induction shows that something *actually is* operative, [and] abduction merely suggests that something *may be*" (Peirce, 1901/1992, p. 216). It should be noted that over the years Peirce modified his views on the three types of scientific logic, so reading other of Peirce's writings may lead to different conclusions. However in general, Glaser and Peirce utilized the same types of logic—induction, deduction, and abduction. Since Peirce was more interested in experimental science, the order and practical procedures are somewhat different between Peirce's and Glaser's methods. For both Glaser and Peirce, though, the three types of logic are integrated to form the scientific method.

Induction. Peirce defined induction as "generalization from a number of cases of which something is true, and inference that the same thing is true of a whole class" (Peirce, 1898/1992, p. 189). Glaser seems to accept the same definition of induction, but unlike Peirce, Glaser's scientific method begins with induction. In grounded theory, explanation is grounded in empirical data and clarified through the process of constant comparison. Glaser (1965) coined the term *constant comparative method,* which he proposed as a key intellectual strategy of grounded theory analysis. Both Glaser and Peirce see the heart of the scientific method as the ability to cluster items by similarity. Clustering can result in formal concepts and can be used for generating hypotheses (Burch, 2008; Glaser, 1978). In grounded theory, theory emerges as the analyst goes back and forth in an iterative process constantly comparing the empiric data (Glaser, 1965, 1998). This method increases formal abstraction and corrects for poor data as it brings each concept into closer grounding (Glaser, 1965, 1998, 1999). It is at this point that Glaser and Peirce diverge slightly. Since Peirce was interested in experimentation rather than theory development, he usually, though not always, viewed induction as a method to test theory through comparison with empiric data (Houser & Kloesel, 1992a, 1992b). For grounded theorists, though, induction occurs as empiric observations lead to generalization and conceptualization as a basis of theory development. Even though Peirce and Glaser have slightly different twists on the placement of induction in the scientific process, the logic itself is the same: abstract inferences flow from empirical data.

Abduction. According to Peirce, "Abduction ... is motivated by the feeling that a theory is needed to explain the surprising facts. Abduction seeks a theory. .... In abduction the consideration of the facts suggests the hypotheses" (Peirce, 1901/1992, p. 106). Peirce also used the term *hypothesis*, when he wrote, "a hypothesis, then, has to be adopted, which is likely in itself and renders the facts likely (Peirce, 1901/1992, p. 106). In other words, the logic of abduction stimulates the act of hypothesizing. In grounded theory, as concepts and processes emerge the theorist proposes tentative hypotheses, i.e., *abduction.* Abduction is fundamental to grounded theory since related hypotheses are joined together to form theories. For both Glaser and Peirce abduction creates hypotheses about what actually is going on (Burch, 2008; Glaser, 1978, 1998). Glaser and Peirce both valued abduction as the only way to uncover new knowledge (Glaser, 1978; Peirce, 1901/1992). Glaser's ideas about emergence are very similar to Peirce's words: "Abduction makes its start from the facts, without, at the outset, having any particular theory in view, though it is motivated by the feeling that a theory is needed to explain the surprising facts" (Peirce, 1901/1992, p. 106). Abduction is always inference to some explanation or to something that makes routine some information that has previously been surprising in that we would not have routinely expected it, given the then-current state of knowledge (Burch, 2008). Thus, for both Glaser and Peirce, concepts emerge from empirical data, which suggest hypotheses and subsequently culminate in theory.

Deduction. Glaser and Peirce defined deduction in similar ways Peirce defined deduction as the application of general rules to particular cases (Peirce, 1898/1992, p. 187). In deduction, one draws conclusions about what observable phenomena should be expected if hypotheses are correct (Burch, 2008). Peirce concluded that "deduction proves that something *must* be" (Peirce, 1901/1992, p. 216). Glaser and Peirce both viewed deductive inference as conclusions about other things that must occur if the hypothesis is assumed to be true. For Glaser, deduction serves to complete a theory. As a theory begins to emerge (through a process of abduction), the theorist may notice gaps in the theory. Through deductive reasoning, the theorist will make inferences about the proper direction of subsequent data gathering. Glaser discusses deduction thus: "Conceptual elaboration ... is the systematic deduction from the emerging theory of the theoretical possibilities and probabilities for elaborating the theory as to explanations and interpretations. These ... guide the researcher back to locations and comparative groups in the field to discover more ideas and connections from data. The data constantly check deductions that lead nowhere, as the analyst takes his directions from emerging relevancies" (Glaser, 1978, p. 40). Thus, for Glaser, deduction points to the most appropriate avenue for further investigation and subsequently fills out a theory and packs it full of relevant facts. Peirce's definition of deduction is congruent with Glaser, but his placement of deduction in the scientific method is different. Peirce would have the experimental scientist utilize deduction to draw predictions from a hypothesis (Peirce, 1901/1992).

Glaser and Peirce both utilized deduction, induction, and abduction as iterative components of the scientific process and both viewed the scientific method as one that should be essentially public, reproducible in its activities, and self-correcting. Both believed that no matter where different investigators may begin, if they closely follow the method, their results will eventually converge toward the same result and that further study will tend to correct the results. This ideal point of convergence is what Peirce meant by *the final opinion* (Burch, 2008) and what Glaser views as the ideal end to modifiable and self-correcting theories. It is at this point that the ontology, epistemology, and methodology espoused by Glaser and Peirce culminate.

Conclusion

With a thoughtful understanding of this highly integrated philosophical foundation, grounded theorists can better understand the process and product of grounded theory studies. This paper proposes that the philosophical framework established by Charles Sanders Peirce fits the classic grounded theory method and can be rationally used to undergird its processes. Classic grounded theory is highly consistent with Charles Sanders Peirce's philosophy of pragmatism, his epistemological and ontological assumptions, and the correlative principles of methodology. Use of the broad and integrated ideas of Peirce can serve as basic philosophical assumptions of the classic grounded theory method for students and experienced grounded theorists and can prevent erosion and misinterpretation of the method.

References

Annells, M. (1996). Grounded theory method: Philosophical perspectives, paradigm of inquiry, and postmodernism. *Qualitative Health Research, 6*(3), 397-393.

Benoliel, J. Q. (1996). Grounded theory and nursing knowledge. *Qualitative Health Research, 6*(3), 406-427.

Blumer, H. (1969). *Qualitative research methods for the social sciences.* Englewood Cliffs, NJ: Prentice-Hall.

Boychuk-Duchscher, J. E., & Morgan, D. (2004). Grounded theory: Reflections on the emergence vs. forcing debate. *Journal of Advanced Nursing, 48*(6), 605-612.

Burch, R. (2008). Charles Sanders Peirce. In E. N. Zalta (Ed.), *The Sanford Encyclopedia of Philosophy* (Online ed.).

Charmaz, K. (2000). Grounded theory: Objectivist and constructivist methods. In N. K. Denzin & Y. S. Lincoln (Eds.), *Handbook of Qualitative Research* (2nd ed., pp. 509-535). Thousand Oaks, CA: Sage.

Cutcliffe, J. R. (2000). Methodological issues in grounded theory. *Journal of Advanced Nursing, 31*(6), 1476-1484.

Glaser, B. G. (1965). The constant comparative method of qualitative analysis. *Social problems, 12*, 10.

Glaser, B. G. (1978). *Theoretical sensitivity: Advances in the methodology of grounded theory.* Mill Valley, CA: Sociology Press.

Glaser, B. G. (1992). *Emergence vs forcing: Basics of grounded theory analysis.* Mill Valley, CA: Sociology Press.

Glaser, B. G. (1993). *Examples of grounded theory: A reader.* Mill Valley, CA: Sociology Press.

Glaser, B. G. (1998). *Doing grounded theory: Issues and discussion.* Mill Valley, CA: Sociology Press.

Glaser, B. G. (1999). The future of grounded theory. [serial online]. *Qualitative Health Research, 9*(6), 10.

Glaser, B. G. (2001). *The grounded theory perspective: Conceptualization contrasted with description.* Mill Valley, CA: Sociology Press.

Glaser, B. G. (2002). Conceptualization: On theory and theorizing using grounded theory. *International Journal of Qualitative Methods, 1*(2), 1-31.

Glaser, B. G. (2002, rev. 2007). Constructivist grounded theory. [Online Journal]. *Forum Qualitative Social Research, 3*(3).

Glaser, B. G. (2005a). *The grounded theory perspective 3: Theoretical coding.* Mill Valley CA: Sociology Press.

Glaser, B. G. (2005b). *The roots of grounded theory.* Paper presented at the 3rd International Qualitative Research convention, Johor Bahru, Malaysia.

Glaser, B. G., & Strauss, A. L. (1966). The purpose and credibility of qualitative research. *Nursing Research, 15*(1), 56-61.

Glaser, B. G., & Strauss, A. L. (1967). *The discovery of grounded theory: Strategies for qualitative research*. Chicago,: Aldine Pub. Co.

Glaser, B. G., & Tarozai, M. (2007). Forty years after Discovery: Grounded theory worldwide. *The Grounded Theory Review: An International Journal, Special Issue*, 21-41.

Greckhamer, T., & Koro-Ljungberg, M. (2005). The erosion of a method: Examples from grounded theory. *International Journal of Qualitative Studies in Education, 18*(6), 729-750.

Haig, B. K. (1995). Grounded theory as a scientific method. *Philosophy of Education Yearbook*. Retrieved from http://www.ed.uiuc.edu/EPS/PES-Yearbook/95_docs/haig.html

Honderich, T. (Ed.). (1995). *The Oxford companion to philosophy*. New York, NY: Oxford University Press.

Houser, N., & Kloesel, C. (Eds.). (1992a). *The essential Peirce: Selected philosophical writings* (Vol. 2). Bloomington, IN: Indiana University Press.

Houser, N., & Kloesel, C. (Eds.). (1992b). *The essential Peirce: Selected philosophical writings* (Vol. 1). Bloomington, IN: Indiana University Press.

James, W. (1898). *The will to believe and other essays in popular philosophy*. New York, NY: Longmans, Green, and Co.

Kinach, B. M. (1995). Grounded theory as scientific method: Haig-inspired reflections on educational research methodology. *Philosophy of Education Yearbook* Retrieved from http://www.ed.uiuc.edu/EPS/PES-Yearbook/95_docs/kinach.html

Kushner, K. E., & Morrow, R. (2003). Grounded theory, feminist theory, critical theory: Toward theoretical triangulation. *Advances in Nursing Science, 26*(1), 30-43.

Lazarsfeld, P. F. (1972). *Qualitative analysis: Historical and critical essays*. Boston, MA: Allyn and Bacon.

Lomborg, K., & Kirkevold, M. (2003). Truth and validity in grounded theory: A reconsidered realist interpretation of the criteria: fit, work, relevance and modifiability. *Nursing Philosophy, 4*(3), 189-200.

McCann, T. V., & Clark, E. (2003). Grounded theory in nursing research: Part 1--Methodology. *Nursing Research, 11*(2), 7-18.

Merton, R. K. (1949/1968). *Social theory and social structure* (Enlarged ed.). New York, NY: Free Press.

Merton, R. K. (1994). *A life of learning*. Paper presented at the Charles Homer Haskins Lecture for 1994, American Council of Learned Societies. Lecture retrieved from http://www.acls.org/Publications/OP/Haskins/1994_RobertKMerton.pdf

Mills, J., Bonner, A., & Francis, K. (2006). The development of constructivist grounded theory. *International Journal of Qualitative Methods, 5*(1), 1-10.

Peirce, C. S. (1868). On a new list of categories. *Journal of Speculative Philosophy, 2*, 103-114.

Peirce, C. S. (1871). Fraser's The Works of George Berkeley: A critical review by Charles Peirce. *North American Review*, 449-472.

Peirce, C. S. (1878). How to make our ideas clear. *Popular Science Monthly, 12*, 286-302.

Peirce, C. S. (1883/1992). Design and chance. In N. Houser & C. Kloesel (Eds.), *The Essential Peirce: Selected Philosophical Writings* (Vol. 1). Bloomington, IN: Indiana University Press.

Peirce, C. S. (1886/1992). One, two, three: Kantian categories. In N. Houser & C. Kloesel (Eds.), *The Essential Peirce: Selected Philosophical Writings* (Vol. 1). Bloomington, IN: Indiana University Press. (Reprinted from: 1992).

Peirce, C. S. (1894/1992). What is a sign. In N. Houser & C. Kloesel (Eds.), *The Essential Peirce: Selected Philosophical Writings* (Vol. 2). Bloomington, IN: Indiana University Press. (Reprinted from: 1894).

Peirce, C. S. (1898/1992). The first rule of logic. In N. Houser & C. Kloesel (Eds.), *The Essential Peirce: Selected Philosophical Writings* (Vol. 2). Bloomington, IN: Indiana University Press.

Peirce, C. S. (1901/1992). On the logic of drawing history from ancient documents. In N. Houser & C. Kloesel (Eds.), *The Essential Peirce: Selected Philosophical Writings* (Vol. 2). Bloomington, IN: Indiana University Press.

Peirce, C. S. (1905). Issues of pragmaticism. *The Monist, 15*, 481-499.

Peirce, C. S. (1906/1992). Letters to Lady Welby. In N. Houser & C. Kloesel (Eds.), *The Essential Peirce: Selected Philosophical Writings* (Vol. 2). Bloomington, IL: University of Indiana Press. (Reprinted from: 1992).

Peirce, C. S. (1955). *Philosophical writings of Peirce.* Mineola, NY: Dover Publications.

Peirce, C. S. (nd/1992). Pragmatism. In N. Houser & C. Kloesel (Eds.), *The Essential Peirce: Selected Philosophical Writings* (Vol. 2). Bloomington, IN: Indiana University Press.

Redding, P. (2003). Early American pragmatism. Retrieved from http://teaching.arts.usyd.edu.au/philosophy/phil3015/EarlyAmerPrag.html

Reed, P. G., & Runquist, J. J. (2007). Reformulation of a methodological concept in grounded theory. *Nursing Science Quarterly, 20*(2), 118-122.

Seaman, J. (2008). Adopting a grounded theory approach to cultural-historical research: Conflicting methodologies or complementary methods? *International Journal of Qualitative Methods, 7*(1). Retrieved from http://ejournals.library.ualberta.ca/index.php/IJQM/article/view/1616/1145

13

THE AUTONOMOUS CREATIVITY OF BARNEY G. GLASER: EARLY INFLUENCES IN THE EMERGENCE OF CLASSIC GROUNDED THEORY METHODOLOGY

Judith A. Holton
Mount Allison University

In this chapter, I discuss early influences that have shaped the course of Barney Glaser's scholarship and its manifestation in the development of classic grounded theory (GT). Glaser has frequently acknowledged the influence of both Paul F. Lazarsfeld and Robert Merton on his conceptual ideation of GT. This chapter explores other early influences as well including sociologists Richard LaPiere, Hans Zetterberg, Herbert Hyman, Alvin Gouldner, David Reisman, Daniel Bell, and Edward Shils. I suggest that these and other influences in Glaser's early life stimulated an autonomous creativity that would progress and develop throughout his scholarly career and foster his determined advancement of GT. Indeed, many of the methodological principles that distinguish GT from other research paradigms can be glimpsed in the work of these early influences.[1]

I contend that autonomous creativity enabled Glaser to transcend the predominant positivist tradition by effectively rejecting that paradigm's premise of preconceived theoretical bases for hypothesis testing and verification in favor of developing theory from empirical data. In so doing, Glaser had begun to lay the foundation for the development of GT while still a student at Columbia University where he worked to capture essential ideas of methodological innovation in the work of Lazarsfeld, Merton, Zetterberg, and contemporaries. This ability to conceptually transcend the dominant view would enable him to approach Anselm Strauss's qualitative paradigm with its rich possibilities for data without imposing the strictures of positivism, an intellectual stance that not only facilitated the emergence of GT but also supports Glaser's contention of grounded theory as a distinct paradigm and a general methodology for research open to any epistemological perspective and using any type of data (Glaser, 2003, 2005a).

I look forward and seldom look or dwell back. So I live more on what I am doing rather than what I did. My orientation is future (still).

–Barney G. Glaser[2]

Early Influences

Glaser's association with the Columbia School is well known and frequently cited as a significant epistemological influence on classic grounded theory (GT). Glaser, however, also acknowledges influences as early as his childhood with having a significant impact on his intellectual development. A privileged yet perhaps rather solitary child from a San Francisco society family, Glaser recalls spending much of his time in the company of household staff. While meeting needs for social development and support, the environment was perhaps not always sufficient stimulus for an overly bright and actively curious child yet may have enabled a sense of autonomy to emerge much earlier than perhaps would have been the norm in many American families. This early autonomy may well have given him a perspective that encouraged observation and questioning rather than simple acceptance of a status quo. Autonomy also left him with time to reflect and process what he observed, time to make sense of what was going on in his social world. Solitude led him into the world of books. He became a voracious reader, picking up ideas from extensive and broad-ranging sources in both classic and contemporary literature, an engagement that may well have provided fertile ground for his early conceptual sensitivity and a developing fascination with the abstract nature of social patterning that would become a key characteristic of the classic grounded theorist's stance.

Glaser speaks of being "turned on to sociology" while doing his first degree at Stanford (Glaser, personal conversation, August 22-08). Sociology would give him a perspective for processing his curiosity and innate desire to understand and explain social behavior. His interest in sociology was particularly piqued by Stanford sociologist Richard LaPiere (1899-1986), who is perhaps most remembered for his scathing attack on Freudian psychoanalysis (LaPiere, 1959). Described in a memorial tribute as "an original … [t]hemes in his life included personal responsibility and self reliance, and he illustrated them by protean displays of competence" (Dornbusch, Berger, Shaw & Snyder, 1986), LaPiere may well have served as an early influence in Glaser's autonomy development.

While some of his contemporaries were disinclined to accept his unorthodox stance (Coser, 1960), LaPiere's critique was characteristic of a growing movement in psychology that challenged the Freudian emphasis on talk-based analysis in favor of more holistic, awareness-based approaches perhaps best exemplified by the emergence of the gestalt therapy movement in the 1960s. The central idea of gestalt therapy, that of personal independence, would seem to resonate strongly with Glaser's developing autonomy. On a meta-cognitive level, one could draw some interesting parallels between the contemporaneous emergence of grounded theory and gestalt therapy, particularly the California-based "way of life" version advanced by Fritz and Laura Perls (Perls, Hefferline & Goodman, 1951), but such elaboration is beyond the boundaries of the current chapter.

While at Stanford (from 1948-1952), Glaser was encouraged to spend some time studying abroad--to take the traditional year to travel after graduation (Stern, 2009b). He chose Paris and spent a year studying at the Sorbonne where his focus would again be literature, contemporary French literature. Here he was trained in the process of literary analysis known as *explication de texte*, a detailed examination of a piece of literature for the purpose of abstracting underlying structures, designs or patterns as evidenced in repetitions, polarities, contradictions, etc. and in so doing to elicit what exactly an author is saying. Explication added a new dimension to Glaser's voracious reading habit and would later offer him an important technique for conceptual clarity in the latent structure pattern analysis that is foundational to GT methodology. Glaser's now famous questions for open coding of data (Glaser, 1978, p. 57) can be seen to have come directly from the explication de texte tradition.

Columbia

Following his year in Paris, Glaser spent an additional two years in Europe in military service before returning to the United States in 1955 to begin his PhD studies at Columbia University. His time at Columbia was set against a post-war backdrop with its focus on scientific progress. Methodology in the social sciences was striving to achieve scientific status and the field of sociology was caught up in the quest for theory verification resultant of advanced statistical methods for data analysis (Wilson, 1940). While qualitative research was regularly undertaken, it was generally in service to designing quantitative studies for theory verification; and despite its obvious value in this surrogate role, there continued to be considerable anxiety over the perceived inadequacy of qualitative methods for theory generation (Blumer, 1940).

In his own words, it was "pure coincidence" that his teachers at Columbia included famous men of sociology – most prominently, Merton, Lazarsfeld, and Zetterberg (Glaser, personal conversation, August 22-08). Glaser explains his decision to attend Columbia as serendipitous, as his being not especially in pursuit of learning at the feet of "great men" but simply open to another source of ideas. "New York sounded like an interesting place to study and live" (Glaser, personal conversation, August 22-08). If interesting was the quest, it would seem that he had chosen well. Hans Zetterberg, who would become Glaser's dissertation supervisor, has described the Sociology department at Columbia as "a hot house of ideas and personalities in the 1950s and 1961 when Barney got his PhD" (Zetterberg, personal email communication, September 9-08). Writing at the time, Zetterberg would comment on a methodological shift already apparent at Columbia suggesting that "[s]ome fortunate members of the generation now being trained in sociology will be the first ever to orient themselves from the very start of their careers toward actual theory construction" (Zetterberg, 1954, p.21).

Another of the Columbia "greats," Robert K. Merton (1910-2003), would write:

> the most important thing to come out of Columbia sociology back then were the students ... successive large cohorts of students, including a striking number of brilliant students, brightened the Columbia scene in the I940s and '50s, owing in no small part to the end of World War II and the ensuing G.I. Bill. They did much to produce the intellectual excitement that then brought us a continuing flow of new talent in the '60s and '70s ... The continuing array of talented and vibrant graduate students provided an evocative environment — unsympathetic observers might say, a virtual turmoil of ideas — for each of them and for their teachers as well. Each cohort had its own contingent of especially evocative students, recognized as first among equals by both their fellow students and their teachers. (Merton, 1998, p.193)

Glaser has described the PhD program at Columbia as, first and foremost, about developing one's autonomy and originality, making a contribution and using one's power wisely (Glaser, 2005b). He describes being "steeped in Paul's elaboration analysis and use of secondary data and focusing on unobserved variables (discovery) and Merton's approach to theory, especially his theoretical coding models for substantive data" (Glaser, personal email communication with Hans Zetterberg, September 12-08). He speaks of Lazarsfeld, Merton and Zetterberg as models of independence, creativity and autonomy who were to influence his intellectual development at Columbia, but emphasizes that he "wanted their ideas not their supervision ... I used their ideas; I didn't work on their projects ... [my] dissertation was about my autonomy and their ideas" (Glaser, 2005b). This focus on autonomy was, no doubt, influenced by Merton's (1957) work on social structure in scientific careers. Indeed, the role of autonomy in a scientific career would figure significantly in his doctoral research (Glaser, 1961; 1963a, b; 1964a, b).

Merton

From Merton, Glaser learned to read for ideas, underlining and noting concepts as they emerged from careful reading of text (Glaser, 1998, pp. 29-30), adding to his earlier training in explication de texte and further developing his natural ability for recognizing and naming emergent ideational patterns--a conceptual brilliance that continues to amaze those who attend his GT seminars. Other methodological ideas from Merton's work that would later find their place in GT were snowball sampling as a precursor to theoretical sampling, serendipity as a precursor to emergence, and the interweaving of data collection and analysis, albeit in Merton's case, the data collection was preplanned (Glaser & Strauss, 1967, p.49); and, as Glaser points out, "Merton was preoccupied with how verifications through research feed back into and modify theory. Thus, he was concerned with grounded modifying of theory, not grounded generating of theory"(p.2).

Glaser saw Merton's concept of career recognition as "juicy," a concept with grab. While Merton (1957) had developed the concept, Glaser saw this work as largely logical speculation, too abstract with little or no substantive grounding. He suggests that Merton got the concept right, but not the process; that Merton did not see that *local* and *cosmopolitan* were, in fact, both dimensions of the same person; that his conceptualization was not empirically grounded through constant comparison of conceptual indicators (Glaser, personal conversation, August 22-08). In his own dissertation, Glaser (1961) carefully analyzed secondary data to produce the emergent and grounded concept of *modest* recognition as the career goal of the majority of scientists, not the sense of all or nothing as theorized in Merton's concept. Modest recognition enables the average scientist to get a career, pay the bills, and feel acknowledged for his contribution without any realistic expectation of career achievement on the scale of a Darwin (Glaser, 1998, p. 83). The power of empirical grounding was evident; as Glaser notes, "... I did my dissertation totally on my own on secondary data ... It passed easily. My supervisor Hans Zetterberg was delighted, PFL [Lazarsfeld] was delighted with the core variable and the development of new method analytic techniques and Robert K Merton was confounded since it cast grave doubt on his famous paper, *Recognition in Science*" (Glaser, 2005c).

This experience so early in his career left Glaser with the strong conviction that earned relevance was essential to conceptual analysis. It would become a hallmark of GT analysis (Glaser, 2005c). "I became confirmed in the notion that the data were the way things were. The data were often not the way people would like to see things theoretically. Theory had to fit to be worthwhile" (Glaser, 1998, p. 30). Glaser's private speculation was that the notion of comparative failure may have been a projection of Merton's own struggle for professional recognition but it was not the norm: "It was clear to me that he was a comparative failure, so I wrote that paper published in Science and I sent out over 2000 reprints all over the world" (Glaser, personal conversation, August 23-08; Glaser, personal email communication with Hans Zetterberg, September 12-08).

While Glaser may have rejected Merton's ego and propensity for preconception, Merton's influence on Glaser's methodological development is unmistakable and is perhaps most apparent through Merton's emphasis on conceptual integration as essential to theory development (Merton, 1968, p.143). This notion resonates with Glaser's later insistence on theoretical coding as essential in organizing and integrating a grounded theory (Glaser, 2005), an ability that came naturally to Glaser. He recalls another Columbia sociologist, Bernard Barber (1938-1988), remarking, "Barney, you really know how to integrate and sort ideas." In later recounting this conversation, Glaser admits that it came as a revelation to him to discover that something he did so naturally was perhaps a gift unique to him; he'd assumed everyone else did likewise (Glaser, personal conversation, August 22-08).

Glaser comments that while Merton taught others to think conceptually, he did not embrace their concepts and his were too abstract to have any relevance despite his calling them middle range theories. "Merton suggested that you didn't need to think up something new to be innovative, you simply needed to find new relationships between existing ideas" (Glaser, personal conversation, August 22-08). He gives Merton only a nod of recognition in having a "brief flicker of light, to admit to emergence" (Glaser, 2005c). For Glaser, however, approaching data open to conceptual emergence would become second nature; no received theoretical framework was required to "understand" the data. Conceptual abstraction came naturally after years of simply picking up ideas from wherever; ideas that work to explain latent social patterns of behavior; patterns that are not forced but that emerge through preconscious processing of conceptual ideation.

Lazarsfeld

Glaser cites Barton's (2001) paper on Lazarsfeld as capturing the intellectual history of GT (Glaser, personal notation on copy of Barton's paper). "Analyzing the logic of research operations to clarify concepts" (Barton, 2001, p. 246) was key to Lazarsfeld's life and to his own (Glaser, 2005c). Lazarsfeld, widely acknowledged for his scholarly development of research methodology, advocated studying the "'methods" of various good theoretical studies to see how they had been developed, to ascertain what makes a really good study so convincing. He taught analysis via "explication," emphasizing the importance of reading for conceptual clarity, for exactly what is being said not what one might interpret from the text; then seeking interchangeable indicators to confirm and elaborate emergent conceptualizations. An acknowledged Francophile, he readily recognized Glaser's training and capability in explication de texte; its empirical objectivity would later become a cornerstone of classic GT.

Lazarsfeld's focus on methodology over methods and data over ideology--his quantomania (Glaser, 2008, p.4)-- would inspire Glaser's focus on theory discovery over verification and systematize Glaser's inherent propensity to identify and integrate ideas, capturing the autonomous creativity of the developing theorist and leveraging his brilliance in conceptual clarity; his reading for ideas, for patterns and structures; his picking up ideas that work (that are empirically grounded) and conceptually developing them. Lazarsfeld's interest in identifying unobserved variables through latent structure analysis and his focus on the minority of cases in a joint distribution would later emerge in GT's focus on latent patterns for studying complex social processes rather than isolated variables. From Lazarsfeld, Glaser learned to "look for the unexpected, the unanticipated" (Glaser, personal conversation, August 22-08). This focus on the serendipitous may well have fed a latent resistance to the dominant quest for theory verification, a subtle backlash to the topic narrowing, energy draining and frequently null hypothesis outcomes of the "scientific method." Serendipity by contrast was exciting and rewarding!

Glaser saw the practical relevance and value in theory generation and recognized the immense waste of resources in the vast caches of untapped empirical data collected at schools like Columbia as preparatory to undertaking large scale survey work. Observing Lazarsfeld's index formation process, Glaser saw the stripped down, summing up quality as a loss. One had an index but without meaning. He took issue with the privileging of theory testing over theory generation, particularly the pretense of disguising inductive hypothesis generation in a cloak of verification as Lazarsfeld's students were encouraged to do. His idea was rather than simply summing up indicators, to instead compare indicator to indicator thereby generating conceptual properties and dimensions – an index with meaning! (Glaser, 1998, p. 24). He would go on to develop the constant comparative process as a key aspect of the data analysis in the dying study and of GT methodology (Glaser, 1965; Glaser & Strauss, 1967). Despite Lazarsfeld's advising students of the poor return on investing time in verification of logically deduced hypotheses, Glaser recognized that Lazarsfeld was too deeply engrained in the positivist tradition to embrace theoretical emergence (Glaser, 1998, p. 23).

Many methodological ideas that would later become key elements of GT were readily evident in Lazarsfeld's writings of the time (Glaser, 2008). For example, in his famous study, The Academic Mind (Lazarsfeld & Thielens, 1958), Lazarsfeld draws on Kant (in Polanyi, 1951, Preface, p.v) in describing extensive data collection and analysis as the "building materials" upon which a well-integrated theory can be generated (Lazarsfeld & Thielens, 1958, p.160). Here, one can glimpse GT's iterative and tandem processes for data collection and analysis and the constant comparative method. Glaser also attributes his famous GT dictum, "All is data," to Lazarsfeld's influence (Glaser, 2005c) and identifies several other analytic techniques from Lazarsfeld's inductive quantitative analysis that would become important elements in GT, including property space analysis, scale theory, qualitative math, elaboration analysis, content analysis, contextual analysis, substruction, reason analysis, secondary analysis and multi-attitude distributions (Glaser, 1998, pp. 27-29; 2008, pp. 4-5, 7). Lazarsfeld's quest to pursue methodological innovation along with his need to pursue various substantive avenues of research (to provide funding for the Office of Radio Research, later to become the Bureau of Applied Social Research, at Columbia) led to his viewing the field of research from a more abstract – or general – perspective. Glaser would adopt a similar perspective whereby the abstraction and conceptualization of methods and concepts rendered the substantive areas of research irrelevant (Glaser, personal notation on copy of Barton's 2001 paper).

He has specifically acknowledged Lazarsfeld's influence as "seeding" him with four important methodological contributions to the development of

GT: index formation, interchangeability of indicators, constant comparative analysis, and core variable analysis (Glaser, 2005c). The first two come directly from Lazarsfeld's work (Lazarsfeld & Thielsens, 1958, pp.402-407) while Glaser discovered and developed constant comparative analysis on his own, noting that Lazarsfeld "missed this one ... The power of this procedure to generate theory is phenomenal. What a theoretical yield of discovery. What a miss. The constant comparative technique became the influential analytic procedure of GT to generate and discover theory" (Glaser, 2005c). With regard to core variable analysis, he notes:

> Lazarsfeld and Thielens (1958) proved the core variable analysis model has great yield. I used it in my dissertation with the recognition index and it literally opened up the data to a plethora of findings about the quest and consequences of recognition. I transferred the analytic notion of core variable to qualitative data and did the book on the core variable 'awareness of dying' (Glaser and Strauss, 1965a) ... Thus, I made core variable analysis the key to generating GT. (Glaser, 2005c)

Both Glaser and Zetterberg have suggested that, like many of their contemporaries, Merton and Lazarsfeld were well known for using graduate students to work on their areas of interest. In Glaser's opinion, the great men of the time like Parsons and Merton, "... played 'theoretical capitalist' to the mass of 'proletariat' testers, by training young sociologists to test their teachers' work but not to imitate it" (Glaser & Strauss, 1967, pp.10-11). "Merton controlled everyone, undermining their confidence and straight-jacketing creativity" (Glaser, private conversation, August 22-08). Zetterberg comments that Merton was a "demanding thesis advisor, thoroughly vetting manuscripts, making superb editorial suggestions – and , it must be added, taking an inordinate long time before returning the texts to the authors," but that his specialty, the sociology of science, "resulted in amazingly few dissertations at Columbia" (Zetterberg, personal email, September 9-08). He describes Lazarsfeld as "sponsor[ing] dissertations that dealt with methodological problems that interested him... He decided the topics, not the graduate student" (Zetterberg, personal email, September 9-08).

Glaser recounts how, during his time at Columbia, Merton had a research grant with six PhD candidates to work on his theory of professions but none got their degrees. The study had been framed using Merton's conjectured theory as the conceptual framework for the six research studies and consequently produced only independent correlations, thus no findings. The absence of findings was then attributed to 'poor data' and the students left unable to achieve their degrees (Glaser, 1998, 2005c). Such experiences may well have left Glaser frustrated with the social structural limitations of academia and their numbing effect on individual creativity and intelligence, perhaps explaining why he continues to caution PhD candidates to be aware of the social structural power of supervisors and their departments; why he bemoans the current tendency of so many supervisors to control and own student work, even insisting on joint authorship of papers from student research; and why he continues to focus his seminars on helping PhD students to get their degrees. No doubt he witnessed the agony and disappointment of those unfortunate six at Columbia and possibly others who, without autonomy and a methodology to guide them through emergence to discovery, forfeited huge stakes in their career potential.

While "skipping and dipping" for methodological ideas and absorbing several of Lazarsfeld's novel perspectives on theory generation, Glaser would continue to maintain intellectual autonomy. Commenting on his time at Columbia, he would later comment that "no one at Columbia got elaboration analysis. I did and Paul Lazarsfeld knew this but I was not a minion" (Glaser, personal conversation, August 22-08). Glaser describes how he "'dipped" into Lazarsfeld's notion that crude indexes were as effective as those perfected through latent structure analysis:

> So the latter was a waste and expensive. With this notion, I was off and running and further developed the analytic techniques of consistency analysis [that] I used with his elaboration analysis model and mine of theoretical saturation. The interchangeability of indicators and theoretical saturation subsequently became prime ingredients of GT procedures for generating substantive theory. (Glaser, 2005c)

This propensity to control attributed to Merton and Lazarsfeld would have been antithetical to Glaser's strongly developed autonomy and perhaps explains his choice of Zetterberg as his dissertation supervisor. Glaser comments:

> Throughout my whole training I resisted the efforts of both Lazarsfeld and Merton to co-opt me to work for them ... I had no time for them personally, just their ideas. It was clear in Merton's writings on the sociology of science that the key to creativity was to study ideas with autonomous freedom in order to put them together by seeing the connections at will, hopefully, for maximum yield and creativity. (Glaser, 2005c)

Indeed, Glaser offers an interesting incident that clearly illustrates his autonomous stance. He describes taking notes in Merton's classes on role theory, then writing up the notes and submitting them to the American Sociological Review for peer review. "I did it and was amazed at what others thought anonymously of RKM's theory ... I was told it was reified gibberish" (Glaser, 2005b). Whether the reviewers were passing judgment on the master's theory or the student's notes is unclear but the boldness of the submission goes without saying!

Zetterberg

The quiet-spoken Hans Zetterberg viewed sociology as not simply scientific but also humanistic (Zetterberg, 1965). Glaser was particularly taken with Zetterberg's focus on the practical value of social theory and the importance of empirical research as the basis for theory development. He cites Zetterberg's famous museum study (Zetterberg, 1962) as especially significant in providing "the only real content" to the ideational evolution of GT, "as he was relating concepts to concepts to produce practical theory, not conjectured theory" (Glaser, personal conversation, August 22-08). Writing to Zetterberg, Glaser comments:

> the central sources of my generating a methodology of grounded theory [were] your two books, on theory and verification and social theory and social practice. I was impressed [with] how your data and theory increased attendance at the museum. I thought why shouldn't all sociological theory have such practical use. And to be practical it had to be grounded. (Glaser, personal email communication with Zetterberg, September 12-08)

This notion of the practical value of sociology was certainly au courant during Glaser's years at Columbia. As Schulze (1967) comments, "An increasing postwar concern with social problems, the failure of applied psychology, the concurrent rise in sociological respectability and the mounting foundation and federal expenditures on social action programs have combined to create a large number of opportunities for applied sociologists" (p. 1027). Certainly, Zetterberg's perspective of the social scientist as consultant and of social theory as applied to the resolution of practitioner issues resonated with Glaser's concern for the practical value of sociology and can be seen mirrored in the later articulation of criteria for evaluating a good grounded theory; i.e., a theory that fits, works, has relevance and is readily modifiable on the basis of new data (Glaser, 1978).

It is interesting to note, however, that Zetterberg attributes the possible origins of Glaser's "practical approach to grounded theory" not to his own work but rather to Lazarsfeld's "reason analysis" (Zetterberg, personal email, September 9-08). Lazarsfeld had developed the approach in consumer survey self-report research to account for, or justify, consumer purchases. Zetterberg describes it, "At the core of this talk is the researcher's 'accounting scheme' that listens to what the actor preferred, liked or disliked about his previous choice or situation, what he preferred, liked or disliked about his prospective choice or situation, and equally important, what kind of trigger event caused him to change course to the latter alternative" (Zetterberg, 2010). So, here too, perhaps was a hint of that now famous GT dictum, the importance of finding out "what's really going on" in the area under study.

Zetterberg also refers to a long-standing debate at Columbia between those favoring the advancement of sociology through theory, led by Merton, and those favoring its advancement through methodology, led by Lazarsfeld, noting that he sided with the theorists, not because he had rejected the arguments of the methodologists, but because he had fully accepted them, as "even the most celebrated research projects in our field left something to be desired from a methodological point of view ... trivial conclusions with efforts toward maximum precision" (Zetterberg, 1954, p. vii-viii). He railed against conjectured theory and called instead for "research grounded theories" (p. 20), albeit, like Lazarsfeld, his primary concern was with theory verification. Zetterberg admonished the definitional and descriptive nature of contemporary sociology and called for more effort at integrating and explaining the complexity of what is a substantively human science, a move past simply setting forth findings and simple propositional statements in favor of integrating propositions into systems, or theories. "Propositions are the central elements; definitions are auxiliary" (Zetterberg, 1954, p. 22). The classic grounded theorist may perhaps see in Zetterberg's distinction between ordinary propositions of low informative value and the more abstract theoretical propositions of high informative value a glimpse of Glaser's later advice to think theoretically and write substantively (Glaser, 1998, p. 197).

Zetterberg rejected the focus of his contemporaries on taxonomies stating that "a concern with taxonomy and descriptive studies does not furnish explanations" (p. 26). A similar call is frequently echoed by classic grounded theorists in response to the numerous "remodelled" versions of grounded theory that produce at best conceptual description (Glaser, 2003). Despite his contention that the main task of the sociological theorist is "the discovery of general propositions" (p. 28), Zetterberg's primary interest remained with the systematic verification of social theory: "Theories summarize and inspire, not descriptive studies, but verificational studies – studies construed to test specific hypotheses" (pp.28-29). Indeed, Zetterberg's 1954 book may well have provided a model for Glaser in terms of his own later explication of grounded theory methodology, perhaps particularly with respect to his preeminent methodological guide, *Theoretical Sensitivity* (Glaser, 1978).

Zetterberg comments that his "simple ambition as Barney's advisor was that [the] thesis should be as good as the sum total of what Merton, [Herbert] Hyman, and Lazarsfeld could have expected from their doctoral students ... Barney worked independently, ambitiously, carefully, and without delays and excuses ... Best of all, his thesis met the cumulative criteria of Merton, Hyman, and Lazarsfeld; in other words, it was very, very good" (Zetterberg, personal email communication, September 9-08). In so doing, Glaser had successfully managed a "conflicting truce at Columbia: an agreement not to conflict the theoretical vs. the empirical side of the department but rather to combine the best of both approaches" (Glaser, 1998, p.31).

Zetterberg, who had established his own publishing company – Bedminster Press – would also influence Glaser in the empirical value of self publishing (Zetterberg, 1961). As Glaser has often said at his GT seminars, how else is an author to know who is using his ideas, exactly what they are buying, the purchase intent, where they're located, etc? In other words, he wanted to discover the latent pattern in the publication and reception of his work from which he would know what needed to be published next. The result is a collection of methodological and substantive monographs and papers spanning a period of close to fifty years; a collection arguably unequalled by any other scholar in advancing a general research methodology.

Sociology Press is therefore another manifestation of Glaser's creative autonomy. While self publishing is non-normative in academia, Glaser dismisses references to Sociology Press as a "vanity press" (Stern, 2009a, p. 59), noting its considerable value in advancing GT scholarship. In admonishing Phyllis Stern for her reference, he wrote:

> Wow. You neglected the various very positive consequences of Sociology Press. It pays for stipends, seminars, web sites, keeps books in print that sell all over the world, prints books that would not be printed, travel grants, pays for the GT review journal, which has taken off and [is]peer reviewed. Prints other's GTs in our readers ... This is not vanity, it is many areas of academic contribution and you demean it ... My books sell voluminously, no matter how old they are. People need them. The negative properties of routine publishers are a long list, especially dropping stale dated titles. (Glaser, personal email correspondence with Stern, May 26 & 27-09)

Other Contemporary Influences

Herbert Hyman (1918-1985), who Zetterberg describes as having written "the bible for secondary analysis," was also at Columbia during Glaser's time there. Zetterberg attributes Hyman with formalizing the art of secondary analysis at Columbia and suggests that "Hyman's access to national surveys benefited Columbia graduate students in search of data for theses" (Zetterberg, personal email, September 9-08). Glaser, however, doesn't attribute his interest and pursuit of secondary analysis to Hyman, instead noting that the interest was "in the air" at the time (Glaser, personal email, September 12-08).

Those who have studied with Glaser and can attest to his frequent calls to staying open when studying any situation or problem (Glaser, 2005) can appreciate them in light of his expressed frustration as a student in always being told what to study; in the privileging of preconceived professional interests of the faculty: "Connecting theory to data had to occur, it seemed to me, in order to get the attention it deserved from users instead of being ignored because it was not relevant or not connected to reality. Induction required data; deduction requires only logic and a fertile, intelligent mind that is closed to empirical happenings" (Glaser, 1998, p. 31).

While the call to openness is certainly reminiscent of Lazarsfeld's reason analysis, Glaser also recalls how meeting Alvin Gouldner (1920-1980) impressed upon him just how far off a preconceived interest can be from what is really going on in a research setting.[3] He recounts Gouldner's using the example of a study in which a student was sent out to study risk-taking behavior among steeplejacks in New York City. The student was frustrated by the difficulty he had in getting the "jacks" to talk about risk taking in their line of work until one day, he observed them drawing straws before beginning a job and he was certain they were using the process to allocate the potential risk. To his surprise, the tactic had nothing to do with risk; rather they were drawing straws for the best vantage points in terms of window peeping while on the job! The preconceived professional concern with risk-taking was displaced with a focus on what was really taking place – a study in *strategic positioning*. Indeed, this transcending perspective has become a hallmark in Barney Glaser's approach to life, indicative of a persistent quest to abstract ideas without being encumbered by particularistic detail, whether in research or in real-life issues and problems. Those who have worked closely with him can attest to the many times we've been implored to take a transcending view when presented with a puzzling or troubling predicament.

There were, of course, others who were to influence Glaser's progressing autonomy. He notes reading the work of eminent sociologist David Riesman (1909-2002) who, with Nathan Glazer and Reuel Denny, wrote *The Lonely Crowd: A study of the changing American character* (1953). In Glaser's eyes, this book, which coined the phrases "inner-directed" and "other-directed," lauded the notion of individual autonomy (Glaser, personal conversation, August 22-08). Another influence was sociologist Daniel Bell (1919-), whose book, The *End of Ideology* (1962), was cited by the Times Literary Supplement as one of the 100 most influential books since the end of World War II. Bell taught at Columbia University during Glaser's student days and was awarded an honorary degree in 1960. He is perhaps best known for his concept of *endism*, the idea that both history and ideology have been rendered insignificant through the triumph of Western democracy and capitalism. While Bell's ideas were criticized for being overly abstract and without sufficient empirical support, for Glaser, he served as another model for intellectual autonomy. Glaser may well have felt an affiliation with Bell's rejection of the preconceived privileging of the status quo.

Glaser has mentioned Edward Shils (1910-1995) as an early influence as well. Shils was a prolific writer and "an avid reader of nineteenth-century French, English and Russian literature and of nineteenth-century political, philosophical and belletristic writings" (Floud, 1996, p.86). An unorthodox academic, Shils had spurned graduate school in favor of practical experience as a social worker. He would, however, later return to academia with a voracious passion and productivity for science over empty intellectualism (Dacre, 1996). In speaking of Shils and Reisman, Glaser emphasizes how autonomy enabled each to escape the resistance and subsequent marginalization and organizational shunning that their progressive creativity sparked, "The danger of progressive creativity is a take-over of academic norms but for the autonomous and independently resourced, it can be an escape affording the liberty to continue the progressive creativity curve" (Glaser, personal conversation, August 22-08).

Return to San Francisco

Having completed his PhD studies in 1961, Glaser moved back to the San Francisco Bay area and quickly commenced what would become a prolific publishing record (Glaser, 1961, 1963a, b, 1964a, b, 1965, 1966; & Strauss, 1965a, b, 1966, 1967). His dissertation, *Organizational Scientists: Their professional careers*, was among those early publications (Glaser, 1964a). Here, his timing was no doubt fortuitous and his topic well chosen, resonating as it did with the fervent preoccupation with scientific progress – especially the space race. The foreword was written by Anselm Strauss, with whom he had now initiated what would become one of the most widely acknowledged collaborations in social science.

Anselm Strauss (1916-1996) had moved from Chicago to the University of California at San Francisco (UCSF) in 1960 where he had begun to form "a group of colleagues who sociologically investigated an array of issues and topics pertaining to health & illness ... students or organizations, occupations...." (Maines, 1991, p. 4). Strauss came steeped in qualitative methods and a symbolic interactionist perspective. Indeed, the 1960s had evidenced considerable momentum from the Chicago School to displace the privileged status that had been accorded to quantitative methodologies (e.g. variable analysis) as producing "disconnected findings" absent of the context that characterizes a sociological variable as "an intricate and inner-moving complex" (Blumer, 1967, pp. 85, 92). Glaser joined Strauss to commence work on what would become their famous study of dying in American hospitals (Glaser & Strauss, 1965). The study relied largely on observational data captured through field notes and afforded distinct yet complementary roles for the new research partners. While Strauss led the field work, Glaser's focus was methodological as he applied his attention to conceptual coding of the data collected. What emerged fairly quickly in the process was his recognition that the study was that of another index – an awareness index – methodologically similar to Lazarsfeld's apprehension index (Lazarsfeld & Thielens, 1958) and his own recognition index (Glaser, 1961, 1964a, b).

The Awareness study (Glaser & Strauss, 1965) would become another marker along the path of GT's evolutionary emergence. Its uniqueness for Glaser was not in its methodology, as he was simply continuing to refine methods conceived through his doctoral work; rather, it was in the study's offering him a first opportunity to work extensively with qualitative empirical data. Glaser's 1965 paper, "The Constant Comparative Method of Qualitative Analysis," published two years prior to *Discovery* (Glaser & Strauss, 1967) shows clearly his integrative ideation. In it, he set down his procedures for taking explicit coding strategies from quantitative methodology and combining these with methods for generative theory development while at the same time distinguishing the constant comparative method from the more conventional approach to qualitative analysis through analytic induction. Other papers based on his doctoral work also evidence his earlier development and use of methodological techniques that he had learned from Lazarsfeld and which would become significant elements in the development of GT (Glaser, 1961, 1963a, b, 1964a, b).

Thus, Awareness marked not a methodological eureka as has often been suggested, but simply another arena for Glaser's elaboration and refinement of methods initially formulated during his days at Columbia. His training under Lazarsfeld, who had privileged neither quantitative or qualitative methods but embraced elements of both (Glaser, 1998, p.29), had enabled Glaser to transcend the emerging objectivist/interpretivist divide in research methodology characteristic of the 1960s. For him, the process was the same – just reading carefully the data and conceptually labeling the ideas, comparing them for interchangeable indicators to theoretical saturation and eventual integration. "The process was the same: the data was a little softer, but no less rigorously dealt with" (Glaser, 1998, p. 27).

While teaching graduate students at UCSF, Glaser started to see the generality of analysis. During the dying study, he had not consciously captured the emerging GT process through methodological notes but after the significant interest in their published study (Glaser & Strauss, 1965), he suggested to Strauss that they should publish their methodology (Glaser, 1998, pp. 21-22). Strauss agreed and encouraged him to proceed. Glaser completed the bulk of the book while Strauss was on sabbatical in Europe. Thus, the methodological capture was retrospective, based on his evolving articulation of the procedures that he had been preconsciously processing from his days at Columbia.

Strauss had not been trained in methodology. Maines (1991), however, suggests that Strauss embraced Glaser's methodological perspective and Corbin (1991) has acknowledged that, " … though Anselm had been struggling with the problem of method through the earlier years of career, it is doubtful that he would ever have elucidated his analytic procedures so explicitly had he not worked with Barney Glaser" (Corbin, 1991, p. 26). While according to Glaser, Strauss did contribute three chapters to *Discovery*, more generally Strauss' writings would stay close to his interests in exploring issues of identity, temporality, action and interaction, social order and social structure. Glaser has commented, "He couldn't get beyond his favorite TC's" (Glaser, personal email, June 9-08). Despite the perspectives offered by

Maines (1991) and Corbin (1991), it would seem that Strauss had not embraced the emergent methodology sufficiently to incorporate it into his own work and much of his writing is more appropriately characterized as conceptual description. As such, perhaps the real 'discovery' was Strauss's discovery of Glaser's work and its immense value in helping him to begin to resolve his own methodological frustration with qualitative data's propensity to overwhelm.[4] Just as the uniqueness of Glaser and Strauss's discovered concept of awareness context as applied to dying in hospitals had most certainly raised the profiles of both researchers in the emerging field of medical sociology, so too did *Discovery* serve as a breakthrough for them in the newly emerging field of qualitative research which suffered from a lack of rigorous or systematic methodological procedures.

Perhaps the continuing evolution of methodology that Corbin (1991) describes in Strauss's later work is really a manifestation of his search to achieve methodological clarity, to reconcile his earlier training in context-bound analysis with Glaser's conceptual abstraction. Still, old habits do die hard and Strauss's trained need for a structure to aid analysis as well as his uneasiness in staying open to emergence would be forfeited to a persistent forcing of preconceived theoretical frameworks. This need would drive his efforts to more precisely capture and elaborate a tightly structured guide to qualitative data analysis (Strauss, 1987; & Corbin, 1990). These efforts also marked a clear methodological divergence in the famous collaboration.

Glaser would continue to apply GT methodology to a wide range of studies including topics of everyday life interest such as contracting the building of a house and safe investing (Glaser, 1969, 1972). For Glaser, it was a natural evolution in applying and refining the methodology as practical sociology. In his writing and in seminars, Glaser continues to underscore the substantial power of GT and frequently advocates that this power deserves to be applied to those areas of life that matter most – relationships, parenting, careers, health and wellness, etc.

The Ideational Struggle for Grounded Theory

While *Awareness of Dying* (Glaser & Strauss, 1965) was well received by the health professions and the popular press, *The Discovery of Grounded Theory* (Glaser & Strauss, 1967) met with some perhaps to be expected resistance from the traditional research community. Loubser (1968) described it as a manifesto – "… a strange mixture of polemic for a certain type of approach, cookbook prescriptions of how-to-do-it yourself, and an attempt to present and share with colleagues a strategy which the authors found useful in their own research" (p. 774). The critique reflected perhaps the very kind of preformed and rigid methodological dogma that Glaser and Strauss had set out to challenge. Scott (1971) appears to have largely dismissed *Discovery* commenting that it "should appeal to many youngsters in the field as an alternative to some of the newer sociologies" (p. 336). One gets the distinct impression from his rather tongue-in-cheek review, that he did not share the same concerns as Glaser and Strauss with the dominant focus of the time on theory verification as the "true" science. Scott's perceptions of the book may well have been influenced by its marketing. Certainly, Glaser has frequently commented at seminars that the initial marketing was completely wrong; the publisher had promoted it as an introductory, undergraduate text.

Some will point to examples of language used by Glaser and Strauss to support an assertion that GT is rooted in symbolic interactionism. If, for example, one reads their early work (Glaser and Strauss, 1965a, 1966), one readily sees the focus on qualitative data in references to social worlds, bias, and the "real life" character of fieldwork (1965a, p.9; 1966, p.59); yet, within the same papers, one finds various references to informed *detachment*, *going native*, *outsider status*, *evidence*, and hypothesis *verification* (1966, p.57, 59), concepts less in sync with qualitative methodology. Certainly, since the publication of *Discovery*, considerable confusion and debate have ensued in academic circles as to the philosophical stance of GT. Despite its genesis within the positivist paradigm of 1950s scientific research, the early ideational genesis of GT methodology clearly transcended positivism's verificational focus. Zetterberg has suggested that Glaser's grounded theorist stance was not "visible" until he had encountered Strauss's symbolic interactionism, but he goes on to add that "a close re-reading with hindsight of *Organizational Scientists* and *Discovery* might change that conclusion" (Zetterberg, personal email communication, September 9-08).

Indeed, the transcendence of ontological and epistemological camps is perhaps best evidenced in Glaser's own comments regarding the Chicago School influence that Anselm Strauss brought to their work together.

> Through Anselm, I started learning the social construction of realities by symbolic interaction making meanings through self indications to self and others. I learned that man was a meaning making animal. Thus there was, it seemed to me no need to force meaning on a participant, but rather a need to listen to his genuine meanings, to grasp his perspectives, to study his concerns and to study his motivational drivers. (Glaser, 1998, p.32)

This statement, however, does not render GT as symbolic interactionism as some have suggested; rather, here Glaser is simply acknowledging his introduction to working with a new kind of data, a new epistemological perspective. However, in GT "all is data" (Glaser, 2001, p. 145) and while socially constructed data is valid data, it is not the only valid data; it is not to be privileged over other types of data for theory generation. Rather, the "impressionistic influence" (Glaser, 1998, p. 32) of socially constructed data is mitigated through GT's foundational procedures of constant comparative analysis and theoretical sampling that enable the emergence of multivariate latent social structural patterns from whatever the data and from whatever perspective the researcher may bring to the data.

While Charmaz (2000) attributes this perspective confusion to a lack of explicit procedures set down in *Discovery* (p. 524), my contention is that the apparent confusion simply signified a gestational period in the emergence of the classic methodology, a time in which Glaser was bringing to his work with Strauss many ideas regarding theory development that he had previously formulated and applied in his own dissertation research. As such, the process followed a natural learning curve and the perceived confusion was a normal and necessary stage in the development and explication of the methodology; the discovery was not a *eureka* cutting point but rather a gradual and cumulative capturing of a normal ideational progression — an evolutionary emergence.

The publication of *Basics of Grounded Theory Analysis: Emergence vs. Forcing* (Glaser, 1992) most certainly signaled a watershed moment in the famous collaboration. The book would become a landmark publication in the development of GT. *Basics* was written after Glaser had made several attempts to convince Strauss to withdraw his earlier book co-authored with Juliet Corbin (Strauss & Corbin, 1990). Rather than withdraw, Strauss encouraged Glaser to publish his objections. Having done so, however, Glaser would become a virtual outcast at UCSF and therewith, he "lost a whole network at the Medical Centre" (Glaser, personal conversation, August 22-08). The inability— or at least unwillingness—of his UCSF colleagues to take a transcending view of his objections to the Strauss and Corbin book meant that Glaser would revert to his familiar stance as an "autonomous person in an autonomous context" (Glaser, personal conversation, August 22-08), pushing on in clarifying and advancing grounded theory as originally conceived — the classic methodology.

In reference to *Basics* (Glaser, 1992), Glaser would later comment, "I discovered that I was the originator of GT as a discovery method and that Anselm did not have a clue about these ideas of emergence" (Glaser, 2005b). Yet, socially vested fictions being what they are, GT's association with the qualitative paradigm continues to dominate perspectives and much of the scholarly discourse surrounding this most frequently cited social science research method. It is a mark of Glaser's essential humbleness as well as the joy he had experienced in his collaboration with Strauss (Glaser, 1995) that he acknowledged Strauss, the senior and more established scholar at the time, as having contributed equally to the writing of *Discovery*, and by inference to having contributed equally to the emergence and development of the methodology. Katja Mruck, Editor of *Forum Qualitative Sozialforschung/ Forum: Qualitative Social Research*, has suggested to Glaser that Strauss and Corbin should be held accountable for the initial undermining of his contribution to the *Discovery* legacy, "... a part of the problem has been that Strauss in his *Qualitative Analysis for Social Scientists* used extensively *Theoretical Sensitivity* by not citing you directly. So after the book had been translated into [G]erman, many concepts you introduced [in] 1978 were cited in following [G]erman texts as Strauss" (Mruck, email correspondence to Glaser, November 30, 2007).

Glaser's career trajectory would suggest that his interest was not in personal advancement; but rather in the scholarly pursuit of ideas. He speaks of GT as preserving the traditional scholarly value of creativity, a foundational pillar in preserving the autonomy of scholarly pursuit (Glaser, 2008, p. 8). He took these notions of autonomy and creativity as fundamental to any intellectual pursuit. Creative autonomy had freed him from the need to pursue the traditional trappings of academic recognition and personal wealth had facilitated a level of financial independence that further bolstered his autonomous stance. While the fact that during his entire career at UCSF, Glaser was never tenured but always "on soft money" (Glaser, personal email communication, March 28-09) kept him on the periphery of academia's inner sanctum, it did facilitate his autonomous creativity. The pursuit was that of ideas, not the accolades of peers. Only after Strauss' rather careless slighting of Glaser's substantial - indeed primary - contribution to the classic methodology did Glaser 'go public' in denouncing him (Glaser, 1992). Even then, his criticism was confined to asserting an absence of good scholarship all the while continuing to acknowledge and respect the man he considered to be a close personal friend.

A Legacy Intact

Glaser has described his publication of *Doing Quantitative Grounded Theory* (2007) as "a return to my sociology roots, which I cherish all my life" (Glaser, personal communication with Hans Zetterberg, September 12-08). While the philosophical disputation regarding GT continues to surface regularly in doctoral theses, conference presentations and publications, Glaser persists in dismissing it as a rhetorical wrestle (Glaser, 1998) and, instead, continues to position GT as a general methodology that can accommodate any epistemological perspective and any kind of data (Glaser & Strauss, 1967, p. 18, 21; 1998, p. 43; 2003, p. 99). The stance is consistent with his transcending perspective in relation to data collection and analysis. This epistemologically flexible perspective was first set out in *Discovery* where Glaser suggests that the ontological perspective of the individual researcher (and not that dictated by a specific methodology) together with his theoretical sensitivity "will help him to see relevant data and abstract significant categories from his scrutiny of the data" (Glaser & Strauss, 1967, p. 3).

While his early writing (pre-1992) illustrates a broad reading and integration of ideas from the work of others, his later works have focused primarily on developing his own ideas and methodology. This is not to suggest that his later work is in any way insular in focus, far from it. Each later publication has been extensively grounded in the empirical experiences of the increasing numbers of students who seek his direction and guidance.

Bryant and Charmaz (2007) have suggested that Glaser has become "far more amenable" (p. 4) to the possibility of alternative conceptions of GT. It is hard to fathom on what basis they make this presumption other than perhaps mistaking Glaser's acknowledgement of GT's remodeling as his giving approval. While the inferred amenability might assist in legitimizing such remodeling efforts, nothing could be further from the truth than his acceptance of the same (Glaser, 2009). In true Glaser fashion, however, he has chosen to transcend the persistent remodeling efforts. Rather than arguing the particular merits of one perspective over another, Glaser employs a transcending perspective. His *Jargonizing* book (Glaser, 2009) is a response to Bryant and Charmaz (2007) in which he offers readers a grounded explanation of the latent pattern in GT jargonizing and suggests how the jargonizing assists qualitative researchers to explicate their multiple methods and perspectives by giving them a language imbued with established legitimacy and acknowledged rigor. While writing the book, Glaser adopted his typical transcending perspective to suggest that "if the vocabulary is the main contribution [of classic GT methodology], maybe it can be the first step into properly using the methodology" (Glaser, personal conversation, August 22-08).

Undeterred by efforts to remodel or "evolve" grounded theory, Glaser has continued to advance GT methodology by elaborating its fundamental elements and processes through his writing[5], his online Grounded Theory Institute (www.groundedtheory.com) and the international peer-reviewed Grounded Theory Review (www.groundedtheoryreview.com). In addition, he regularly hosts his GT troubleshooting seminars in both Europe and the USA and sponsors additional seminars throughout the world led by Fellows of the Grounded Theory Institute, a testament to his amazing energy and continuing commitment to a life's work.

References

Barton, A. H. (2001). Paul Lazarsfeld as institutional inventor. *International Journal of Public Opinion Research*, 13, 3, 245-269.

Bell, D. (1962). *The End of Ideology: On the exhaustion of political ideas in the fifties*. Glencoe, IL: Free Press.

Blumer, H. (1940). The Problem of the Concept in Social Psychology. *The American Journal of Sociology*. Vol.45, no.5, pp. 707-719.

Blumer, H. (1967). Sociological Analysis and the "Variable", in Manis, J.G. & Meltzer, B.N. Eds. (1967). *Symbolic Interaction: A reader in social psychology*, Boston: Allyn and Bacon.

Bryant, A., & Charmaz, K. (Eds.). (2007). *The Sage Handbook of Grounded Theory*. London: Sage Publications

Charmaz, K. (2000). Grounded theory: objectivist and constructivist methods. In N. K. Denzin & Y. S. Lincoln (Eds.), *Handbook of Qualitative Research (2nd. ed.)*. Thousand Oaks: Sage Publications, Inc.

Corbin, J. (1991). Anselm Strauss: An intellectual biography. In Maines, D. R. Ed. (1991) *Social organization & social process: Essays in honor of Anselm Strauss*, pp. 24-25.

Coser, L.A. (1960). Book Review: The Freudian Ethic, *The Annals of the American Academy of Political and Social Science*, vol.328, pp.211-212.

Dacre, L. (1996). *Edward Shils (1910-1995)*. Minerva 34: 85-93.

Dornbusch, S.M., Berger, J., Shaw, E.S., Snyder, R.K. (1986). *Memorial Resolution: Richard T. LaPiere (1899-1986)*. Stanford University. Retrieved from: http://histsoc.stanford.edu/pdfmem/LaPiereR.pdf

Floud, Jean (1996). *Edward Shils (1910-1995)*. Minerva 34: 85-93.

Glaser, B.G. (1961). Some Functions of Recognition in a Research Organization: Unpublished doctoral dissertation, Columbia University.

Glaser, B. G. (1963a). Attraction, Autonomy, and Reciprocity in the Scientist-Supervisor Relationship, *Administrative Science Quarterly*, vol.8, no.3,pp. 379-398.

Glaser, B. G. (1963b). The Impact of Differential Promotion Systems on Careers, *AIEEE Transactions on Engineering Management*, vol. EM-10, no.1, pp. 21-24.

Glaser, B.G. (1964a). *Organizational Scientists: Their professional careers*. Bobbs Merrill: New York.

Glaser, B.G. (1964b). Comparative Failure in Science. *Science*, vol.143, no. 3610, pp. 1012-1014.

Glaser, B. G. (1965). The Constant Comparative Method of Qualitative Analysis, *Social Problems*, vol.12, pp. 436-445

Glaser, B. G. (1966). Disclosure of Terminal Illness, *Journal of Health and Human Behavior*, vol.7, no.2, pp. 83-91

Glaser, B.G. (1972). *Experts versus Laymen: A study of the patsy and the subcontractor*. Mill Valley, CA: Sociology Press.

Glaser, B.G. (1978). *Theoretical Sensitivity: Advances in the methodology of grounded theory*. Mill Valley, CA: Sociology Press.

Glaser, B.G. (1992). *Basics of Grounded Theory Analysis: Emergence vs. forcing*. Mill Valley, CA: Sociology Press.

Glaser, B.G. (1995). In Honor of Anselm Strauss: Collaboration, in *Grounded Theory, 1984-1994*, B.G. Glaser, Ed. Mill Valley: Sociology Press, pp.103-109.

Glaser, B. G. (1998). *Doing Grounded Theory: Issues and discussions*. Mill Valley, CA: Sociology Press.

Glaser, B. G. (2001). *The Grounded Theory Perspective: Conceptualization contrasted with description*. Mill Valley, CA: Sociology Press.

Glaser, B. G. (2003). *The Grounded Theory Perspective II: Description's remodeling of grounded theory methodology*. Mill Valley, CA: Sociology Press.

Glaser, B. G. (2005a). *The Grounded Theory Perspective III: Theoretical coding*. Mill Valley, CA: Sociology Press.

Glaser, B.G. (2005b). On Being a Sociologist, Invited Address to the Sociology Department, Columbia University, January, 2005.

Glaser, B.G. (2005c). The Roots of Grounded Theory, Keynote address to the 3rd International Qualitative Research Convention, Johur Bahru, Malaysia, August 23.

Glaser, B. G. (2007). *Doing Formal Grounded Theory*. Mill Valley, CA: Sociology Press.

Glaser, B. G. (2008). *Doing Quantitative Grounded Theory*. Mill Valley, CA: Sociology Press.

Glaser, B. G. (2009). *Jargonizing: Using the Grounded Theory Vocabulary*. Mill Valley, CA: Sociology Press.

Glaser, B.G. & Crabtree, D.J. (1969). *Second Deeds Of Trust: How to Make Money Safely.* Mill Valley, CA: Sociology Press.

Glaser, B. G., & Strauss, A. L. (1965a). *Awareness of Dying.* Chicago: Aldine Publishing Company.

Glaser, B. G., & Strauss, A. L. (1965b). Discovery of substantive theory: A basic strategy underlying qualitative research. *American Behavioral Scientist*, vol.8, no.6, pp. 5-12.

Glaser, B. G., & Strauss, A. L. (1966). The purpose and credibility of qualitative research. *Nursing Research*, vol.15, no.1, pp.56-61.

Glaser, B. G., & Strauss, A. L. (1967). *The Discovery of Grounded Theory: Strategies for Qualitative Research.* Hawthorne, NY: Aldine de Gruyter.

LaPiere, R. (1959). *The Freudian Ethic: An analysis of the subversion of American character.* New York: Duell, Sloan & Pearce.

Lazarsfeld, P.F. & Thielens Jr., W. (1958). *Academic Mind: Social scientists in a time of crisis.* Glencoe, IL: The Free Press.

Loubser, J.J. (1968). Book Review: The Discovery of Grounded Theory: Strategies for qualitative research, *The American Journal of Sociology*, vol.73, no.6, pp. 773-774.

Maines, David R. Ed. (1991) *Social organization & social process: Essays in honor of Anselm Strauss.* – Corbin, Juliet, Anselm Strauss: An intellectual biography. New York: Aldine de Gruyter

Merton, R. K. (1957) Priorities in Scientific Discovery, *American Sociological Review*, Vol. 22, p. 635-659

Merton, R. K. (1968). *Social Theory and Social Structure.* New York: The Free Press.

Merton, R. K. (1998) "Le style de recherche de Lazarsfeld", in Jacques Lautman & Bernard-Pierre Lécuyer (Eds), Paul Lazarsfeld (1901-1976): *La sociologie de Vienne à New York.* Paris: Editions L'Harmattan.

Perls, F., Hefferline, R., & Goodman, P. (1951) *Gestalt therapy: Excitement and growth in the human personality.* New York, NY: Julian.

Polanyi, M. (1951). *The Logic of Liberty.* Chicago: University of Chicago Press.

Reisman, D., Glazer, N. & Denney, R. (1953). *The Lonely Crowd: A study of the changing American character*, Garden City, NY: Doubleday Anchor.

Schulze, R. (1967). Book review. Gouldner, A.W. & Miller, S.M. (1965). Applied Sociology: Opportunities and Problems. *American Sociological Review*, 32, 6, 1027-1028.

Scott, J.C. (1971). Book Review: The Discovery of Grounded Theory: Strategies for qualitative research. *American Sociological Review*, vol.36, no.2, pp.335-336.

Stern, P.M. (2009a). "Glaserian grounded theory", In Morse, J.M., Stern, P.N.; Corbin, J.; Bowers, B.; Charmaz, K. & Clarke, A. E., Eds., *Developing Grounded Theory: The second generation*, pp.55-85. Walnut Creek, CA: Left Coast Press.

Stern, P.M. (2009b). "In the beginning Glaser and Strauss created grounded theory", In Morse, J.M., Stern, P.N.; Corbin, J.; Bowers, B.; Charmaz, K. & Clarke, A. E., Eds., *Developing Grounded Theory: The second generation*, pp.23-29. Walnut Creek, CA: Left Coast Press.

Strauss, A.L. (1987). *Qualitative analysis for social scientists.* New York: Cambridge University Press.

Strauss, A.L. & Corbin, J. (1990). *Basics of Qualitative Research: Grounded theory procedures and techniques.* London: Sage.

Wilson, E.B. (1940). Methodology in the natural sciences and the social sciences. *The American Journal of Sociology*, 45, 5, 655-668.

Zetterberg, H.L. (1954). *On Theory and Verification in Sociology*. Stockholm: Almquist & Wiksell.

Zetterberg, H.L. (1961). *To Publish Books by Scholars for Scholars*. Totowa, NJ: Bedminster Press.

Zetterberg, H.L. (1962). *Social Theory and Social Practice*. Edison, NJ: Transaction Publishers.

Zetterberg, H.L. (1965). *On Theory and Verification in Sociology. 3rd Ed.* Totowa, NJ: Bedminster Press.

Zetterberg, H. L. (2010) *Volume 3. The Many-Splendored Society: Fuelled by symbols*. Scotts Valley, CA: CreateSpace.

Endnotes

[1] While much of this paper has been sourced through published work by Barney Glaser and contemporaries, I am indebted to Hans L. Zetterberg for his recollections of the era as shared through our email exchanges and, of course, to Barney with whom I have been privileged to share a close collaboration over the past seven years. Where the recollections and observations in this paper are his, I have cited them as such; otherwise, the perspectives offered are my own as developed through the benefit of our interactions.

[2] Glaser, personal email communication, June 9-09.

[3] Glaser would later acknowledge Gouldner's assistance in the preparation of his paper, *Attraction, Autonomy, and Reciprocity in the Scientist-Supervisor Relationship* (Glaser, 1963a).

[4] Corbin (1991) has acknowledged Strauss' inability to analyze collected data (from his earlier studies on daydreaming and on lonely people), suggesting that his frustration was in how to penetrate the surface of data for what is really going on, what procedures to use to identify and label concepts and how to relate concepts in theoretical formulations: "I find him occasionally talking around a subject, rather than attacking it head-on, perhaps because he himself is not yet clear about what he is trying to say.." (p.36).

[5] Published and available online through Sociology Press (www.sociologypress.com).

14

GROUNDED GLASER

Evert Gummesson
Professor, Stockholm University

Before you find out yourself, let me share a secret with you. This tribute to Barney Glaser and GT is about me first. It is a cover for my ego-centricity and a vehicle for me to develop my own relationship to science, even to life itself.

Once I have admitted this, isn't my credibility going to fade? Perhaps, but to my defense I claim that I acknowledge the naked truth – which academics usually don't. The academic protocol prescribes more subtle ways of promoting oneself. One of the subtlest and safest is using the halo effect of academic celebs and their brands to gain credibility. It is done by referencing articles from so called international top journals (meaning US journals), written by those who publish a lot and whose names therefore are widespread and give many hits on Google. They are the poster names of the social sciences industry just like Johnny Rotten and Sid Vicious in Sex Pistols are in the rock music industry and Tom Cruise and Julia Roberts in the movie industry. The more a reference is cited the more it is cited, especially in these days of rankings and Internet searches. If you do not list the established references in your script the double-blind reviewers will be deeply upset and even reject your script without a chance for revision. You must show that you will carry the burden of received "wisdom" forward. References become name-dropping. The instruction to the authors is the same as the instruction from the Chief of Police in the classic movie *Casablanca*. When some mischief had been discovered by the Chief his instruction was: "Round up the usual suspects!"

If scientists were as rational as they brag they are, they should of course use the references which are relevant to their publications whoever wrote them – recognized scientists or not – in whatever form they are published and in whatever language they are written.

Being self-centered is not wrong by nature. It is liberating to be able to share one's personal interests, conclusions, opinions and recommendations without feeling the restrictions imposed by journal reviewers, examiners, and grants and promotion committees. The risk is on me and there is no protective net, only a free fall. I may appeal to the reader, be judged as conceited with hang-ups and idiosyncrasies and the worst of all, being stamped as non-scientific. I am caught with my pants down and there is no one to hide behind. But I also stand the chance to be considered creative, straightforward, unique and even brilliant.

Ego-centricity is a great driver, perhaps the greatest of them all. But it is not the same as being just selfish. You may be on to something from which others can benefit, too. In that case self-interest coincides with common interests and the two can live happily ever after. Barney Glaser is on to something through GT. Solidly based, its methodology and applications keep evolving. But to say that someone is grounded may not sound uplifting. Like an eagle that can't get off the ground. With Barney Glaser it is different. His research is grounded in real world data and his life is grounded on what the data teaches him. That makes him fly. It may sound contradictory but life is contradictory, ambiguous and complex.

I will never learn to master GT to the extent I would like. In my relationship with Barney Glaser I am clearly Dr. Watson and he is Sherlock Holmes. But GT has taken me part of the way. Among other things it has made me aware of the importance of substantive and genuine data, the beauty of emergence through an inductive stance, the perils of received categories, and the inescapable pilgrimage from description to conceptualization. Here I will juxtapose this knowledge with other knowledge and experiences.

Data Sources and Persona

From where do we get the data that may teach us about reality and make us see what we didn't see before? The conventional answer is: from the outside. But a single brain supposedly holds as many nodes and links (yes, I see the brain as a network) as a galaxy in the sky. The brain just reflects the complexity and diversity of the universe. In the language of fractal geometry the same shapes and patterns repeat themselves in smaller and smaller bits. Only in physical terms is my brain smaller than a galaxy; the content is as rich.

The universe of anybody's brain is a source of data that needs to be picked. Just as data can be grounded in external interviews and observations, data can be grounded in internal interviews and observations through introspection. Seeing us all in a network context, the following happens. If I interview you I enter your internal universe, which of course is an external universe for me. If you interview me you enter my internal universe, which is an external universe for you. So our universes connect. We live in an era of individualism – quite recent in the history of mankind – but we also acknowledge that we are all connected. These days I am repeatedly told that we live in a global world, even in a global village.

At an early stage I was taught that genuine science is objective. Scientists like to think that they deal with a fact-based reality and make major contributions to their discipline, perhaps to the whole of society. They strip the alchemism of practitioners, politicians and common people to the bare bones. Some hope to become immortal, get the Nobel Prize or at least get a campus building named after them.

To begin with I was fooled to believe in the supremacy of objectivity. I gradually felt that objectivity deprives the scientist of his/her persona and that the personal qualities of a researcher are important. I plucked up courage to acknowledge my feeling that science is both objective and subjective and both detached and involved. The evidence? It's simple: solid facts are hard to come by. What for example are the facts in the current financial crisis? And what is the reality behind President Obama's eloquence? I have to trust my own experience and intuition. But I have to do it with an open, reflective and systematic mind, letting knowledge emerge and not force myself on it. At the end of the day, I have to take responsibility and trust myself more than I trust others. With accumulated experience and non-stop curiosity it may end up in wisdom. Don't we all want to be wise guys?

In its fundamentalist version, Western science is detached from the human being and lives its own sublime life. Research is rigorous, strictly following rational logic on the highway to the truth about reality. It represents the crown of knowledge. All other "knowledge" is fake or at least inferior. If the researcher gets involved, he/she will be corrupted by such irrational dimensions of life as emotions, sympathy or antipathy, and can't see the real reality. Reliability means that other researchers can replicate and verify your research by getting roughly the same results. Can science handle this? Yes, of course – if social life is simple and static. Some social phenomena are probably simple and static but more typically they are complex and moving, sizzling and buzzing like an erupting volcano. Further the notion of an independent and a dependent variable that establish an unambiguous cause and effect relationship is too simplistic. In the dynamism and interaction of life, A sometimes causes B, but sometimes B causes A, sometimes A and B cause each other, and sometimes they get tired of each other and temporarily break their relationship or perhaps divorce for good. What is even more aggravating: A and B are not alone. They live in a network of complex and contextually dependent influences that are just ignored.

I thought first that science must be the Immaculate Conception and that Jesus is a metaphor for Western science. Many sciences think that they are the Savior of the Universe and God's gift to humanity and some of their representatives see themselves as Gods. Medicine is the uncontested epitome and the contemporary Jesus, and MDs and pharmaceutical companies are his disciples. Jesus loved metaphors but I am sure not in this case. Science is the religion of our times and universities are its cathedrals and professors its cardinals and priests. But can the sparkle of life really be ignited without interactive involvement; can science be the outcome of the Immaculate Conception?

In the "dark" ages (as opposed to our "enlightened" times) master and disciple, mentor and young protégé, were learning institutions. The good news is that being with a master will give the disciple both explicit and tacit knowledge; the bad news is that the disciple may become just a clone like the gene-manipulated Scottish sheep Dolly, or an impersonator and a preacher of the master's words. I have seen how such disciples in maintaining their master's reputation eventually become imprisoned by it, not being able to develop anything themselves. They keep asking "What would my master have said or done?" After the years pass by, the legacy of the master can turn into a burden and a hindrance. At the same time, keeping the master's knowledge alive is essential; it may otherwise be diluted, misunderstood, or just forgotten. Jesus preached love but its followers found motives for starting wars; Marx preached equality and solidarity and out came dictatorships. GT may become just another qualitative and inductive approach, action research becomes a questionnaire survey, and statistical surveys hide behind numbers that pose as facts whereas in reality they only represent an accumulated set of approximations and uncertainties.

The Terrorism of Images and Proxies

A famous painting by Belgian artist René Magritte shows a traditional smoking pipe, the type a bearded and weather-beaten fisherman would smoke. The painting is called *The Treachery of Images* and under the pipe is written "This is not a pipe." It may stand out as strange at first sight but of course it is not a pipe; it is an image of a pipe. This imaginary pipe is reduced to two dimensions; it can't be touched or be turned around; and it can't be lit so there is no smoke and smell of tobacco.

Conclusion: The image of the pipe is poor in data whereas the real pipe is rich in data.

Today's ubiquitous and invasive media bombard us with images, voices, texts, music and sound effects. Only with difficulty can we separate the factual from the fictional, information from noise, and the genuine from the spin-doctors' seductive stories.

Today an image scores higher than the real thing. In business schools and marketing, brand management has acquired divine status. A brand is a symbol and its image is shaped in the eye of the beholder. The trick is to communicate a story that charges the brand with positive associations and then to maintain the brand. This is done through increasingly more subtle tricks. If a product is good or not, the brand gospel claims, is just a perception among consumers. If consumers believe that the mile of fancy yoghurts in a supermarket will keep them slim and stimulate bowel movements through benign bacteria, they will buy it – even if only a few, if any, of the products will fulfill their "promise". Just keep claiming it! Make it credible! Keep up the brand but not necessarily the true value of the product. Appearances are everything.

An elderly eye-doctor, whom I consider not only a man of knowledge but of wisdom, said to me: "Young doctors don't know how to make a physical diagnosis. They rely on technology, test results and images. They send patients to x-ray as routine." X-ray or more advanced digital tomography deliver images that can be presented and accepted as evidence in court. A doctor's professional assessment based on ocular inspection, sensitivity to the patient's signals, tacit knowledge and observed patters from continuous clinical work, has less credibility than a picture or a number from a test. We live in a society where images, words and numbers – which are there to mirror reality but often do it poorly – are mistaken for reality. Hands-on action and results become less important. The rituals of science dictate the world view.

The Terrorism of Received Categories

Terrorism is not only physical and sensational. Unobtrusively it sneaks into our lives and we innocently let ourselves act as promoters of received categories. GT is an enemy of received, mainstream categories that have emerged God knows from where. Some categories may have been relevant in a specific time and place context. They sometimes become so well-established that they imprison us for life. Some categories are overriding dichotomies that turn life into opposites with ensuing conflicts. Everything becomes black or white. They are usually a heritage from long ago; they may be just folklore, misunderstandings and witchcraft. I have challenged some of them: goods vs. services, suppliers vs. customers, quantitative methods vs. qualitative methods, and social sciences vs. natural sciences vs. the humanities (Gummesson 2006, 2007). People feel secure in them; everybody knows everybody's place. These terrorist categories are often badly and sloppily defined, if defined at all. They become so established that they stay forever like old castles along the river Rhine, even if they are ruins. The categories are promoted in textbooks to which young students are exposed centuries and decades after they have been shown to be misleading.

For example, Barney Glaser emphasizes that GT is a general methodology and not just qualitative as it is conventionally categorized. This is, however, somewhat nebulous, but categories are often just that. They become Procrustean Beds where the content is stretched or reduced to fit the bed meaning that the bed has empty space at the same time as limbs hang out from it. The original book by Glaser and Strauss (1967), *The Discovery of Grounded Theory,* carries the subtitle "Strategies for qualitative research." For the Strauss and Corbin (1990) book it is the other way around: *Basics of Qualitative Research* with the subtitle "Grounded Theory Procedures and Techniques" and the third edition (Corbin and Strauss 2008) has removed the subtitle from the cover; only qualitative research is left to catch the eye. One of Barney Glaser's latest books is *Doing Quantitative Grounded Theory* (2008). The qualitative-quantitative dichotomy is used to create hostility in science for no rational reason whatsoever. These categories represent different languages with different symbols, different structures and different strengths and weaknesses, but they are equally dependent on subjective assumptions, judgment calls and approximations. They both require someone to draw conclusions from the results and to take action. The qualitative/quantitative divide causes a pointless competition about which is universally superior; it is a pseudo-ranking. Carpenters are smarter than professors; they would not declare civil war between the hammer and the saw or try to rank the two in order of importance and relevance.

Walk Your Talk

At some point of time during my incessant struggle to understand science, I discovered that it helped to get to know the person who had made a seminal contribution. I have gained more understanding about GT being with Barney Glaser in a discussion in his kitchen, on a hike in the Californian Mountains, and together watching Michael Moore's documentary *Fahrenheit 9/11* in a Mill Valley movie theatre on the 4th of July, than reading his books. This is not a depreciation of the value of his books; they are important supplements that can add missing links and dispel the mist in my mind. It means that the written word has less impact on me than actual deeds.

Barney Glaser and I first met in 1984. One thing struck me: he quickly grasped what I tried to say without knowing my discipline, marketing and management. He wanted to bring the dialog forward. He demonstrated theoretical sensitivity through his actions and not just through articles and lectures, in contrast to one of my professors who responded to everything with a scornful smile expressing his misgivings about my mental capacity and concluding that what I said was completely unfounded. He was always predictable and reliable, but he was a dead end.

Barney Glaser lives GT; it's part and parcel of his persona. How many politicians fulfill their promises and how many of them live the life they demand voters to live? How many economists know how to handle money – just look at the crisis of the global financial system! How many lawyers live by the law? Don't they just cynically take advantage of their technical knowledge of a fragmented and incoherent legal "system" instead of the wisdom inherent in a truly ethical and coherent legal system? How many physicians lead a healthy life? Barney Glaser became a world-renowned sociologist, but he used his scientific method to start a building company, a financial business, and a publishing house. And they are all successful. His windsurfing skills are the outcome of a GT study; his mini-GTs help him to quickly get to the point in all walks of life. Doing what he preaches, he personifies GT methodology.

Personally I am good at transferring my experiences – "empirical" originally meant experiential – to abstractions and generalizations. I am not as good, sometimes even useless, in transferring the general and abstract to a concrete personal reality. For other people it may be different, so let's capitalize on our differences. My general conclusion is that any science and any methodology is not only intellect but also flesh and blood, and it walks on two legs. Paradigms are two-legged.

Changing Role Models: From Major Bigglesworth to Dr. Glaser

Through the passages of our lives we scrap some heroes and pick up others. A hero from my early life was Major James "Biggles" Bigglesworth. He is the object of 104 books by Captain W.E. Johns. Biggles was a flier who successfully fought every war and terrorist for a few decades. Both Biggles and Glaser are guys with the mission to make this world a better place. They use different tools. Biggles had his aircraft and gun and Glaser fights with words. But the titles of the Biggles books could very well fit Glaser; just put Glaser there instead of Biggles in, for example, *Biggles Investigates* and *Biggles Sorts It Out.*

They both go places: *Biggles Flies North, Biggles Flies South, Biggles Flies East, Biggles Flies West* – there is even *Biggles at the World's End.* After having given up his chair at Berkley, Barney Glaser became an academic recluse in the woods on Mill Valley and entered the business world. From an academic horizon he stayed at the world's end for a long period. In 1994 I invited Barney to Stockholm to lecture on GT. "This is the first time in twenty-five years that I leave the US," he said. Since then he has followed in Biggles' footsteps and flown north, south, east and west.

There may be one Biggles title, though, where the Glaser name would not fit: *Biggles Goes Alone.* Barney Glaser wouldn't do that; he comes with an entourage of his wife, children and their spouses, and admiring disciples, his "groundies." In this way one gets to know him not only in the context of universities and personal one-to-one contact but also in a family and friends context.

Orthodox GT or GT Light?

The publication of Strauss and Corbin's book in 1990 brought Barney Glaser and me closer. Although dedicated to him "with admiration and appreciation," it upset him. In 1992, he published a book against it, *Basics of Grounded Theory Analysis* subtitled "Emergence vs. Forcing," and he sent it to me. It was published by Sociology Press, which is his own publishing house – where else? For many years I used both books in doctoral courses to show that methodology is not given by God; even co-authors and close friends can disagree.

It was generally thought that long-time buddies Anselm and Barney became enemies and I fell into the same preconceived trap. They didn't. I can't tell the whole story behind the books here. Had it occurred in Hollywood and the movie industry it would have made *The National Enquirer*, but who wants to read about social scientists. Anselm Strauss died in 1996. I never got to meet him, which I regret.

The event brought to the fore how strictly a methodology should be applied. When I first met Barney Glaser in 1984 he was disturbed by the fact that many claimed they had done a GT study but had only applied some of the GT strategies and not done it well. GT got criticized and Barney Glaser thought it unfair. During the past decades GT has become a whole range of applications from orthodox and classic GT to GT light. A recent example of a consistently orthodox GT application in my discipline management and business is the book by Christiansen (2007). There is also GT-inspired research – I am in this category – where GT has contributed with certain strategies and procedures, but this is not full-fledged GT and should not pose as such. It is light one-calorie-only GT.

Premium Brands

Has the brand of GT been diluted by the various uses or have they strengthened the brand? A brand, especially a premium brand, becomes a story that guides people. If the brand represents true content and not just a superficial image, a well-known brand is a unique platform from which to speak. Once I stayed in The Midland Hotel in Manchester, UK. At the entrance was a plaque saying that this is where Mr. Rolls first met Mr. Royce. In London I found another plaque on a house – I think it was an old hospital, but I lost track – saying that this is the place where Sherlock Holmes first met Dr. Watson. The two plaques illustrate the difference between fact and fiction but also the difficulty of separating the two. Rolls-Royce is for real but Sherlock Holmes and Dr. Watson are fictional characters from short stories. But even as such they are in a way real. They can be read about in books and be seen on screens and stages. The actors are real people who pose as proxies for fictional characters who in turn may be the synthesis or abstraction of substantive knowledge. But they *are* not the people; they are impersonations and look-alikes.

As a bonus Sherlock Holmes stories can be read as cases on methodology. Holmes was passionate about data. He observed and interviewed and compared data. He did not find a grain of tobacco or a cigar butt on the scene of a crime; he found a specific make of tobacco. He knew where it was manufactured, in which stores it could be bought and what kind of people smoked it.

Glaser & Strauss and GT are just as well-known as brands in the social sciences industry as are Rolls-Royce and Sherlock Holmes & Dr Watson in the auto and detective novels industries. But are there any plaques commemorating great social scientists? I would like to see a plaque saying: "This is where Mr. Strauss first met Mr. Glaser." I'm sure Barney Glaser knows where it occurred. Now that GT has entered a respectable age and survived the slings and arrows of outrageous fortune, I think it is time to commemorate it with a plaque.

Academic Ritual or Usable Results

I have also been taught that the hallmark of science is the systematic application of approved methods. If you follow these you can't go wrong; nor can you be criticized and jeopardize your smooth road to academic tenure. You are safe; you are one of the boys and girls. Medicine claims it is the science of sciences today; it counts itself as part of "life science." But how then is it possible that one day fat is dangerous and the next day healthy? For a long time fat has been claimed to be the cause of obesity, but now the culprit is carbohydrates. Most likely it is neither. The ludicrously simplistic and fragmented "explanation" that one single factor causes one single universal outcome is based on nothing but immature assumptions. The human being – body and soul – is a complex system living in complex systems. The outcome is caused by a constellation of factors – or in network theory terms of nodes, links and interactions in scale-free networks – that ultimately determine a patient's well-being. Science cannot just be referred to image, ritualism and hubris triggered by some occasional and contextless results. To a large extent relevance and validity have evaporated – but that would never be admitted and no one is ever held responsible. The ritual – the methods and techniques – are paramount. The researcher is reduced to a technician let be skilled as such, a scientific bureaucrat who can never be blamed: "I followed approved scientific procedure!" If patients do not get well from the medication the doctor ordered and even get sicker, the doctor is benchmarked against his/her compliance with approved, mainstream knowledge and not against the ability to cure patients. Technicalities have pushed real life into a dark corner.

Unfortunately professors who want to make a straight university career get blindfolded by academic jargon, fancy research techniques and historical theory that they succumb to the increasing force of the US career systems that is imposed on the world. The world's No. 1 management guru Peter Drucker (1988) has warned that "...we are prone, both in academia and in management to...mistake the surface gloss of brilliance for the essence of performance. But it is so easy to fall for sophistry – to mistake clever techniques for understanding, footnotes for scholarship and fashions for truth..."

So the methodologies we use must be good, which sounds trivial. GT is good methodology, perhaps the best. I also favor case study research, action research and network theory. These are not in competition with GT; they can live in symbiosis.

I gradually understood that science is not necessarily the highway to enlightenment. It could just as well be the bureaucrat's dirt road to an academic pseudo-life. True scientists are innovators and entrepreneurs and break the more-of-the-same and me-to deadlock in a discipline.

A Stopover on a Never-Ending Journey

I have neither written this to be applauded by Barney, nor be criticized by him. I have written it as an expression of where I am and how I feel about it. If I cannot come to peace with my own scientific thinking and behavior how can I expect anyone to listen to me? Although an individual has some core qualities, the individual is many-bodied; he/she is different to every other individual. My perception of Barney Glaser and GT and my projection of myself on them create my reality. We live in a context – or a network of relationships within which we interact – and that is where my persona is shaped. We become through social interaction; as stand alones and recluses we are nothing to nobody except perhaps to ourselves.

One thing that Barney Glaser and I have in common is the desire to be genuine, walk our talk, promote what we truly believe in, and protest against what we believe is flawed, and not bend to power pressure or temporary hypes. But I don't know if my contributions to social sciences will have any impact or live on. Perhaps my major contribution is reflected in what Barney Glaser wrote in two of his books which he has presented me with over the years. In *Doing Grounded Theory: Issues and Discussions* (1998) he writes, "To Evert who started me on the worldwide trek to this book" and in *Doing Quantitative Grounded Theory* (2008), "For Evert awakening me up 20 years ago." Together with the late professor of educational sciences Sven Styrborn, I took the initiative to nominate Barney Glaser to the Stockholm University Board for an Honorary Doctorate. He was awarded the doctorate in 1999, not an easy title to get. Among other honorary doctors in Stockholm in social sciences are Amartya Sen, University of Cambridge, UK, and later Nobel Laureate, and Mark Granovetter, sociologist and famed network specialist from Stanford University, USA.

It so happened that the same year I was selected by the Social Sciences Faculty at Stockholm University to be the one to officially confer the insignia to new doctors at the annual celebration in the Stockholm City Hall. I put the laurel on Barney Glaser's head, handed him his diploma and doctor's ring, and made a short presentation of him. It felt good.

I keep wondering why Barney Glaser has not received more honorary doctorates.

The original idea of scientific work – to find the truth, acquire knowledge and wisdom, and make earth a better place – have been devoured by a gargantuan bureaucratic, ritualistic, political and financial apparatus. The parts of this apparatus are controlled by power groups within the sciences themselves, by grants and promotion committees, evaluation and certification bodies, corporations, politicians, and government institutions. To keep up the brand of science has become more important than the content, and advertising agencies, public relations consultants, lobbyists, and reporters are busy creating an image of science moving ahead faster and faster. The knowledge society, the knowledge-based company, the knowledge worker, the professional, and evidence-based medicine are some of the hyped expressions. Knowledge and techniques are necessary but they are meaningless unless they transcend into wisdom.

As I warned you initially, this was going to be an unabashedly ego-centric account, acknowledging the human being behind science. In that aspect I think I have been both consistent and successful. I wish more scholars would feel free to let a bit of their persona out and stop hiding behind an established elite. You should of course learn from others, but you should not be cloned by them. After all, the personality of the scientist is his/her most important research instrument.

This is the end of my ego-trip – for now. Science is a journey and the journey is the goal. There is no end station.

References

Christiansen, Ó. (2007). *Opportunizing: How companies create, identify, seize and exploit situation s to sustain their survival or growth. An orthodox grounded theory approach.* Torshavn: Faroe Islands University.

Corbin, J. and Strauss, A. (2008). *Basics of Qualitative Research.* Thousand Oaks, CA: Sage (3rd ed.).

Drucker, P. (1988). Teaching the Work of Management. *New Management Magazine*, 6(2).

Glaser, Barney G. (1992), B*asics of Grounded Theory Analysis: Emergence vs. Forcing.* Mill Valley, CA: Sociology Press.

Glaser, B. G. (1998). *Doing Grounded Theory: Issues and Discussions.* Mill Valley, CA: Sociology Press.

Glaser, B. G. (2008). *Doing Quantitative Grounded Theory.* Mill Valley, CA: Sociology Press.

Glaser, B. G. and Strauss, A. L. (1967). *The Discovery of Grounded Theory.* New York: Aldine.

Gummesson, E. (2006). Qualitative Research in Management: Addressing Complexity, Context and Persona. *Management Decision,* 44(2), 167-179.

Gummesson, E. (2007). Exit Services Marketing – Enter Service Marketing. *Journal of Customer Behaviour*, 6(2), 113-141.

Strauss, A. and Corbin, J. (1990), *Basics of Qualitative Research.* Newbury Park, CA: Sage (1st ed.)

15

LIVING THE IDEAS

A BIOGRAPHICAL INTERVIEW WITH BARNEY G. GLASER

Astrid Gynnild
University of Bergen

How does a person become himself? That is one of the questions that has intrigued Barney Glaser both professionally and personally throughout most of his life. In this interview, Glaser talks about experiences that have had crucial influence on his extraordinary career in social science research. For the first time, Glaser reveals that the inspiration for the grounded theory principle of staying close to the data, the true source of the idea, is psychoanalysis. He also discusses how he has used grounded theory successfully in all areas of life, and how any obstacle can be overcome and positively redirected through the use of aikido, jujitsu with swords, in an abstract sense.

"Who-o-o is it?"

A thin, shivery voice slips out through a narrow opening of the door. It sounds like an old grandma who is afraid of letting anybody in, and for a moment I am wondering whether I have got a wrong apartment number. Then suddenly, the security locker is removed and there stands Barney Glaser in a timeless sweat shirt and a pair of well-worn college pants, with a big grin on his face.

After a short pause, his dark voice is back to normal and he continues as if nothing had happened:

"Well hello there, come on in! I'm just about to have breakfast, do you want some?"

It is a late Friday morning, 14 floors above street level in Manhattan, and another grounded theory seminar is over. Barney Glaser is preparing scrambled eggs and a cup of tea in his New York apartment, where he stays during his annual GT seminars.

The co-founder of grounded theory, now in his eighties, appears to be a little tired but happy and energetic. While padding slowly around in the newly refurnished apartment, he is playing practical jokes, thereby doing a necessary warm-up for the conversation to come. Glaser brings his breakfast over to the table, and is a bit surprised when I ask if it is OK that the conversation is tape recorded. After all, he does warn against tape recording in most of his books. The argument is that taping is time consuming and ties the researcher up in details that are redundant for grounded theory generation. And yet, he says, he respects that sometimes other levels of descriptive accuracy are needed. So fire loose.

OK then, since we're talking during a meal, when doing GT seminars, you always invite participants out for dinner, both before and after the seminar itself. Why this social focus on food when teaching grounded theory to PhD candidates?
Why? It's just obvious. To binary deconstruct to the max to keep them up.

Did you do anything similar as a professor at the UCSF?
Totally. We dressed down, let everybody talk, express themselves, go around the room, whatever it does to take the professor-student edge off. Because I live with ideas, not status or identity in that sense. My identity is someone with ideas, not someone with position and power. Ideas are far more powerful than positions.

Where did you get the inspiration from?
Me. It's doing Barney. I was the youngest professor at the UC. I was 47 or 48. And that was great stature and I never monopolized. I never used stature and position in approach on how I dealt with people. Let me put it this way: I studied sociology and science. Do scientists become famous because of their stature? No. They become famous for their *ideas*.

Leaving UCSF

Two decades have passed since Barney Glaser said goodbye to his career at the university to continue building his own research and business universes from his base in Mill Valley, California. He left shortly after the controversial book The Basics of Grounded Theory *appeared in 1992, at the age of 62.*

How was it just before and just after you left the University of California in San Francisco?
As far as I remember, we were always on soft money. And then I think they ran out of money. It was sort of a gradual progression, with a lot of factors. Business was doing well. I published *Basics* and everybody was unhappy. Anselm was running out of soft money. It wasn't even quitting, so much as that of a progression. I don't think I noticed it that much.

But it did change your career?
Well, I didn't teach any more, but I was still writing, wasn't I? It wasn't a big split or big disjunction, or juncture, more just a progression of doing more of one thing and less of another. I think *Basics* turned a lot of people against me.

It wasn't that they threw me out, but all of a sudden there was no more research money. That also progressed.

More than 40 years after *Discovery*, and nearly 20 years after you left institutional academia, you're running more grounded theory seminars than ever. Almost half of your books are written after the age of 60. You are helping an increasing number of PhD candidates getting their degrees. And yet many aspects of your personality and life are little known. If you were to describe yourself in one sentence - who are you behind the concepts, Barney Glaser?

Me. I hate to tell you where it comes from, but I think I've already told you. Didn't I?

What do you mean?
How do people become themselves?

He looks up from his tea cup.

My me is bigger than any identity. Where did it come from, just doing me? I own my own press, I own 40 houses and where do they all come from? I'm just doing me and I teach by just being me. I'm not talking about being a CEO or all that. So just being me and whoever I am always transcends social identity, although most people just hang on to their identity because they don't know who they are. And identity is given to them by society. They just do what mobilizes it.

Suddenly, Barney Glaser's face lights up. He starts giggling and then waving his right arm like an excited little boy: "Look at that guy outside the window - Good Lord!"

It is like a scene from a movie; three guys watching us from the outside, on the 14th floor in Manhattan. They are sitting on a plank moving upwards in slow motion. Window washers. They are waving back and smiling, too. Barney Glaser bursts out in laughter. In the next moment he is into reflection again:

I mean, how do people *disbecome* themselves- as opposed to positions, jargon expects or whatever?

How have you become yourself?
Well it's an important child-rearing that most people can't have. There's three factors. Four factors. One is: My parents never said do this and do that; they never evaluated. I was always on my own. But we were wealthy and we had a staff, and the staff always took care of me based on what I thought we should be doing or whatever. One day I said: 'Mom, I'm five, aren't I supposed to go to kindergarten?' She said she didn't know, so she sent me over with a chauf-

feur. We had upstairs maid, downstairs maid, cook, chauffeurs, cleaning maid. They all, you might say, obeyed me with judgment, but I was always on my own. And that only happens positively for intelligent people. That's how I raised my kids. You never tell them what to do. You always say I trust your judgment. And then they bring their activities to you to discuss how their judgment worked. So I was on my own at the age of five. My parents hardly knew that I went to college.

You didn't have much contact with your parents, then?
I made dinner with them every day. I had a lot of contact with them. They were always there, but what I'm saying is they never evaluated; they never requested I do this, do that. I was always left to be myself and I never disapproved of it. It made me very productive. Most people are always wondering if they are going to be evaluated properly or doing what their mother wants to do or something. They evaluate themselves and worry if they are going to be OK, to be approved. I never had this problem. And I was very intelligent so I read constantly. If there was anything, I talked with the staff.

What did you read?
Whatever. I read constantly.

There is a long silence.

Psychoanalysis

And then, to further become myself, I've been to three psychoanalyses.

As he is expressing this one sentence, Glaser speaks very slowly. He seems to be diving back into a far gone past. There is a long silence.

...at what age?
22, 24 and 26. They all took several years pushing to the fore.

Why did you do it?
To become more me. I went to psychoanalysis because I lived in Germany for a year and I was going around with a German girl. We were living together and I brought her back to America and she ditched me and I was devastated. Probably lucky.

So then you wanted psychoanalysis?
Yes, because it really hurt me. I've always learned a lot about myself.

What was the most important thing you learned from the periods of psychoanalysis?
One guess.

Be more you?
No. Getting the exact data to see what it is. Well, it was used in grounded theory, but no one knows it. The true devotion to data, the true source came from psychoanalysis. Honestly. The result is me, putting a lot of different ideas into one package. In psychoanalysis you learn to look at exactly what is and just to be yourself.

Some people would see it as a lonely way of getting to know yourself. I mean, psychoanalysts are not known for talking.
It doesn't always work like that. There's interactions. I'm a kind of guy who easily gets along with people; it's a kind of waste to be alone. So you need to find ways to be alone. I spent hours trying to look at the data exactly as it is. And I learned a lot about relationships.

He takes another break before he presents an example of what he has learned about relationships and how to handle them, particularly when one has been betrayed.

The first that happens then is to go and say 'thank you, anything I can do to help?' Because…who needs someone who cheats on you? Help the person to get away from you.

If people do something badly to you, you should help them first and then let them go?
Yeah, help the other person; hug him and get rid of him.

How has that experience been helpful for you in your life?
It's just data. Most people get very upset. They think they are cheated upon. They get angry and so on. But they are the luckiest people on earth.

It resonates with Dalai Lama's saying, you should always love the people who kick the legs under you, because they are the ones who really help you to move on in your life?
It's just an example of being me.

Suddenly, again, Glaser is putting on his most cheerful face while waving to the new friends outside the window. This time, they are moving downwards in slow motion. In the next second, Glaser is becoming serious again, speaking very slowly:

But I trust the patterns. People can't do anything else but their patterns.

So that means then when you're in charge of building houses…
…so I don't need to control them. I'm running a mega virtual organization. A big one. But I know what each person does. Most of the time I don't say anything because I know they just run their patterns.

Does that mean that you do theories on people close to you, too, or can you tell just by getting to know them?
Mainly by getting to know them.

Collaborating with Anselm Strauss

Were you able to foresee your relationship with Anselm Strauss, too?
No. But I discovered his pattern as life went on. He always linked up with other people who wrote his books.

Are you saying that part of his pattern was that he couldn't resist being under the influence of others?
He was a full professor and people wanted his OK. He OK'ed everybody and never really wanted to control anybody. He OK'ed. He used that power. And he wrote books with them. But they wrote most of those. I mean, he wrote his own books, a few. And I was sucked into the pattern and I didn't realize it at the time; I was delighted to write a book.

I notice that all your books are dedicated "For Anselm"; you're always thanking him for what you did together and how he inspired you. In the festschrift for him, shortly after the *Basics* book, you emphasized the good collaboration.
He was a lovely guy. We were close friends. He was the one who told me to write *Basics.* He wasn't domineering or anything. He just let everybody go in whichever way they wanted so he could go his way.

You were friends in private all the time?
Oh yeah, absolutely. I talked to him a few days before he died.

Many academics have been quite concerned about your professional breakup from Anselm Strauss in the *Basics of Grounded Theory* in 1992. How has that been?
What do you mean?

Do you get tired by other people's curiosity about this issue?
Well, I wrote it to keep the classic GT on track. Anselm said 'if you don't like it, write a book about it.' But you know, I read a lot and...you know the rule of getting read, don't you?

Get cited?
No. *Dissonance* creates lots of reading because everybody is trying to figure it out.

Did you see the dissonance between you and Strauss as good for your career and your books?
Up to a point. I knew a lot of people would work to figure it out for years to come. That's what they do when people write something meaningful. If you want to write a popular book, write something with dissonance so they could try to find out what you're really saying.

GT businesses as non-profit

After he separated professionally from Anselm Strauss and left University of California, San Francisco, in the early nineties, Barney Glaser continued and expanded his four different businesses: Sociology Press and the Grounded Theory Institute, house rentals in Hawaii, and a loan transaction company. The businesses were all developed as a result of grounded theories in the field, the Sociology Press and The GT Institute as non-profits. He is also involved in the growth of a world wide network of grounded theorists online.

Why do you run Sociology Press as a non-profit? And why did you choose to publish your own books?
My first books were all published by Aldine. But I got tired of the exploitation, you know, the lack of contact with the readers, curbing of the distribution of the ideas, the recognition mares. I guess I was hard fired with *Basics* so I published examples so people would see what a grounded theory really looks like. Everything I've done has been published immediately, and Sociology Press does what the presses don't do. Try to put in a contract that you want the name and address of everybody who buys a book so you can ask them what they are writing and thinking. When people call to order a book, I give them a stroke. 'Oh, am I talking with Dr. Glaser?' And I say 'Yes, this is Barney.' I mean, just that stroke. And you see, I can watch where the books are being used, I can tell where there's a class. Getting all kinds of data you won't get when using a publisher. You just get a summary at the end of the year.

What do you use the data for?
Knowing where grounded theory is pitching on. Someone called me from New Zealand and wanted to buy a book. He started talking to me and invited me to New Zealand. I mean, I can see what students are buying from what universities. I learn an awful lot just by looking at it. And Sociology Press on the Internet helps selling it, and PayPal helps selling it. But copyright is bullshit in our world, there's no money in them. It's not like intellectual capital, which is very significant - corporate secrets.

Your, so far, next latest book, *Jargonizing: Using the vocabulary of grounded theory*, dissects the content in the Handbook of Grounded Theory published by Sage. What responses did you get from the authors whose work you analyzed?
Not much. They wouldn't understand well. But lots of people are buying it. They'll read it in a lot of different ways. That's what you have to learn. You have no idea of how people are going to take it. There's no control. Just like your career cycling, people start seeing all kinds of different things in it. But there's a lot of grab in the ideas no matter.

So you ask how can I do so many things? I just do me. And most of the time I trust their patterns and most of the time I don't have to tell anybody what to do. I mean, I have the bank in my pocket. I have a fabulous manager over in Hawaii, who manages all the real estate, you know. I do the booking, renting, but the manager calls me, and I tell her I love her and she says 'I love you, too.' And I have a guy who runs around and takes care of all my houses, I have about thirty, and I never tell him what to do. He runs around, always painting, fixing, whatever, he says 'Barney is this OK with you?' and I say 'I trust on your judgment.' You just let them run their patterns and I say just keep it on the even. So I do me and doing me is watching people's patterns. Once I see them they can't do anything else, unless they're deceiving. And I haven't run into anybody who deceived me.

Even people who deceive run their patterns. I'm a very moral man. I don't use it against them and…you know.

Doing aikido with words

The way you use your own and other people's energy reminds me of the principles in jujitsu. Instead of trying to stand up and conquer the enemy you transfer the other person's energy and power for your own benefit.
You just said the word! I do aikido [jujitsu with swords] all the time, every day. You grab it and pull it faster, and you don't block. Only it's their patterns. I mean, you know, that seminar we had is elitist beyond belief to be open and flexible and getting what's going on. But people need the structure to keep body and soul together. And they're one-track patterners. I mean, like I have my own wire officer at the bank. I call and say 'how am I?' and he tells me everything. But I have to ask. And I have tile officers doing all kinds of things for me. It's very easy.

Does that mean that it's not possible for anyone to hurt you?
Of course it is. But what you use is any negative pattern against them.

How do you do that?
Any way you can. What I'm telling you is very youthifying. Because there's no wear and tear. So you never tell your kids what to do again except one thing –

get their PhD because it brings a social value they'll never get any other way. So I never told my kids to go to college. Lila went to Yale, Jill and Bonnie went to Berkeley.

Daily Life

You were talking about the simplicity of letting people run their patterns. But according to your daughter Jill, you are the most hardworking man she's ever met. I'm curious about your daily life schedules.
Well, I've had a trainer for ten years. I've told her one thing: If I don't hate you, you're not doing your job. She's a very smart lady and she laughs constantly. But she keeps me in good shape. I'm going to start ski exercises soon. She comes at seven in the morning, three days a week. And then I go down to the office. It's a house and I have to be alone because I can't stand other people 'cause I work and think faster. And I'm running around, doing all kinds of things, packing books, making phone calls. And then I go to the post office and I come back. Sometimes I bike-ride, sometimes I walk. But most of the time I drive. I check in with Carolyn at three to see what's for dinner. She's like me, she can't stand long descriptions.

Glaser adds that Carolyn is his "fourth lady." He has been married three times. His first wife died in the seventies. Glaser says that he has taken a dive for his women throughout the years, but it never got him off the productive track.

There was a long time span between *Theoretical Sensitivity* in 1978 and *Basics* in 1992 where you didn't produce many books, though. Since 1998, at the age of 69, you've written a book every one or two years. What caused this sudden turn in productivity?
You believe in the universe?

It depends on what you mean by believing in the universe. I get a lot of associations but they might be different from yours.
Well, yours are all right. I mean, you hit the nail on the head. Isn't it obvious? When did I start writing a lot of books?

After 1998.
And what happened in 1998?

I know that your wife had an accident.
Absolutely. And one more thing. I got the honorary doctorate at the University of Stockholm. So all of a sudden I was confronted with a lot of time and a lot of motivation to make the honorary doctorate worthy of my work, or make me worthy of the honorary doctorate. I could go to earn the value that was given and start correcting all the bullshit. In other words, Carolyn had an accident and couldn't do much. Did it stop my life? No, it opened up another

section of my life, which gave me immense amounts of time to write books. She couldn't do a lot and she loved the idea that I was still being busy, and that her accident didn't threaten my life. I would say that 90 percent of men would have gotten divorced to get out and get active. And I see upon it as really having a lot of time. Interesting, huh? In other words, that's what I mean about the universe. The universe presents you with things that you run with.

What happens and when is not accidental - is that what you mean?
I don't know if it's not accidental; that gets to be a little spooky. It happened. I mean, I wouldn't wish it to happen so I could write books. But before that she was here every month for a week and I was travelling back and forth to New York. And I wasn't having the time I needed. I need a lot of alone time.

When do you usually get the alone time you need?
In the office. In the morning. At night, after dinner. It didn't wreck my life. I changed it and I found a way to improve it. She never curved my own activities, she was delighted. We would stay together. She's very smart. She would talk a lot. She thinks the jargonizing book is fabulous.

It seems that you have a good relationship.
Yeah, I mean, the universe opens up for me all the time. Most of the time, I just wait. For example, when she had her accident…

Suddenly, Barney Glaser gets up and is heading toward the bookshelf that covers a whole wall in the apartment. He is looking for a book written by a man who, according to Glaser, saved his life during the first, difficult phase of reorganizing life after the event known as "the accident." Glaser's wife Carolyn was seriously wounded in a car accident while on visit to India. This friend John loves library work and told Glaser he would go out to find all accessible articles on spousal care of the SCI patient, spinal cord intervention.

The papers were marvellous and he sent them all to me. He saved my life. Isn't that the universe opening up? Interesting, huh? One of the biggest things it said was: First of all, you don't engage in dependence care. In other words, make the other person more independent, if not you have to take care of them over and over again. Then we got semi-workers. There was a lot of good information in those articles on how to take care of a spouse with an SCI problem without killing yourself.

In another case the universe opened up, too. At the age of 40 I was always very sick. I couldn't figure out what it was. And I had a very close friend, a psychoanalyst, very intelligent. We started doing some work together on diet research. This was in the early seventies and he was testing among patients the effect of diet on people's thinking attitudes. He had all his colleagues trying his diets. We analyzed it, not in terms of foods, which was a waste of time, but in terms of the chemicals that go into metabolism. Like acids, acids in alcoholized dosages and levels. They're getting into the body levels and I was trying all these diets. His friends were becoming upset because they couldn't go to social dinners and eat anything. We had several people who couldn't continue. When they broke the diet they became suicidal. And he started thinking it was too dangerous to sort of have death on his hands. His colleagues started saying he was crazy, that he only had his exams in psychoanalysis. In any case, he cured me and I started to diet at the age of 40. And I've been in perfect health and sanity ever since. I was really sick.

What was it?
Too much sugar and too much acids. We had it all worked out based on simulation. There were only two things, acids and alcohol. So now I can eat like a horse as long as I eat the right stuff. And I'm always going to my doctor for a physical checkup. The last time we spent 15 minutes talking about his skiing trips. And I said 'what about me?' And he said, 'Oh, I only deal with sick people." You know, so that's the reason why I am so healthy.

...and I've never drunk. I tried to drink once when I was 16 but that made me sick. See, that's another thing. If you start telling your kid what to do, they won't like it. So you never tell them what to do, you work with what they are doing. Then I like myself, see that's yet another thing. I like my self a hell of a lot, so why would I do something that do me misery? Interesting, huh? I'm very valuable and I'm creating that value. People are doing all kinds of things, suicide, over drinking, smoking, whatever, physical stress.

When you see such people, what do you think then?
I'm my own best friend. It's not a question of self protection. It's a question of living in a way that...indicates you like yourself. You can go to a party and people really don't like themselves. Most people don't like themselves unless somebody says they are OK for the last five minutes and then they need another positive evaluation.

You are concerned about being your own best friend. If you didn't go to psychoanalysis far back, do you think your career would have been just as successful? You would not have been the same person.
Aaah - no. But it happened. It's not an either or. No I'd probably be dead, you know, the doctor saying 'you're healthy and I don't know what to do with you.' It's the path. It's never the reason you want to do something; it's the reason why you stay doing it. A person I know well thought I was a Buddhist. Now I can say that what distressed him was that 'Oh you mean that Buddhists do it, too'? See most people would think that what I'm telling you has to come out of a spiritual legislation. But Buddhists tend to let you be yourself, I guess, as long as you obey. So that's from parental control to spiritual control.

So that's why you've never been interested in any religion?
No, to me it's always control. They want you to join up.

Speaking of control. What does "imbuement" mean to you?
That's one of the main uses of grounded theory. It imbues your way of thinking. Do you think in terms of concepts as supposed to description? One time one of my women went skiing with me and I said 'follow me, don't fall off'. She laughed and she came back up and she got hurt and she went on skiing. I started thinking she was supernormalizing. People supernormalize until they really get hurt, which is part of the theory. It imbues your way of thinking. She got to emergency. All these ideas of imbuement. Your paper imbued my way of thinking, too. It just becomes a part of the way you think. For example if my partner is a control freak and I know one of his patterns is to stay in control. Since I'm not a controlee it'll be total control or compartmentalized. So if I really don't want him involved in something, I really take away, I compartmentalize it and then he's left alone. But I think, in terms of compartmentalization...

Barney Glaser interrupts himself and adds, "interesting." Suddenly his telephone starts ringing and he gets up. In the phone: "Hi, I'm deep into an interview about something very interesting. Yeah. It's about me. Boring, do you say? I'll call you when we're done."

Glaser is more or less constantly engaged in, in his own terms, reversal humor, which often turns out to be a friendly form of self irony. "So you know, there's things." Glaser is now talking about people who are trying to delineate classic grounded theory.

Acceptance and demarcation

From the stories you're telling about other people right now, it appears that you fully accept them as persons although you might dislike their actions?
Well I don't know about acceptance. I think they are OK. But I only accept people as far as I am capable of. I mean, I'm only human, and I do have needs. Or I think I had them once. I accept people for what they are whether I need it or not. So there's always accepting them and dealing with them according to my own needs.

Your work to demarcate grounded theory in research indicates that it's hard to copyright a concept. What solutions do you see to that dilemma in the future?
I don't see there's a dilemma at all. Look, I've studied methodology. Methodology is something you want to tell someone how to do it. And the person you tell doesn't want to be told how to do it. They want to do their own. There's no ownership. There's no possession. I studied the sociology of science. There are just ideas. The fact that grounded theory jargon is being so used beyond the method is just a simple fact. It is based on how good

grounded theory is because it in itself is a grounded theory. So sure, I'll try and correct it and get people straight but it's dealing with the way the world it is. It'll be like dictatorship, it's awful and always going on, and genocide is awful, but it goes on. There's no stopping to it. Grounded theory is only for people who are very intelligent and can conceptualize. But people have to do research. So I think we are improving on capturing people.

You know, after a seminar people come up and say that they saw it and have to do it. They get grabbed and they want to do it their way. The imbuing answer is that there's nothing more boring than description. Ah. You can't generalize description anyway. Or you can, but that's for...does that answer your question? There's nothing else to do, just acknowledge the world the way it is. It's socially structured fiction that rules the spread of QDA. And I don't need this departmental based... I mean, classic grounded theory does rule. It has its own grab. That's why I want to start The Department of Advanced Grounded Theory Training in the Institute. That's a place to do it and there are a lot of people that would help each other. Did that answer your question? It's not going stop but it can be explained and people like you and the rest will understand its explanation. So is it unfortunate that grounded theory methodology is so grabby in its concepts? It's a consequence. It can't be stopped.

Have your investments in the remodeling wrestle been at a high personal cost, or has it felt good?

I don't know if those variables matter. It's just, for the people who follow me it makes it clear it is remodelling. Now that I've realized jargoning is not going to stop, I see that jargoning brings people in the classical grounded theory mode. When they start doing it they discover it goes nowhere. So it's just ideas, trying to keep it straight for those who want it straight. More and more people are using the jargon for qualitative work, but a lot more people are becoming classical grounded theorists. I think. So let's see what happens. I mean, I certainly have no problem getting a group together who want to learn to do it right.

Reviewing

You have always been critical towards the review system of academic publishers, though. But wouldn't a general lack of reviewing put the quality of an academic text at risk?

No absolutely not. No-no-no. Remember I took over my life at five. None ever said yes or no. I just did. How can someone tell you should do this or that? You have no idea of how people are going to see your book. You have to live with what you have written. So if some others could tell you beforehand, they don't let you live with what you've written and be totally responsible for it. So you want someone to give you an OK, a stroke? You need a stroke from them?

A review process might be viewed not as accidental assessment or control but as a way of contributing to the further development of ideas. By improving the ideas, the quality of academic work is usually enhanced, too.
If people have a good reviewer, sure. But that's different.

Why is that different?
When I wrote *Basics*, I showed it to people. Lots of people said 'Don't publish it, you'll get into trouble', 'you're wrong'. Then I knew I'd hit. Very few people can critique me. The only one I let is Judy, but she never changes anything. She helps if I'm tired. In other words, I'm worried about ideas, not English, and I'm very good at integrating ideas. See, I used to teach reading, and reading is like looking at a house: It's constructive. It's done by a human being. You just have to figure out the structure, the patterns, and you can read a book in a half hour. I think that's the answer to the review question. You built it. How do they know? It's almost like saying that *Discovery* was published because the word qualitative was in the title and it would sell.

Well, when talking about building houses you pointed out that your workers were so good that they even improved some doors that you had put in. Isn't it a fact that many times, different people can add and improve elements and parts of a house and they all contribute to making it better, and it's still your house?
Yeah, you can improve the doors. But I wouldn't mind. Mine worked. But I'm not sure it works that way in writing. It depends on how much they want to fuck with what you really want to put over your patterns, your construction patterns. You put the idea in the first part of the paragraph or last part of the paragraph. You put it in the introduction to the chapter or in sections. You know, there's all kinds of ways to construct. I look at all these ways and I figure it out. Like when I constructed *Jargonizing*, I did it in a new way because I didn't have time to do it the usual way. It was a write up, since there's nothing like writing in our world. To me I need data to write up. So no, I don't think reviewers are helpful. I think you can choose one who you know and will review a certain way. When I wrote *Doing Grounded Theory*, Phyllis Stern got furious at me and wanted it re-edited. She said I had so many English errors it's gonna kill it. So I had it re-edited and re-published, but I didn't let them change an idea. In another book I wrote, I gave it to an English editor and he wrecked it and I put it all back.

It's pretentious for two people to tell you this is good or this is bad. I learned it as a student when doing papers. Some guy gives you an A and some gives you a C. What right does he have? When I taught, I required my students to get it published in journals to get peer reviewed, because how would I know? You know how I got so many articles published immediately?

I studied the journal, figured out the format and wrote it that way. You need to fit into it.

Your emphasis on full research autonomy does not quite fit in with the goals of multidisciplinary, international research collaboration, which has become one of the main issues in academia.
That's interesting. My son's going through that in a field that is highly formed in terms of writing. I mean, could someone have come up to me a few years ago and say to me, 'Gosh, you should write a book about jargonizing!'. I didn't even know jargonizing existed by that time. Jargonizing has got more power than a 200-page book if anybody really understands it. You know, I know the secrets for densifying your writing. But most reviewers can't stand dense. But the students I've written for will understand.

So what's the secret?
You write up books. You don't write them and you sort all kinds of ideas. In that way you can put five ideas in a paragraph. As opposed to the chapter having only one idea and describing it for ten years.

At several seminars you have warned that when doing theoretical coding, one should only think of the social process and not so much of the psychological aspects?
Well lots of people have a psychological background and tend to talk about people's feelings. But we do social action. We study what is going on. All people go through basic social processes, they relate to people. It's not all a psychological system. It's a psychological-social system. I always said I used to give lectures on situational mental illness at the psychological center, mental illness that comes to people because of the situation.

Are you saying that grounded theorists should stick to the social process – since psychological processes are integrated in the social ones?
Abstractly, I guess, yes. Like fronting normality. One fronts normality to feel OK. It's done in terms of people. That's the only thing. Psychology puts all the pressure on people; it's all their fault. I saw people as cultural beings. It's not their fault, it's their roles, status passages, status sets.

After a short pause…

Helping people to get their self
You know what I'm thinking the core variable is? I know how wonderful it is to have one's self. I want to give people their sense of being themselves. You know, in academia, the first thing professors do is taking away their self by possessing them, by learning what they want them to do. GT is a wonderful way to help people getting themselves. It gives them creativity, their inde-

pendence, their autonomy, their contribution, their self satisfaction and their motivational joy. That's why in one sense reviewing, once again, is potentially taking somebody's self away, as if they know better. Now in a field where there are five people and everything has a model, the persons don't have themselves. It is all about how good the model is. And you'll find competiveness, jealousy, subversion. I think that's why everybody gets so excited at the seminars. You can talk, you can interrupt, you can do anything anytime. Interesting, huh? And that's my joy. I don't want your self, I want *you* to have it. In terms of science I always say autonomy, creativity and originality. But in terms of human beings, it's them, their product. It's not how well they have developed a preformed thought. On the other hand you can't do it to too many people because we need a bunch of people who just obey.

Glaser looks intensely at me.

I'm serious. I'm talking about society. So we can trust they won't be too original, they are going to do the same thing over and over again. People who do grounded theory is an elite group. We're allowing ourselves to be whoever we are. It's very important and it's my joy. I think that's why people love it; classic grounded theory gives them more of themselves. The less preconceived the more open. It's always asymptomatic. You might say, in my life I've been involved in many, many situations which are really difficult. But I was so much myself that it didn't bother me. I was so busy saying I'm me and that's the other person. I became objective. Yeah, it's not me. I was just always into my own thing. I never broke my pace, I just did what I thought was right.

The future of grounded theory

Many people, including yourself, have claimed that grounded theory was ahead of its time. In the globalized world, where things change very fast, how do you see the future of grounded theory?

From my own point of view, my own body, as opposed to the pass of time et cetera, I see it as just giving these seminars because I want the product. I want the transition from candidate to PhD, because that's where the social value changes. In terms of my own, you know, human existence and its limits, I just want to go on doing these seminars helping people getting PhDs. And it's working. As far as where grounded theory is going in the future, I guess the Institute is a structure that allows future development, like a department. But others have to do it. And others are very busy, too. So I guess the future is gross GT institute, but also it's informally seeping into more and more departments as a viable method for doing dissertations.

My seeing it as the future, as people get PhDs, it will continue to be informally absorbed and formally it can be more developed more on the institute. People have the time and the incarnation. And the other things are: By getting people PhDs they get their careers. Their social value gets up, and they may not get a chance to really do it again, but at least they've gotten their careers. They've had a chance to do it and then they can have a life, and they can pay insurances, mortgages, have kids. In that sense grounded theory is life producing by advancing careers. So that's the future of it. I think the informal seeping into the departments as we're getting more PhDs and maybe some more formal structuring up. I see that almost as epiphenomenon - all the things that's going on around something that's going on no matter. The actual job is producing dissertations and then publications. But even just dissertations. You know, I see it as a fact of life until it isn't a fact of life. If some people want it more structured, more universal, fine.

There will always be people who want to do that.
Yeah, because that's their format. And it's good. I guess. I don't want to engage although I support it. I think it's a good vehicle for it. I see the main feature as the informal absorption of classic grounded theory in so many areas as it captures people. And I'm also amazed on how departments legitimize grounded theory disseminations based on people coming to the seminar getting trained by somebody off campus. Interesting, huh?

It is soon noon. Several people are patiently waiting for Glaser's company, but he does not appear to be in any hurry. Even though he constantly speaks quite slowly and very explicitly, his voice is getting a peculiar strength and pace when I ask him:

What is your most important message to future GTers?
Conceptualize! Get people off, off the substantive to a concept.

Anything you would like to say that we haven't talked about?
No. You've covered the source. The true devotion to data exactly what it is came from psychoanalysis. No one has ever called me on it because most people don't go through psychoanalysis. But that's what we do. Get people to call exactly what it is. Honestly. And you know, lots of this is me putting a lot of different ideas into one package.

Part IV

Advancing Grounded Theory

16

GENERATING FORMAL THEORY

Barney G. Glaser
Grounded Theory Institute

We now turn to procedures for generating FGT (formal grounded theory) from a SGT (substantive grounded theory) core category.[1] They are the same as the procedures for generating SGT with one major change, that is theoretical sampling is different. The same procedures are constantly coding for categories and their properties, analyzing each day by constant conceptual comparisons and successive delimiting based on the core category as its general implications are pursued. No new core category is discovered.

The difference in theoretical sampling is multiple. In SGT one samples within a substantive chosen site or population. In doing FGT, one samples widely in other substantive sites and populations both within and outside the substantive area in order to make the theory more general, as one constantly compares adding new properties and categories to the core category being generalized. One can sample new data being collected from new sites, within existing studies and in documents, or do a whole new study in another area. One samples already collected date, existing research and theory in the literature and do so in all combinations. One can even write up abstractly a SGT. The ranging of theoretical sampling depends on availability of data, the literature and time and resources of the researcher. Theoretical sampling for FGT is "wherever."

The researcher must keep in mind the goal of FGT: that is a theory based on the general implications of the core category. It is not to find another, new core category in new data. The chosen core category is the controller for keeping the generating and theoretical sampling within some conceptual bounds.

The researcher must be skilled at conceptualizing by comparisons, otherwise it will not be worth the "travail." SGT generation is easier as it is so close to the data being explicitly collected to saturation. FGT loses this so immediate control and delimitation, as it is infinitely expandable. Time and resources and available comparisons are largely involved in deciding on unending completeness as opposed to the tidy theoretical completeness experienced with SGT.

---

[1] This chapter was previously published in Glaser, B. (2007). *Doing Formal Grounded Theory: A Proposal.* Mill Valley, CA: Sociology Press.

Existing Writings
Barney Glaser and Anselm Strauss have written several chapters on generating formal theory:

- DISCOVERY OF GT

Chapter IV, from Substantive to Formal Theory

- THEORITICAL SENSITIVITY:

Chapter 9, Generating Formal Theory

- STATUS PASSAGE: A FORMAL THEORY

Chapter 9, Generating Formal Theory

- MORE GT METHODOLOGY; A READER

Chapter 13, Generating Formal Theory
Chapter 17, Awareness Context and Grounded Formal Theory
Chapter 18, Discovering New Theory from Previous Theory
Chapter 19, Time Structural Process and Status Passage.]

Here I will add to these writings with more recent thought, precision and direction on generating a FGT of a core category.

General Properties of an FGT
Let's be clear. Doing FGT is not a test of the SGT, nor is it a correction of the SGT (see Buroway et al, Ethnography Unbound, Univ. of Calif Press, 1991, p. 10-12). Theoretical sampling for FGT is neither for verification nor for credibility. It is not for confirmation of accuracy or validity. It is simply an extension, by modification, of the core category theory as the constant conceptual comparative method is used in different substantive areas.
New categories were not missed in the SGT; they were just not relevant for that substantive study. New categories do not change meanings of the theory, they just extend and modify it and give a breeder generalization. (see Dey, op cit, p 62, who insists that meanings are changed.) As one colleague wrote me: "a theorist would have to determine whether one indicator was sufficient to warrant the modification or whether further sampling might be needed to ascertain the significance of the emergent indicator." (Judith Holton: email 10/15/05) This is one goal of FGT that is to extend meanings. The general meaning of how the core category continually resolves the main concern in the SGT is extended. For example, a FGT of cautionary control still has the resolutions of safety in mind.

The generation procedures for FGT can shift relevance and completeness in reference to the initial SFT as theoretical sampling swings wide in other areas. But bear in mind that the generating of FGT progressively modifies and extends to a more complete general application. The FGT becomes more dense thus slower to read, than the base SGT as one compares it to field and manuscript data and finds new indicators that vary the original categories and their properties. Saturation of new indicators delimits the boundaries of the emerging FGT.

As the generation of FGT continues, the progression increasingly develops broad generalizations that apply in many substantive areas, which generalizations become integrated more abstractly. As saturation occurs and contexts change, one can see the abstract application to many new areas. But the available data will delimit the increase to new boundaries and a level of completion and to saturation of concepts. Like SGT, a FGT is only as good as it can get within the given resources of the researcher.

When saturation occurs within given resources, it is then time to start sorting one's memos for writing up the theory. Keep in mind that the general implications of a core category can be apparently infinite. For example one can see credentializing, cultivating, discounting media awareness and worsening progressions virtually everywhere. The FGT researcher has to stay within the boundaries of available research data and research studies to sample. This is important even though completeness and saturation may seem provisional. The used data must be for or of research and the rigor of GT procedures adhered to. Once the research data is found the FGT can go to the limits of the data.

There is no open coding transcending to selective coding as in generating SGT, when generating FGT, since the core category was found and chosen already. Again, it is just using available data to extend its general implications. Thus doing an in quotes "formal theory" without the core category general implications of a SGT or straight from data, is clearly outside the purview of this book. It is a different entity. A FGT starts from a SGT's core variable as the automatic, stepping stone or springboard to the FGT. For example my theory on the comparative failure of scientists easily can become a formal theory on comparative failure. Starting from data only for a FGT would not have gotten to this core category. (See Examples of GT, chapter 22).

Sorting one's memo bank from FGT generations does not necessarily change the theoretical code used to model the SGT though it might (GT Perspective III). The theorist should be alert to the emergence of a new model or TC when sorting. But he should not change the original TC until virtually and strongly indicated by sorting. In Awareness (p282) we said: "The formulation of formal theory often requires guidance from explicit models more than does substantive theory, if only because the great level of abstraction of its concepts requires integration according to such models." The existing TC of the SGT should work well unless a new TC emerges during sorting. A new TC should not be imposed before emergence as an "ought" to work.

In addition remember that sorting for the writing-up is only a slice of the moment as FGT is so easily, continually modified. Especially so it is easily modified, rightly or wrongly, in the reader's head as he/she sees more general implications grounded in personal experience and interest. Sorting for an integrated set of conceptual hypotheses only solidifies the theory for the writing moment. And the writing moment typically only taps about a third of what was generated for reasons of space, high density, tight integration and the theorist's sanity. It is best to read FGT quick overall and then go back to a slower pace read or study on currently relevant parts to the reader. Density comes from fewer examples used in favor of getting down more FGT conceptual hypotheses and their generalizations. In the next chapter on uses of FGT, I tackle the question of why bother to do or read FGT.

The presentation of a FGT is best done in prose like a SGT. Formal does not mean writing a list of hypotheses or offering a list of theoretical propositions. The reader can pull out of written context salient hypotheses at will for his own needs. Nor does it mean using a complex diagram. It is a written understanding of how a core category applies extensively and inclusively across a range of substantive areas. It is not the intensive explanation of the details of action within a specific substantive area as SGT is. Yet when applied to another particular area the FGT can become intensive in understanding and explanation.

I have been asked by my students where to publish a FGT. Who publishes such work? Since publication is peer reviewed and so varied in style, writing a FGT for publication is journal driven. The theorist must analyze the "how to" integration of his FGT into the style and requirements of the specific journal of current choice. The theorist must accommodate to the journal peer review and existing readership's preexisting knowledge and requirements without losing a FGT grab and momentum (see Stake, p 96, Denzin and Lincoln: Strategies of Qualitative Research 1998) . Of course publication is the desired end result and necessary contribution.

Doing FGT is "research" age graded. Thus beginners have enough trouble doing SGT. It takes an experienced grounded theorist, thus someone with the experience grounded skill with GT to tempt doing a FGT. Doing FGT does not come first in skill development. Doing SGT does. It is best to do a few SGTs before tempting a FGT. Furthermore FGT is not for dissertations, since the candidate probably does not yet have the skill. Also dissertation committees typically may want more contained grounding data examples from a specific substantive area than FGT has space for. Though FGT is carefully grounded, it may appear less so using data from so many different substantive areas. It can appear airy or speculative. It can appear as being ideological or have strong allegiance to pre-existing "grand theory". The data-theory link is clearly rigorous GT procedures. It is not permeable as Van

Maanen proffers (1979), Denzin and Lincoln (1998). And it is especially so, that the link is not permeable by widely varying interpretations as Van Maanen says, "if the theorist uses the constant conceptual comparative process and not one incident impressionism" (See Doing GT) .

The researcher-theorist should detail where the data came from that he is comparing. The independence of original thought required in a dissertation may not clearly be related to a defined set of facts in a FGT though it is. The grounding argument may be a bit much for the PhD committee to agree to. In contrast, an experienced GT theorist in the middle of a career would be wise to show his/her advanced professional skill and knowledge by doing a FGT and avoiding the attribution to an over claim of existing speculative theory. Remember TCs emerge during sorting; they are not pre-framed in advance using speculative theory, however thoughtful it may be. (See GT Perspective III). Furthermore even if the FGT reads in an "as if" trusting way so could be seen as conjectured. IT IS NOT since it is generated by the rigor of GT procedures.

Directions for Theoretical Sampling

The question arises in deciding to do a FGT where do I go for data to compare? Two advanced GT theorists wrote me: "I am seriously thinking about writing a proposal to develop a formal theory of moral reckoning. Tell me what type of sample/sampling do you think would be effective for this project. Should I think about concentrating on two or three groups? For example would I look at moral reckoning in teachers, social workers, police (or any other type of in the middle group) etc. separately? Would it matter if participants lived in the same general area of the U.S. or should the sample represent a larger geographic area? Different countries?" (Alvita Nathaniel: email 4/12/05).

Antoinette McCallin emailed in May of 2005, "Formal theory development — I am interested in this although I wonder if I have enough experience as a GT researcher to develop just yet. Disadvantage — it sounds sophisticated and wonder if I have the GT skill just yet. Modify the theory and start collecting data for formal theory development to take place in a couple of years when I have had more practice as a GT Researcher."

Both these accomplished researchers have doubts about their ability because they need direction on where to start theoretically sampling. They need a method to start theoretically sampling and comparing. The general implications of their core variable can take them anywhere and everywhere it seems as they see it anywhere and everywhere. They are forestalled without a starting direction.

The method answer to their question is to decide on a path to comparative data and research findings which exists and then once chosen stick with it. This brings the "anywhere" problem under control. It provides the boundaries for pursuing the emerging core category FGT. It provides not only the boundaries but also the saturation end point within the data sources chosen. It takes the FGT that far, and that is all.

The path to choose is not a research question, nor is it pursuing hypotheses for confirmation or correcting. The path is core category driven. Data is sought and chosen for comparisons that enhance and extend the core category with new subcategories and their properties and the resulting, subsequent conceptual modifications. What to compare is guided by the core category and subcore categories of the theory being extended. The theorist will be surprised what can be compared. Apparent lack of comparability can easily become moot at the abstract, conceptual level. For example, a theory of 'expendable' populations exists at all levels of social life, ranging from nations to small local units of family teams, etc. Abstract ideation is the goal of FGT, not comparative description. The kinds of data selection for sampling do not have crucial implications for credibility, as in descriptive generalization. Data selection is crucial only for generating conceptual generalizations for the FGT. In sum, the path stops with saturation of the data source(s) and the resulting FGT extent of completeness.

There is no order to theoretical sampling except the order the theorist applies from a data source(s) to compare the core category to. "Sometimes it seems that a formal theory can "go" just about any way that an analyst desires. (Theoretical Sensitivity Chapter 8). The FGT is generated within the context(s) of the new data sources. That is it for the present effort. The theorist need not swing as wide as Alvita wonders. Triangulaton is irrelevant or moot. Multiple comparisons of data and research finds come from wherever, and usually a data cache. The rule is to follow the core category. The data should be within reach of the theorist's time and available costs and resources. For example using a reader or a journal with several articles on the same subject is efficient.

The theorist must be sure in choosing data sources that his choice is core category driven. Different cores will require devising different orders and paths to data sources. The theorist can use the general implication of the core category to direct him/her to data sources for theoretical sampling research data and other SGTs. The theorist can trust to the relevance of the core category in other substantive areas. For example, credentializing is universally relevant as is cautionary control. However the theorist must be humble if he has a core category that has general implications but no obvious data sources to turn to for comparison. The core area is not ready for a FGT. Thus, doing a FGT is then not ready and it is too hard to do the research in other nonexistent substantive areas as a start. Best to do a FGT when you can with some grounding ease from available data sources.

But it amazes me how many data sources there are just bursting for use in a FGT such as readers, journals, documents, researched newspaper articles or areas of much literature coverage with arrays of articles. For example, see my FGT on organizational careers in which I used 63 articles on careers, or look at our Status Passage formal theory. In short it is wise to choose a theoretical sampling path in an area just bursting with accessible articles. The idea is to do a FGT when the data sources are ready for it. For example there is a reader "Social Theories of Risk" (edited by Krimsky and Golding Praeger 1992) with 15 articles that are, in my view, just begging for comparison to extend the theory of cautionary control.

Another major source of data for generating FGT is secondary analysis of data collected elsewhere for other reasons. It is in the nature of QDA to collect large amounts of data by taping recording, which results in data overload and thus much data that never gets used or analyzed. These piles of unanalyzed data are everywhere on tapes or in type form. It is easy enough to screen them quickly for indicators that bear on a core category. See my two articles on secondary analysis in More Grounded Theory Methodology, Chapters l2a and 12b. See also Janet Heaton, Reworking Qualitative Data, (Sage 2004). She is very concerned in the use of secondary data whether or not the procedures for collecting the data are suitable to the secondary use of them. She is concerned about depth and coverage of the secondary date. Also she is concerned about using data that the secondary analyst has not "been there" to see and hear for themselves. To allay these concerns about quality and knowledge of data, she advises interviewing the primary investigator about the data collection, if he can be found and has the time, which is unlikely. She calls secondary analysis "makeshift" research.

Her concerns take many pages to detail, about 123. From the point of view of this book on FGT and for GT in general, her concerns have merit only for worrisome accuracy in qualitative research. They do not apply to GT or FGT as I detailed at length in GT Perspective. Conceptualization Contrasted with Description and in GT Perspective II: Descriptions Remodeling of GT. The credibility of GT and FGT is based on constant modification, fit, earned relevance and workability of groundedness and the probability in application of conceptual generalizations, as I have said many times.

People collect heaps of data thinking that is what research is, and then do not know what to do with it. They are often delighted that someone will or may do something with it. Large heaps of data are easy to skip and dip for theoretical sampling. One PhD candidate after getting her degree said she still had 27 interviews that have not been screened or typed, and that anyone who wants the data could use it. Caches of secondary analysis (interviews, speeches, collections of letters, journals) are not hard to find and can prove to be veritable gold mines of data for a FGT generation. Scanning them for relevant comparisons is the method of choice.

The theorist doing FGT should be careful when using secondary data of a conceptualizing nature that may not be relevant. Just by assuming it doesn't make data relevant or give it earned relevance. This means be wary of face sheet data or context data that lowers the level of abstraction and relevance of the FGT. Generalization relevance must be earned, not assumed. For example abusive power relationships do not just apply to men abusing women in business or couple/marital relationship. Women also abuse, or employee rights are not just union issues. Mexican women domestics also have a quiet non-combative movement to obtain health and pension benefits. Keep in mind that FGT is abstract of context when generating FGT, but can later be contextualized for application.

Doing FGT puts emphasis on continuous memo writing based on conceptual comparisons, as opposed to the data collection emphasis of doing SGT. The theorist writes memos for him/her self to externalize the emerging conceptual generalizations. Since the memos can be written any which way as beginning formulations, they should not be shown to anyone, except the most trusted colleagues or the collaborators on a team. Others may make the theorist anxious and feel critiqued prematurely before mature memos emerge. To be sure the pressure is and must be grounded. The pressure helps direct the theoretical sampling quest for more data on the possible modifications of the core. For example, the theory of moral reckoning in nursing easily leads to the quest for data, in all organizations where the situational bind of personal values, professional ethics and organizational constraints occurs in conflict and has to be continually resolved. The sampling can lead to modification of the solution stages of moral reckoning that make it more general.

Needless to say FGT requires some, if not a lot of illustrations to give applied imagery. Thus memos should always include some illustrations or notations where they can be found later. Of course the illustrations come from a wider range of substantive areas, and this indicates the generality of the theory and its general application.

As the ideas and concepts congeal and mature in memos, the theorist can see that he/she could never have thought of them before and that the chosen comparative data has been productive. He can see the benefit of choosing groups conceptually, not on the basis of differences and similarities. The apparently non-comparable groups on a descriptive level may be comparable on the conceptual level. Credentializing- for a driver's license can easily be compared to the credentializing of a surgeon. Training can vary from taking a few weeks to 10 years depending on subject and importance though they both have to do with warding off disastrous consequences by qualifying future behavior, so to speak.

As the analysis and theoretical sampling continue, the theorist will start to develop a general perspective on problems and dimensions of the core category that goes beyond his available data for comparisons. The general perspective itself can guide theoretical sampling. It also helps bring the memos to a degree of saturation on problem and dimension within the data set. At this point the theorist can start sorting his memos into an outline for the writing-up his theory into a book or article. The writing can begin with the general perspective, its efforts on people and social units, and then funnel down the writing to that portion of the general perspective on the core he has generated a theory of formal probability generalizations for and from whatever data he could find. Then with humility he mentions central and relevant problem areas that still remain unresearched, that he had no data resources for. Staying modest means staying grounded, that means not slipping into speculation, as one appeals for future research.

Remember the theorist cannot ever cover it all, as general implications spread with constant conceptualization. Some coverage is not left out; it is just in reach at that time. What is not covered is the work of a subsequent theorist. (See Strauss, MGTM, "Discovering new theory from Previous Theory", p. 369-371.) The generalizations of a FGT often bear on continuing processes and activities. Therefore in no sense does the completion of the FGT mean that the use of the SGT and subsequent data and SGTs used for the FGT are complete. They may continue with implications that could continue the FGT. The temporal nature of FGT is continuous.

Choosing Data

Several properties of data obtain when theoretically sampling. Anselm said "Filling in of what has been left out of the extant theory is a useful first step in extending its scope." (MGTM, Chapter 15, p. 371.) This is incorrect. One cannot know what is left out on an emergent basis. Apparent "left out" moves to increase coverage is a false track. One simply goes where a data cache exists. That is all the theorist can do, go wherever and anywhere. What data is available to expand, the general implications of a core is the initial direction to the theoretical sample.

Data volume or the more data the better is always favorable. But it is not necessary for generating FGT. What is necessary are the data, however small, that can be compared with concepts of the core by the interchangeability of indicators. The effort is to get to new conceptual comparisons by having new indicators in the data. A little data goes a long way in FGT generation. Sufficiency in volume will emerge as will the concepts. Thus the more data the better is at the limits of the extant data the theorist has found. Data choice is guided by the core, not by volume. And data choice must be separate from the original SGT, so the result FGT is not the SGT worked up even more. The theorist does not do both theories at once. That is, he does not generate a more complete SGT while generating a FGT, if the latter is his choice. The data must be outside the specific substantive area of the SGT in order to extend general implications. The goal is to generate a FGT that goes beyond the SGT.

The theorist must not be too judgmental about the data and substantive study he is comparing to. The comparative goal of FGT is concepts and hypotheses, not accurate findings. Even "poor" studies of stark description and method inadequacies can yield good data from interviews or observations for conceptual comparisons just as long as it is research data. The studies do not have to have any concepts, although they are too useful for comparison. Tiny topic research with lots of descriptive indicators can be rich for conceptual comparison. The reader should reread my chapter on "All is Data" (GT Perspective I, chapter 11) The main point was "all is data" means that "what is going on has to be figured out what it is or means for conceptualization, not for description."

"Data is always as good as far as it goes and there is always more to keep correcting it by conceptualizing properties of categories." (op cit, p.145.) Types of data are only relevant in the comparison. For example base line data can be compared to proper line data. Process data can be compared to cutting point data. More descriptively focus group data can be compared to private interviews. None is special for FGT, although a SGT is usually done on one type of data such as interviews or documents. (See Janice Morse, "Using GT in Nursing", p.8, for a more restrictive view of data.) Data doubts imbued with worrisome accuracy, as I have said many times, is a property of routine QDA research, not GT conceptualization. And it is even less so for EGT as generalization and extension of the core category increases. (See Kearny, Generating Formal Theory, op cit. p 255).

In addition extending the general implications of a core category emerging in a single case study is just as good as a SGT core from a series of cases within a substantive vicinity. (See Silverman, Selecting a Case, Chapter 8, op cit p.109) . It is the core that guides theoretical sampling, not a descriptive property of the case. For example the study of super normalizing emerged from a single case study of heart attack victims, but has far ranging general implications (see Charmaz,PhD dissertation. The same applies to "cutting back" after a heart attack. (Pat Mullen dissertation). We all cut back when necessary for all kinds of reasons. The benefit of FGT, as opposed to a single case SGT, is that it extends the core category ideas over many cases in different sites or contexts. The original single case can be easily forgotten. Thus the super normalizing of professional athletes is well known and the theory applies easily elsewhere, and heart attack victims are soon forgotten.

Literature Review

The rhetorical question I am often asked is what is the difference between a literature review and doing FGT. The straight forward answer is the standard literature review is to show how a new research fits and integrates into the body of knowledge to which it belongs. The effort is comparative description producing similarities and differences that add to the literature, fill in gaps in it, transcends it and fulfills disciplinary requirements of reference to "hallowed" literature. The literature review for FGT is conceptual comparisons of extant substantive data from wherever, and not necessarily though often useful, from the body of knowledge in the particular field. Theoretical sampling according to the core category will guide the literature review.

Vivian Martin has written in an email to the GT forum "There is a tendency for scholars in general to not boldly go where nobody in their discipline has gone before. Such disinclination is particularly pertinent to discussions about the use of GT to create formal theory. This disinclination for data collection and analysis across different areas of interest that would be needed to develop a formal theory to transcend the substantive area will likely keep such a FGT from development. Yet grounded theory holds out the possibility of helping researchers cross disciplinary walls." How true! The theorist should be bold and go where ever in the literature the core category guidance takes one. For example the core category of Trisha Fritz's dissertation on "untenable accountability" in education, can take one far a field with its general implications for organizational and personal accountability wherever. See Carolyn Weiner's monograph on accountability in hospital management ("The elusive Quest: Accountability in Hospitals," Aldine de Gruyter 2000).

Antoinette McCallin's article on "Grappling with the Literature in GT Study" (Contemporary Nurse 2003, 15.61) deals with the SGT research reading of the literature, on which there is much debate over my dictum do not read the literature before the research to avoid preconception forcing the research. This does not apply to doing FGT. The dictum is irrelevant for FGT. FGT starts with going to literature wherever for conceptual comparisons. FGT uses the literature from the start. And I have above suggested caches of data are to be found in readers and journals which are the 'literature', needless to say.

Similarly Creswell's chapter two on "Review of the literature" (Research Design in Qualitative, Quantitative and Mixed Methods, Sage 2003) is an exhaustive approach to before, during and after talking about varying types of descriptive research. But again this detailing of how to use the literature in descriptive research does not apply to GT conceptual comparisons. He totally misses the conceptual comparison as literature use. However, QDA researchers may find his chapter very helpful, see p. 33-35, particularly.

It is unlikely that the literature review will run across another formal theory on the same core category topic. One PhD candidate, Helen Scott (email Feb 2006) was spooked to find a study she thought was exactly like hers: a study of integrating flexible distance learning into a structured life. This is a flexible tyranny many face today. But upon examination her level of comparison with the extant work was descriptive not conceptual, especially on the core category. Thus the result was good substantive data to conceptually compare to. And another author may have a good conceptual idea buried in the description that helps the generation of the SGT and perhaps an FGT in the future.

Literature reviews and literature indexes in standard QDA articles and computer collated references on an area are also a good source of many articles to go for theoretical sampling. Tedious and overwhelming as they may be to read, the articles themselves can have very useful data for conceptual comparisons. Remember skip and dip for indicators.

These reviews and references can tap the burgeoning literature in a field just waiting, so to speak, for a conceptual formal theory to be generated from. They are a stepping stone for FGT. Thus the cumulative nature and/or synthesis of the research in the field by a conceptual FGT is exhibited.
This is a major use of FGT.

As I have said in Doing GT, where the relevant literature for doing a FGT appears is an open question until the SGT and its core category emerge. The literature that pre-framed a SGT study is easily not the best literature to relate the core to for a FGT. The best literature to sample is the literature to which the core relates emergent. Others (peers of importance) discover this as well as the theorist when seeing general implications elsewhere. Karen Locke in her GT in Management Research, discovered this (p. 124) when she said, "Similarly, I had framed the contribution of my study on comedy in medical settings in terms of the literature on organization emotions. The journal editor, however, suggested that the study also had something to say to the literature on service management and proposed that I write its contribution in terms of an intersection between organization emotion and service management." General implications break through with relevance and change the theoretical sampling of the literature, even when the other person is not trained in GT.

The only claim on theoretical sampling of the literature of an existing SGT is the core category to be extended. It is not problematic if and when the research questions emergent in the SGT are not further useful for theoretical sampling of the literature for a FGT. (See Locke, op cit, p. 21.) While the relevance of a literature can vary widely for a SGT, it can still vary even more widely for doing a FGT, since the boundaries or the original substantive area are breached. It would be apparently ideal if many SGTs existed to compare the core category to, but this is seldom the case as yet in the growth of GT. As Kearny (op cit, p. 235) suggested "If the material for a FGT is to be multiple substantive theories on a shared topic of concern the first steps to collect these substantive GTs". This is an appearance based on descriptive comparisons. If the several conceptual GTs do not exist, there could be several on a descriptive level, misnamed GT. And if they did exist, their core category would in all likelihood be different, so comparisons might default to the substantive data of the compared SGT. For example the two dissertations I supervised on heart attacks lead to diametrically opposed core categories: cutting and super normalizing. Remember GT is conceptual focusing on recurrent resolving of a mail concern, not full descriptive coverage. The descriptive studies are easier to compare, as the comparisons are most often based on substantive indicators, even though comparing concepts is useful also. Thus, for example, heart attack victims ordered to cut back, if the attack is not severe, will super normalize to prove that they are still ok.

Kearny realizes this when she says (op cit, p. 237), "In my formal theorizing experiences to date the GT literature on the phenomena of interest have consisted of only a handful of fully developed grounded theories, accompanied by eight of ten reports that are partial or incomplete but contain useful data or single concepts". In sum, descriptive studies provide much data for generating FGT. It is no loss that few GTs may exist on one area, and then the core would be different anyway. It is best to theoretically sample descriptive studies in many for indicators that bear on the core category.

See our discussion of "Awareness and the Study of Social Interaction" (Awareness of Dying, p. 276) for examples of finding awareness indicators "am diverse substantive groups which quickly lead to the development of general categories" on awareness context. See also Glaser "Organizational Career: a Formal Theory" (MGTM, Sociology Press chapter 14, p. 286). I say". I bring together many articles on careers-63 in total-that fit the category of organizational work careers. This act of itself will initiate much general understanding. I wish to start the generation of a form of grounded theory of organizational careers by initial comparative analysis of these articles." A treasure trove of many articles are surprisingly not hard to find, as popular areas get researched every which way, such as in nursing care or organizational careers.

The literature on doing EGT constantly reiterates our original dictum of looking at several substantive areas for comparisons. For example: "Glaser and Strauss noted the drawbacks of formulating FGT on the basis of data from only one rather than several substantive areas. In a book published three years later (1970) those authors presented a formal theory about status passages that was ...based on data amassed from a multitude of substantive areas" (See "Strategies of Qualitative Inquiry" op cit, p. 178). Descriptively different substantive areas can be conceptually similar and suitable for comparisons as core categories can be seen applying "everywhere". Thus "untenable accountability" inherent in the no child left behind program, can be also seen in armed forces accountability and hospital accountability to insurance companies. Credentializing goes on everywhere. The luxury of having lots of data to compare increases as the theorist goes more and more conceptual.

It is a procedural miss to think that one has to develop concepts from descriptive studies first and then compare the concepts to the core category. This is a false intervening step, GT theorists might not realize. One just compares the substantive data-indicators to the core category theory to expand conceptually its general implication. The comparing generates conceptualizations (concept), whether new categories or their properties. It is not necessary, even derailing, to generate mini GTs from descriptive studies before conceptual comparing. If a concept or two does exist in such studies fine, compare if useful. But keep in mind the fruitful comparing is bringing in more data to bear on the core category theory (Holton, email 10/05).

The Theorist's Resources

The theorist pursuing a FGT can do it at home and in the library with extant literature. Going into the field and doing field notes or tapes is very seldom necessary. The constant comparisons that start the generation of a FGT starts with conceptual memos, not field data collection. Since the data for the FGT is wherever, it is best to boundary the FGT theoretical sampling with a cache of data that readily exists, as I said above in a reader, journals, monographs or a strand of articles on an area. Keep in mind we are extending general implications of a core category by careful generation of additional concepts and their properties, so there is no worry about extensive vs. intensive research data or about internal vs. external validity. It will be what it is based on the research data used. The product will be conceptual generalization.

Unlike SGT data collection by field work interview and observation which can be time consuming and costly, especially when it is based on extended QDA research procedures. FGT research costs vary minimally and are virtually nil beyond the normal scholarly reading, study and note taking. Also there is no rush or time schedule, career demands or publishing deadlines as there is for doing a dissertation. Put simply, doing FGT is low cost and well suited to the more mature, advanced GT skills of theorists in their post PhD career. Besides being inexpensive, doing FGT is also comfortably solo, self-paced work.

Our new PhD Mark Rosenbaum wrote me: (email 2/20/06) "consider my theory on consuming commercial places as practical, gathering and home. Obviously, I could expand it to a formal theory regarding the consumption of all places. Furthermore the time required to read the literature across discipline and to clearly organize thoughts to create formal theory would be enormous." Not thinking descriptive comparison and time consuming SGT work. This book will guide him and others to the inexpensive approach to FGT, especially to doing the necessary literature review. Remember always more can be done, but a modest amount goes a long way in doing FGT, which will be delimited by data sources related to the core category.

However, if the "right" theorists are involved, a team approach of two or three people to doing FGT is possible. They must all be skilled with the core category and knowledgeable of the data bearing the core category and knowledgeable of the data bearing the core category as they talk together or otherwise work together by passing each other data grounded memos on the core. Routine seminars among colleagues with patterned analytic conceptual comparison techniques between them, will work and yield faster generated FGT. (See Doing GT, chapter 15, 1998).

In our FGT on Status Passage, for example, Anselm and I said "Because so much relevant data and theory was "in" us from our previous study and research, the principal mode used to generate theory was to talk out our comparisons in lengthy conversation, and either record the conversation or to take notes. We talked through virtually everything we could remember and studied the literature for more data and theory. These conversations went on almost five days a week for three months. Finally we gave up in exhaustion with the realization we could begin to write up our memos." (Status Passage, p. 192-193). The benefits of collaboration are several. A large amount of materials can be covered. The collaborators keep each other on a conceptual level, when it is easy to get bogged down in details. Also the generating proceeds more quickly as each collaborator conceptually takes off on the other's comments and they reciprocally stimulate each other. Others will interfere to maintain continuity of generation no matter where a colleague starts a session. Redundancy is minimized. Preconception and speculation is minimized as colleagues catch and prevent each other's natural bent to just theorize out of their heads, which is easy to do with the logical possibilities that emerge when doing FGT (See Doing GT, chapter 15).

Pitfalls

There are several pitfalls that a theorist can easily slip into in doing FGT that the reader-theorist should be wary of. A brief list I consider here is falling into description comparison, losing core category focus, rewriting a SGT, immodesty, not using research data sources or using particularistic data experience, speculation additions to rich theory veins, using computers, possessive alternatives deteriorating into the standard literature review, and so forth. Perhaps the reader can think of more. In the close of this chapter I consider these as briefly as possible to get the ideas out and across.

First there is a tendency at ties to drift out of conceptualization back into descriptive comparisons and start detailing similarities and differences and then give descriptive generalizations. The result is calling the product FGT and yet it is only a label. The jargon of GT and FGT runs far ahead of the knowledge of the procedures for doing F'GT, and thus is used for QDA generalizing descriptive research. Also dropping into description starts to bring in the problem of worrisome accuracy, which I have discussed elsewhere as having no place in GT. Developing descriptive "road maps" also has no place in GT (Kearny, op cit, p. 244).

For an example of dropping into description, Margaret Kearny (in her chapter "New Directions in Grounded Formal Theory", op cit, p. 238-244), uses the FGT jargon well, but immediately this chapter slips in a synthesis of many studies to achieve a broad explanatory model of comparative description of similarities and differences of women's adjustment to illness, trauma, addition recovery and experience in violent relationships. This is all clearly broadening descriptive data, not conceptualizing. She does not extend the power of any core category. She sticks clearly to a symbolic interaction view and uses constructionism with her general description. She is concerned about the validity (worrisome accuracy) in using some of the studies. Her goal is to synthesize findings into "a common entity" or a "unified entity" pro and conning each study. The general implications by conceptualizations relating to a core category yielded in a FGT are competing out view in her work. What is similar and what is different across all her chosen studies is the goal "to arrive at a synthesized comprehensive description of the phenomenon under the study". So be it, her work should rather be titled formal descriptive, synthesized elaborations and summary to use her words, not titled FGT.

Another example of slipping into description is the article by Janice Morse et al, on a "Comparative Analysis of Conceptualization and Theories of Caring" (Qualitative Health Research, Chapter 7, Sage 1992). They do start off with a core category which is "caring". They say "caring is even described as the essence of nursing and the central, dominant and unifying feature of nursing." Their "purpose of this article is to review critically the nursing literature pertaining to caring and to explore the implication of various conceptualization of caring have for the discipline of nursing." Their comparative analysis yielded five conceptualizations of caring. This sounds like formal theory, but they slip into descriptions of five conceptualizations, which themselves are not really conceptualization, but routine description and then they compare these five conceptualizations to indicate differences and similarities. They say "To some extent, the complexity of the concept of caring is captured by the diversity of these conceptualizations. Comparisons across the five conceptualizations indicate differences related to the focus, purpose and variability of each thesis". Thus as a specification of a concept their article is excellent, but it is not a formal theory of caring, as lengthy description over. Formal theory was not their intent, yet it was so close to going in that direction. I trust this book will help theorists see this distinction and start tempting to do FGT.

Descriptive formal comparisons leading to lengthy minimal conceptualizations non-theoretically integrated is a genre that is well used in the literature. Another excellent example of descriptive theory is by Mark Granovetter in "Getting a Job" (Univ of Chicago Press 1995, p. 139-177). He compares an extensive literature on informal networks, which bear on his renowned theory of get a job through weak ties. It is worth reading but it is not FGT.

Closely related to slipping back to description from conceptualization is starting to engage in the standard literature review, citing support and non-support for findings. Instead of conceptual comparing, the theorist slowly becomes taken with literature coverage descriptions relating to the core and drops conceptual comparisons. Judith Holton expressed this clearly in an email-memo (10/15/05). Instead of doing FGT "the scholar is comparing the studies for 'findings' that support or challenge his/her 'findings'. The results are more likely to be a long descriptive overview of the argument and misalignments between the 'literature' and the new study's findings, rather than to be integrated explanation of a process concept." Alvesson and Skoldberg in Reflexive Methodology, New Vistas for Qualitative Research also touch on this drift into description when they say: ""Searching for data caches (eg collections of letters, interview, speeches, article series, journals), which can prove to be veritable gold mines, the only risk is that the researcher may become too engrossed in these to the point of possessiveness". (op cit p. 21).

The grab of the literature can overcome the focus on conceptual comparisons. The drift into dropping conceptual comparisons in favor of a literature review is closely associated with the loss of focus on the core category. A literature review will certainly do it, but it can also occur on its own. As the loss of focus occurs, many other possible concepts, not related or relevant to the FGT task at hand, can become exciting and claim attention over the core category. Focus is also lost, as stated above, by reverting to generating more complete SGT on the core category, instead of a FGT. Our focus here is on generating FGT, not on revisiting the SFT to make it more comprehensive within the original substantive area with more data. By the same token, as I have said above, one cannot start with generating a FGT without a SGT yielding a core category to start with. In short one cannot start a formal theory from just data.

Another simple pitfall given the "autonomous at all costs scholar" is dropping the notion and name of FGT in favor of a possession alternative originated by the author. They rename the FGT, whether they do it fully or not, something else like general theory or extended theory to make it their solely own contribution. For example, Diane Vaughn (1992) a thoughtful theorist and excellent researcher has written about an alternative but related approach to producing general theory. She advocates "theory elaboration which consists of taking off from extant theories and developing them further in conjunction with qualitative case analysis." By elaboration she means "the process of refining the theory, model or concepts in order to specify more carefully the circumstances in which it does or does not offer potential for explanation". (See "Strategies of Qualitative Analysis", op cit, p. 175).

Another excellent example of a possession alternative method is called "Cross Case Analysis", which borders on the generation of FGT, as it dances between comparing for differences and similarities and actual generation of "new categories and concepts which the investigators did not anticipate," is Kathleen M Eisenhartdt's article "Building Theories from Case Study Research." (Miles and Huberman, op cit, Sage 2002, Chapter 1, p. 5-36) The heavy focus on both descriptions and "theory with a close fit with the data" and new concepts give her a foot in both camps of comparative focus. She uses the constant comparative method for both goals of description and conceptualization. She emphasizes theoretical sampling, theoretical saturation, selecting lots of cases outside the area of one case and coding, which is a start for FGT, but they are diluted by her requirement to "list the similarities and differences between cases" (p. 16). So near and yet so far from just doing FGT.

All she has to do is leave out comparative descriptions and build on her abilities of comparative conceptualization to just do FGT. But her descriptive training and her possession alternative derail this. So be it, her work is excellently conceived and received. She is right on when she says "Thus (with her method) a strong theory building study presents new, perhaps frame breaking insights.") (p. 32). Closely related to this, combined or not generating of FGT, is the tendency to lack modesty in the FGT presentation to the scholarly public. One can easily with pride of accomplishment start acting like a "grand or great man/woman, leader of the field and purveyor of the preempting last word. This is overblowing of the FGT. It is just a simple offering with NEVER full coverage. Modesty of contribution allows others to cut their competitive, threatened nature and appraise the genuine contribution for what the FGT is — just a contribution, not a takeover.

As the joy and excitement grows over doing a FGT on a core, another simple pitfall is to leave available data for the natural tendency to engage on speculative, groundless conceptualization. That is, to develop the FGT more so with conjecture and speculation. Conceptual comparisons occur in the mind of the theorist, so this natural inclination is always there. The procedure is to stick with the data rigorously. And as an afterthought, the "stick to" requirement includes that FGT cannot be done on a computer, which stores data. It is the creative conceptual comparisons that put the FGT together, not stored retrieval. Strauss's emphasis on personal experience and an analytic lifestyle is fine, and makes the theorist theoretically sensitive, but it is not research data to be used in FGT! It makes for good aside conversations and come theoretical sampling.

The drift into logic-deductive speculation typically has a dignified, respected tone to it, since most general theorists who have achieved some notoriety, have generated theory this way. Fine, this is the way they achieve awe and fame, but it is not grounded formal theory. It is just "super think" divorced from reality, the correction of which spawned the start of GT in 1967. (See Discovery of GT). Also beware of using an abstract model to guide and generate a formal theory. It is an old style for speculative theory building and thoroughly neglects the core category approach and the emergent theoretical codes yielded in sorting. Thus Dey (op cit, p. 227) does not contribute to FGT by saying "Analytic generalizations can be made through comparisons with a theoretical model of a typical case and through the reformulation of received ideas." Fine for other theorists, but this is not FGT generation.

Rewriting substantive theory up a notch can sound like formal theory and gives formal theory implications but it is not FGT. At best it is a FGT waiting to happen by comparisons with new data and simply rides on the general implications of the core category. It can be an adequate step toward generating a formal theory, but it is not a formal theory itself. The abstraction just sounds like a FGT. For example a theory on becoming a nurse can be rewritten as is as a theory of becoming a professional by leaving out substantive words, or even becoming in general, an aspect of socializing. Or a theory of cautionary control among dentists can be rewritten, leaving out references to dentists as four general types of cautionary control. Or a theory of cultivating housewives for milk delivery accounts can be rewritten leaving out substantive reference to milkmen, as a theory of cultivating clients for profit or recreation. In short, by rewriting leaving out the substantive attributions the researcher has raised the conceptual level of his work mechanically. He has not done the research to broaden the scope of his theory to the formal by conceptual comparative analysis of different substantive areas. It cannot fit or work or be relevant very well with many of the conditions, contingencies and contexts in diverse substantive areas to which it could be applied. This last section is but a sample of the myriad of research styles that may or may not pretense to FGT as I have stated in the last chapter. They are still in full play, and I have only added to the styles, not subverted or canceled them. I only offer classical FGT as an option. And it works, with fit, relevance and enduring grab.

In closing I trust this chapter will get the F'GT theorist going and going on the right track to generating FGT. To be sure GT is an experiential methodology as well as a set of rigorous procedures. Thus the reader will probably have to experience the experience of doing FGT to begin to fully understand how it is generated and develop the skill to do it. To try generating FGT the theorist must be able to transcend the data with at least some skill in conceptualization.

17

Reflections on Generating a Formal Grounded Theory

Tom Andrews
School of Nursing and Midwifery
University College Cork, Ireland

Generating a formal grounded theory means going outside of the substantive area to look for new conceptual comparisons (Glaser 1967). It is the conceptual extension of the core category of a substantive grounded theory using the generating procedures of grounded theory (Glaser, 2007) and is the logical next step. There are very few formal grounded theories because they are so difficult to generate or the substantive theory is not sufficiently well developed to warrant one (Glaser, 2007). In this chapter I will discuss my efforts to create a formal theory of worsening progressions. Also I will outline some of the pitfalls and how to avoid them.

General Implications: The Starting Point

Originally I had interviewed nurses and physicians working in general medical and surgical wards where I discovered that nurses rely on soft signs in detecting a worsening progression while physicians rely on hard signs. Nurses' main concern in visualizing worsening progressions is to ensure a successful referral to physicians with the aim of convincing those with the power of intervention to do so. Trusting relations based on mutual trust and respect is pivotal to the whole process. To extend the theory and begin developing a formal grounded theory I decided to interview nurses working elsewhere such as Intensive Care, the Operating Room, the Emergency Department and Neonatal Intensive Care. Despite this, no new categories or properties emerged which confirmed that the original categories were saturated. I then decided to go outside of the substantive area of health and interview university counselors. Once these were complete I theoretically sampled and analyzed the literature. Before starting out on formal theory development it is important that that the general implications of the substantive theory are fully understood.

When my theory was originated, I did not realize its general implications. I simply did not understand what this meant. Like most students learning GT, I was a minus mentee and when requested to write a section discussing this aspect of my theory, I did so in a way that remained context and unit-bound; I discussed deterioration within the context of health rather than deterioration in general. It was only later that the realization grew. My understanding of what is meant by the general implication of a generated theory only developed later with continued reading of GT and comments from people following presentation at conferences. At these events several people came up to me and recounted picking up on subtle indicators of a worsening progression in family members, friends and work colleagues. One person told me that while working as a librarian she had noticed changes in a colleague such as not communicating as much, and picked up that something was wrong. Despite appealing to her immediate supervisor to intervene she could not convince her of the need because it was based on soft signs. Having realized the general implications I began considering generating a formal theory. I was further encouraged to do this by the publication of the latest grounded theory book dealing with formal theory (Glaser 2007).

Glaser and Holton (2007), realizing the general implications of my substantive grounded theory, wrote that we are all in situations where we visualize some form of deterioration based on long-term intuition. It is a combination of experience and knowledge, and then we need to figure out what to do about it if anything. This includes how to manage the social structures involved as well as convincing significant others that there is a worsening progression, as well as how to ensure cooperation of others based on soft signs. Their feedback stimulated me more to think of more general implications and convinced me that a formal Grounded Theory could be generated. However, before contemplating such an endeavor, it is essential to think carefully and work out the general implications of the substantive theory. I suggest that this is the starting point and that writing good memos is central to this.

Gradual extension

In generating a formal theory it is seldom necessary to go into the field to collect data, as is the case for generating a substantive theory since theoretical sampling can be done with a cache of data, such as from journal articles (Glaser, 2007). With this in mind I started off by conducting a comprehensive literature search using such search terms as: parental involvement; deteriorating relationships; relationship communication. I began by extending the general implications of the core category by careful generation of additional concepts and their properties. Initially this was within the field of health but from the perspective of relatives rather than health care professions.

The general implications of a core category have the potential to lead anywhere so it may be tempting to broaden the initial literature search. For example, I broadened the search to include literature not only on worsening personal relations, but also workforce friendships and marital relations. I also included literature on nurse-patient interactions and professional support. A cache of literature helps enormously and is often sufficient, since a little literature goes a long way (Glaser, 2007). Within the field I was studying, no such cache was evident. However the effect of doing this resulted in comparing literature in areas that were too diverse for careful theory development. The alternative is to choose a path to comparative data as a means of containing the diversity, ensuring that the search is not broadened too quickly. One such mechanism is to consider carefully the literature in the same substantive area such as health care in general, and to theoretically sample, using that literature for constant comparison instead of comparing literature in diverse fields such as health care, personal relationships, friendship, and marital relations. In generating a formal theory, it is necessary to sample within a substantive area and gradually broaden out into other substantive areas through constant comparison and memoing. If this is done prematurely, it may lead to inadequate saturation of concepts and sampling literature that is too broad in scope, which has the potential to lead to unbounded reading. Sampling within a substantive area before branching out, provides boundaries and brings order to theoretical sampling. To avoid this, theoretically sample until saturation and then go to another likely source of data to further the general implications of the core category, thus gradually building a formal theory. Also, sampling in a number of different substantive areas simultaneously leads to conceptual description rather than conceptualization.

Another important point to remember is that it is tempting to skip coding, since the literature is rich in concepts. It is tempting to use ready-made concepts rather than engage in the tedium of coding. This may have two consequences: concepts that are not saturated or one indicator concepts; the latter results in too many concepts. Coding for generating a FGT is selective since the core category has already been discovered (Glaser, 2007). The advice is clear; saturate one substantive area before moving on to another one.

Using literature as data

The experience of reading literature outside of the substantive area was a very interesting one for me. I was surprised that the literature about relationships was dominated by research using quantitative methodologies. For example, one study investigated how couples communicate using a communication pattern questionnaire and established that males and females communicate differently (Vogel 1999). The naming of this difference had already been established so this study merely confirmed differences. This raises the challenge of using quantitatively generated data in a GT study where the data is purely of a statistical nature with few concepts and no behavioral indicators. Since these indicators form the basis of generating concepts in GT, experience of reading these types of studies led me to conclude that without any behavioral indicators, it is in fact very difficult to use them since purpose of the literature in this context is to generate new indicators for conceptual comparison (Glaser 2007). Concepts may well come from such studies but it is more doubtful if their indicators can be made explicit. However quantitative studies that test hypotheses have greater potential to generate indicators. For example, a study investigating friends at work generated such hypothesis as: “a higher perception of friendship opportunity with the immediate superior leads to employees’ positive work attitudes” (Song and Olshfski 2008, p154). Interestingly this paper discusses results in terms of "multiple indicators for all latent constructs" (p155). What makes it different from GT is that its focus is on establishing statistical relationships and concept definition, whereas GT's focus is on concept specification (Glaser 1978). Depending on the nature of the quantitative paper, in my experience quantitative data can be used to specify the concept. In other words, it can provide theoretical codes that specify when and under what circumstances the emergent hypotheses hold true.

I compared the literature on communication in situations of marital conflict as well as health care literature. This enabled me to extend the properties of categories already generated from the SGT. For example, I discovered the importance of picking up on communication as an indication of a worsening progression and a new category of communicative distancing, with properties of defensive reactions and communicative withdrawal emerged from the marital conflict literature (Roberts 2000). Purposeful seeking increases the likelihood of soft signs being picked up and contextualized and was discovered in the health care literature (Wuest 2000). The original concept I had was "baselining" but the new wording captured much more effectively how normal progression is established. Its purpose is to establish what is normal for a person so that any deviation from that baseline can be visualized. In contrast, my experience is that reading quantitative research data is not as rich in concepts and is therefore limiting. Reading quantitative research data may not be as effective at generating concepts and their properties because it is mainly concerned with generating facts. Although fundamental to GT, using the literature as data is a skill that needs to develop, since it is usually outside the skill development of a PhD student.

The importance of sampling strategy and memos

I did not differentiate between worsening progressions as a general process and as a professional one, which caused problems later because it guides the conceptual elaboration of the core category. It is important to differentiate between the core category as a general process with general implications and the core as a professional one. For example, the original substantive theory was a professional process, concerned with how health care professions pick up and report a worsening progression. It is important to think the issue through by memoing rather than going ahead and theoretically sampling the literature in other substantive areas. I might have more carefully sampled the literature dealing with how other professionals such as social workers picked up on worsening progressions, although I understood this implicitly since I had originally decided to interview social workers working in a number of different areas such as families, the prison service and mental health. It is imperative that a decision is made at the outset whether to focus on developing the substantive theory as a professional or a more general process. In the absence of a cache of literature it is essential to develop a sampling strategy rather than just sampling the literature in general. It could start with sampling in a similar substantive area to the one that the theory was generated in and then carefully selective sampling for the core category in other areas. This will ensure that the literature search remains focused.

As Glaser always cautions, no part of the Grounded Theory process can be omitted. There might be a tendency, as there was for me, not to memo as frequently as during the development of a substantive theory, but this is a mistake. If memoing is not done as it is supposed to be, it makes it more likely that the theory will lack integration and will be written descriptively rather than conceptually. Memos continue to be central to the development of formal theory just as they are in substantive theory generation. In fact there is increased emphasis on memo writing (Glaser 2007).

Writing conceptually is more challenging in FGT generation when trying to use the literature as data. Research methodology courses and academic training in particular teach students how to use the literature to support points being made in a discussion. To my knowledge, no methodology book deals with using literature as data within qualitative research instead focusing on synthesizing extant findings in a meta-synthesis. However because it is based on the interpretative tradition it tends to rely on descriptive rather than conceptual comparisons with the aim of developing a full understanding of the phenomenon or generating theory at a descriptive level (Zimmer 2006). Therefore looking to the literature for guidance on how to write conceptually is not an option. Qualitative research values "thick description" to describe the research setting, observed interactions and context, making it a very rich source for identifying behavioral indicators, concepts and theoretical codes. In writing up the FGT it is essential to differentiate between using the literature in the traditional sense of supporting what is being written and using it in the GT sense. The literature is used as data not to support the points being made. If used in the latter sense, then writing is likely to be descriptive rather than conceptual. The literature should be used for conceptual and not descriptive comparison.

Conclusion

Generating a FGT is a challenging task, but this should not deter researchers from generating one. Careful generation is ensured by gradually extending the general implications of the substantive theory once the general implications of the core category have been identified. This can be done by having a research strategy, particularly in relation to the literature to be theoretically sampled, which ensures against an unfocused literature search by providing boundaries. Qualitative studies are likely to be more useful for generating concepts and their indicators compared to quantitative ones, although the latter may be useful for hypotheses. No step of the GT process can be skipped and there is increased emphasis on memoing.

References

Anderson, W., Arnold, R., & Angus, D. (2009). Passive decision-making preference is associated with anxiety and depression in relatives of patients in the intensive care unit. *Journal of Critical Care. 24*, 249-254.

Glaser, B. & Strauss, A. (1967). *The Discovery of Grounded Theory.* New York: Aldine De Gruyter.

Glaser, B. (1978). *Theoretical Sensitivity.* Mill Valley: Sociology Press.

Glaser, B. (2007). *Doing Formal Grounded Theory: A Proposal.* Mill Valley, CA: Sociology Press.

Kearney, M. (2001). New directions in Grounded Formal Theory. In R. Schreiber, *Using Grounded Theory in Nursing* (pp. 227-246). New York: Springer Publishing Company.

Roberts, L (2000) Fire and ice in marital communication: hostile and distancing behaviors as predictors of marital distress. *Journal of Marriage and Family.* Vol. 62 No.3 693-707

Sias, P. R. (2004). Narratives of workplace friendship deterioration. *Journal of Social and Personal Relationships.* Vol. 21 (3) 321-340.

Song, S. and Olshfski, D (2008) Friends at work: a comparative study of work attitudes in Seol city government and New Jersey state government. *Administration and Society.* Vol. 40, No.2, 147-169

Verhaeghe, S., Defloor, T., Van Zuuren, F., & Duijnstee, M. (2005). The needs and experiences of family members of adult patients in an intensive care unit: a review of the literature. *Journal of Clinical Nursing* (14), 501-509.

Vogel, D, Webster, S and Heesacker, M (1999) Dating relationships and the demand/withdraw pattern of communication. *Sex Roles.* Vol. 41 No.314, 297-306

Wuest, J (2000) Negotiating with helping systems: an example of Grounded Theory evolving through emergent fit. *Qualitative Health Research.* Vol.10 No.1, 51-70

Zimmer, L. (2006). Qualitative meta-synthesis: A question of dialoguing with texts. *Journal of Advanced Nursing, 53*(3), 311–318.

18

From Theoretical Generation to Verification Using Structural Equation Modeling

Mark S. Rosenbaum
Northern Illinois University

Grounded theorists, especially those pursuing advanced doctoral degrees, often wonder whether they selected the methodology because they favor qualitative data or methods over their quantitative counterparts. Unfortunately, many empiricists fail to realize that grounded theory (GT) can be generated from any type of data (Glaser, 2008). In addition, the qualitative versus quantitative question is a simplified guise for a more substantive question—namely, whether the researcher is engaging in theoretical generation or theoretical verification. Grounded theory is an inductive methodology that results in theory generation, including propositions or hypotheses that can be empirically verified in a subsequent study (Glaser & Strauss, 1967). Thus, grounded theorists typically confront the question whether they are ending their study at a theoretical generation stage, which typifies most GT studies, or at a theoretical verification stage, which requires empirical verification of the original GT theory in a future empirical study.

Although GT ends successfully at a theoretical generation stage, some researchers may opt to employ theoretical triangulation by demonstrating the verifiability of their proposed GT in a subsequent study. Thus, GT researchers might present a separate theoretical generation study and follow-up verification study in different publication outlets or present a combined (triangulated) theoretical generation study and follow-up verification study in the same publication outlet (for a triangulation example, see Rosenbaum, 2006). Indeed, theoretical triangulation, or the simultaneous theoretical generation and verification in a single study, represents the epitome of research prowess and produces a study that pundits should find difficult to challenge using the futile qualitative/generation versus quantitative/verification debate.

The goal of this chapter is to show how researchers can employ GT and structural equation modeling (SEM; for extensive review, see Lei & Wu, 2007), using the IBM SPSS Amos software (Arbuckle, 2007; www.amosdevelopment.com), in a theoretical generation and subsequent verification manner. Many SEM software programs exist, including Mplus, LISREL, EQS, and SAS (Proc Calis). However, the Amos graphical interface simultaneously obtains regression coefficients between hypothesized relationships; thus, the program is a natural choice for SEM researchers engaging in theoretical creation and verification because it mirrors the illustration that some grounded theorists may provide at the end of their research project to visually depict a proposed theory (for illustrative examples, see Glaser, 1978, chapter 4; Glaser, 2005, chapter 2). Glaser and Strauss (1967, p. 32) recommend that GT researchers present their substantive or formal theories to readers by employing a "discussional form," as this type of theoretical presentation suggests that theories are "ever-developing." However, they also state that GT researchers may prefer to present their GT theories as pictorials in publications, with each pictorial representing a "momentary (theoretical) product" (p. 32). In other words, a graphical pictorial suggests that a proposed GT theory is a perfected final product; hence, GT researchers, who provide readers with theoretical illustrations, should alert readers that *theory as process* is ever-evolving and in a state of continued development. Perhaps, one may conclude that via SEM, GT researchers may 'momentarily verify' a 'momentary theoretical product.'

This chapter reveals how extant coding families that have emerged from prior GT studies or from sociological conceptualizations (e.g., Glaser 1978, 2005) can be straightforwardly verified in an additional SEM study, using Amos. To complement Glaser's style of providing illustrative pictorials of proposed coding families, this chapter shows how many of these pictorials can be easily employed using the Amos graphical interface or commands. Although LISREL (www.ssicentral.com), along with other key SEM programs (i.e., Mplus, SAS Proc Calis, and EQS), also offers a graphical interface option, researchers must use keystroke data entry that specifies the hypothesized relationships between and among variables contained in an illustrative representation of a proposed theory. Support of Amos for theoretical verification in GT rests solely on the relative ease with which illustrative coding families can be replicated for theoretical verification at a future time. However, researchers can also obtain the same regression coefficients by employing SEM programs that involve keystroke data entry. Although grounded theorists must develop SEM skills and learn how to use Amos, theoretical triangulation offers both the thrill associated with a conceptual "grab" and the knowledge that theoretical creation can be just as powerful as theoretical verification. Finally, GT researchers, especially those residing in America, will confront dissertation committees, journal editors, and grant committees that favor theoretical verification over theoretical creation. Rather than perceive empirical inclination as a career limitation, this chapter suggests that GT researchers should pursue theoretical creation, and then satisfy pundits with theoretical verification; albeit, momentary at best.

Introduction

Both GT and SEM share their origins in the sociology discipline. GT represents an inductive methodology that was created to generate original theories, including concepts (or constructs) and proposed relationship between or among these concepts; SEM is a deductive methodology that is used to test hypothesized relationships between concepts. Note that in deductive research, concepts are conceptualized as variables. The essential difference between these methods is that GT studies tend to yield concepts and hypothesized relationships in a paragraph (discussional) form, and SEM verifies these concepts and relationships in either tabular or graphical forms; thus, SEM is a natural by-product of a GT study. Perhaps the major obstacle to using SEM is that GT researchers must engage in additional theoretical sampling procedures to verify proposed theories using SEM, learn essential SEM skills, and learn how to use one of the aforementioned SEM programs.

SEM is a general term that describes a large number of statistical models used to evaluate the validity of a proposed model—most notably, the model's concepts and hypothesized relationships—with empirical, quantitative data collected through survey or experimental methodology (Byrne, 2010; Lei & Wu, 2007). In addition, the data can be either primary or secondary. For example, researchers may generate a quantitative GT from an extant data set (Glaser, 2008), while retaining a portion of the same data set for theoretical verification with SEM. However, note that SEM adheres to Glaser's dictum that "all is data," and therefore SEM is essentially a means to simultaneously obtain regression coefficients that are representative in the many coding families that Glaser (1978, 2005) puts forth.

More important, SEM is a prime means for assessing latent variables, which are usually indicated by a confluence of observed, multiple measures. For example, a GT researcher may first propose a conceptual category called "Z"—which is based on its related properties (A, B, and C) that all emerged from qualitative interview data—then collect additional data, and then verify the proposed theory using SEM techniques and Amos software.

Employing Amos for SEM is advantageous because it uses a graphical interface that is nearly the same final offering as that of many GT studies that put forth graphical illustrations. In other words, GT theorists who propose graphical theoretical models can easily apply the same design to the Amos graphical interface to evaluate the "fit" of their proposed models based on the data results. For example, in an Amos graphical interface, researchers could test the Z theory using the diagram shown in Figure 1.

The movement from theoretical generation to verification may not be easy, however, because the SEM verification stage requires additional data, typically a minimum sample size of 150–200 respondents for survey data or an experiment, to produce meaningful results. Therefore, GT researchers who opt to engage in theoretical verification using SEM may confront time and resource constraints because the empirical portion of the study requires generating a GT and then engaging in additional sampling for verification.

Furthermore, unlike the theoretical sampling stage of a GT, which encourages GT researchers to strategically obtain sample variance to broaden the conceptual categories that constitute a proposed theory, sampling for verification requires GT researchers to minimize potential sample variance.

In other words, GT researchers engaging in SEM are evaluating the "fit" of their proposed model with a new data set, and any new unexplained variance in a data set, which may be indicative of new conceptual categories, will lower the "fit" of the proposed GT, possibly to the point at which the proposed GT model is deemed invalid. Despite these limitations, empiricists would find it difficult to refute the theoretical substantiveness of a model that emerges from grounded data and is empirically verified in a follow-up study.

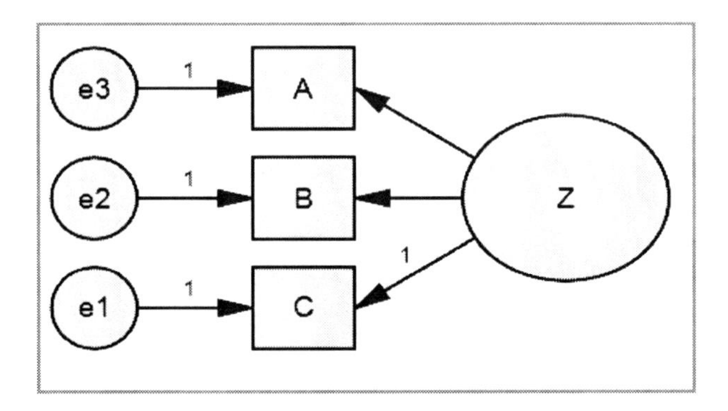

Figure 1: A core category with three indicators

In Figure 1, Z represents a concept, or perhaps a core category (e.g., social support), and A (companionship), B (emotional support), and C (household/chore assistance) are three of the concept's observed related properties, which are supported by qualitative data. After proposing the theoretical model in a GT study, the researcher might use A, B, and C as three questionnaire items, perhaps measured on a 5-point, Likert-scale question (e.g., "strongly agree" to "strongly disagree"). In Amos, latent variables are represented by ellipses, observed variables by rectangular boxes, and hypothesized direct effects by one-sided arrows. The terms e1, e2, and e3 represent error terms or residuals indicating unexplained variance in the dependent variable, and the "1" values are scaling requirements for Amos. After connecting the graphical model to an actual data set, the researcher 'runs' the proposed model, essentially testing the hypothesized fit between the proposed model and the actual data. For example, in Figure 1, item C is hypothesized to be an indicator of Z; however, if the empirical data reveals that this relationship is not a reliable indicator of Z, the model is deemed to have a 'poor fit.'

Because the empirical results could indicate that the proposed hypothetical relationship is not statistically significant, removal of this path, through the Amos modification indexes option, is possible to obtain a better fit with the sample data. Thus, GT researchers may confront a conundrum—that is, whether they should remove a statistically insignificant path, which was hypothesized as substantive according to its origins in inductively collected data. However, the removal of an insignificant path, or paths, likely does not nullify the proposed theoretical model, but rather suggests that the path is not as theoretically substantive or generalizable to a new sample as significant paths. If the empirical results indicate that the proposed theoretical model does not fit the new sample data, GT researchers should consider the empirical study part of a theoretical creation study. That is, if "all is data," poor model fit suggests that the original GT theory requires additional modification as researchers discover other contexts in which hypothesized relationships may differ, thus expanding on the theoretical grab of the original theory. This leads the author to conclude that 'having data' never expunges theory but rather elucidates where changes are warranted.

In the following sections, attention turns to exploring how classic and original GT theoretical codes can be graphically tested in an Amos interface. GT researchers are encouraged to employ both theoretical codes and triangulation in their studies. GT researchers must remain open to the nonforced, nonpreconceived discovery of emergent theoretical codes. Although several classic theoretical codes that have emerged from actual studies or the extant literature are discussed in this chapter, these codes do not represent predisposed packages of coding families that researchers must select. However, GT researchers should be attuned to some of the coding families to help understand emerging patterns in their own data. Although it might be axiomatic that theoretical codes conceptualize how a GT's substantive codes relate to each other, not every GT study needs to have them (Glaser, 2005); however, the conceptual grab of the theory is somewhat attenuated without them.

Causal Model

The causal model prevails in many GT studies. For example, Figure 2 illustrates how the core category, zeta, which may, for example, represent satisfaction with a child's school, is influenced by three conceptual categories—alpha (perceived teacher quality), beta (perceived principal quality), and gamma (perceived parent–teacher interaction). The version on the right-hand side of Figure 2 suggests that none of the categories (alpha, beta, or gamma) is hypothesized to influence each other and that each affects zeta independently. However, in this school example, this is probably false.

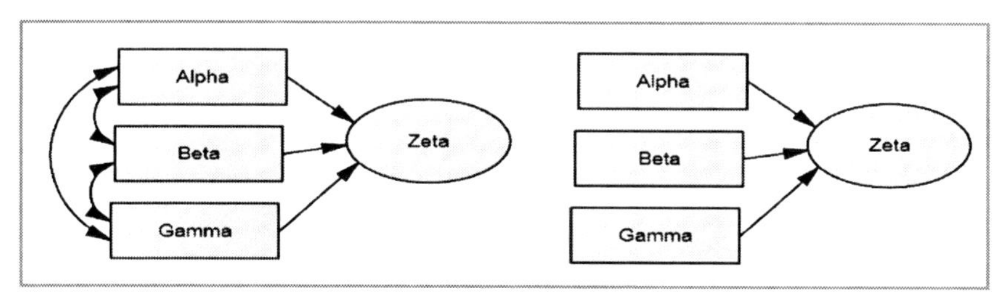

Figure 2: Illustration of how observed variables may correlate or not correlate with each other

The left-hand-side diagram in Figure 2 illustrates a situation in which the three hypothesized conceptual categories are permitted to correlate or, in factor analysis terms, the exogenous, observed variables are considered oblique. In contrast, the proposed right-hand-side diagram is testing a model in which the exogenous variables are considered mutually exclusive or, in factor analysis terms, the observed variables are considered orthogonal. Thus, the simplest GT theory, the offering of a core category with its related conceptual properties, offers GT researchers inherent theoretical codes, which they can verify in SEM.

The Six Cs

Glaser's (1978) "bread-and-butter" theoretical codes—causes, contexts, contingencies, consequences, covariances, and conditions—are often summarized in a causal–consequential model. Figure 3 represents the Amos interface of three proposed causal–consequential models. The first model shows that observed variable A has a hypothesized cause and consequence, as well as residual errors. These errors represent the unexplained variance in the model; more specifically, part of A is caused by an unknown variable and its consequence is a result of more than A itself. The second model represents the addition of an observed condition to explain latent variable A, and the third model represents a covariance between the hypothesized cause (e.g., lack of sleep) and the condition (e.g., reading class level).

GT researchers who want to explore whether the model fits in various contexts should employ the group comparison function in Amos. For example, a context may be three classrooms or three data collection locations, such as the United States, Canada, and Mexico. Using the Amos group comparison function, the researcher could assess model fit in each context and empirically

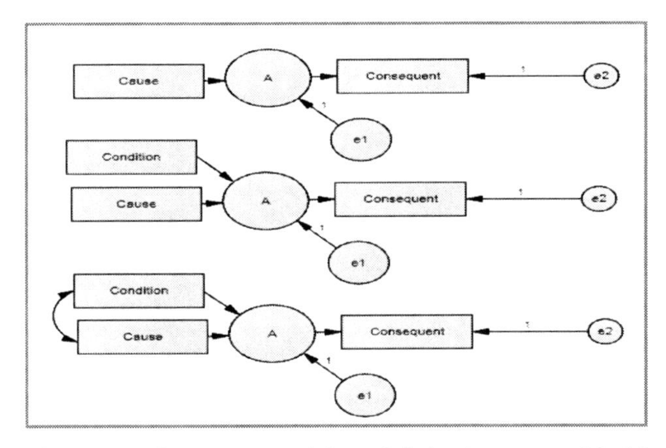

Figure 3: Three causal–consequential models in Amos graphical interface

demonstrate whether the hypothesized relationships, including direct effects and correlations, differ or remain the same. Likewise, the researcher could explore whether the model's residuals differ by context, thus illustrating whether the model is more or less reliable depending on a particular context.

Process Models

Process models represent stages, progressions, and transitions and, at the minimum, must have at least two distinct stages. Figure 4 represents two such depictions. In the top model, a latent variable that emerges in Time 1, as indicated by a, b, and c, later emerges as a latent variable in Time 2, based on the emergence of d, e, and f. The second model depicts a hypothesized relationship between two latent variables and their observed variables.

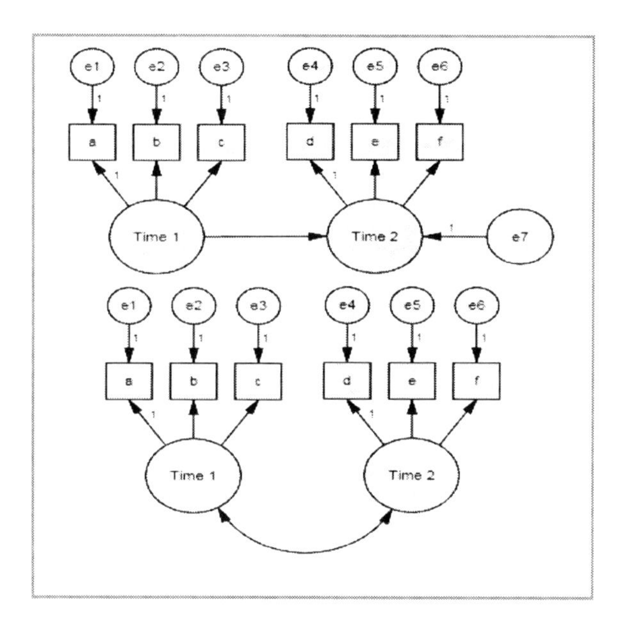

Figure 4: An Amos graphic interface depicting two time stages

Degree Models

Unlike the more simplified causal–consequential models, a degree family model proposes that a latent variable is caused by several antecedents and that its emergence leads to several consequences. For example, Figure 5 illustrates that three mutually exclusive causes result in latent variable A, and its emergence yields three consequences.

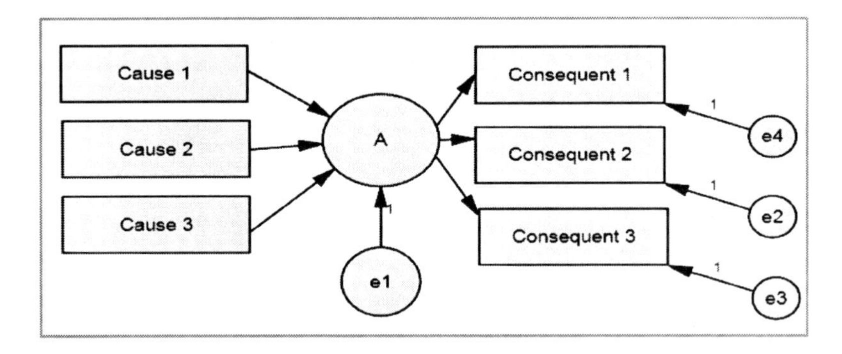

Figure 5. An Amos graphical interface depicting a complex causal–consequential model

Dimension Models

Dimension models depict latent variables that comprise dimensions, elements, pieces, sectors, and so forth. Figure 6 typifies an example of a dimension model, which purports that concept Z can be explained by the confluence of three variables, A, B, and C. Note that concept Z includes three pieces, for example, sectors, elements, and so forth.

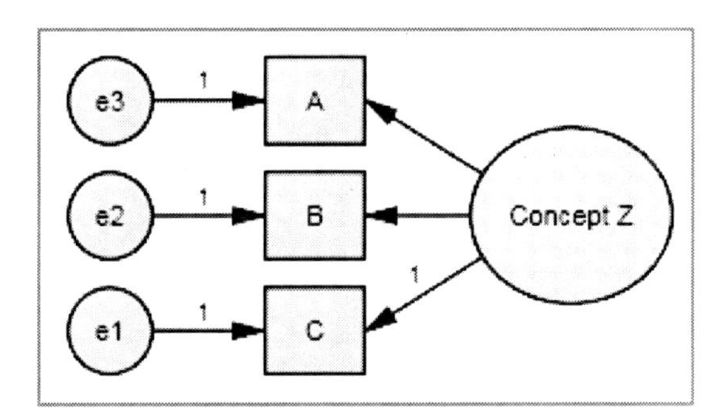

Figure 6: An Amos graphical interface depicting a dimension family

Interactive Models

Interactive models depict mutual effects, reciprocity, mutual trajectories, and so forth, which are easily modeled in Amos. Until this point, all the illustrated direct effects have been one-dimensional; in SEM terms, these unidirectional effects denote recursive models. However, hypothesized relationships may be bidimensional, or nonrecursive. Figure 7 illustrates a model in which variable A causes variable B, which in turn has a direct effect on A; however, both variables A and B have a direct effect on variable C. In this model, variables A and B are interactive, and their interplay has a resultant effect on the emergence of variable C.

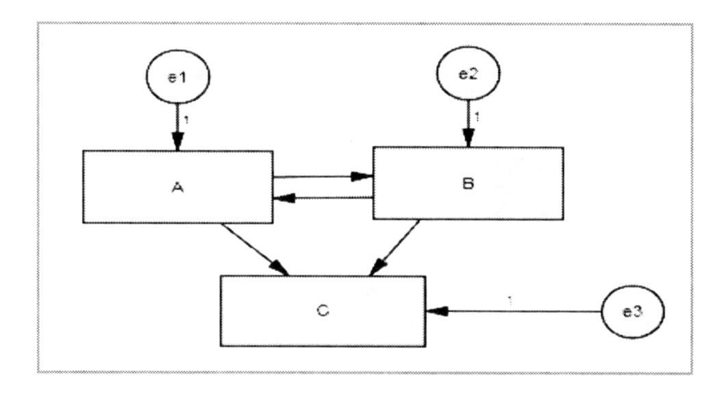

Figure 7: An Amos graphical interface of an interactive model

Other Widely Used Theoretical Codes in Extant GT Studies:

- Strategy models. Strategy models, which illustrate strategies, tactics, mechanisms, and so forth, are really complex structural equation models that depict the hypothesized relationship between two or more conceptual categories. That is, they are simply combinations of the previously mentioned coding families.
- Identity-self models. Any model that reflects a concept of self-image, self-concepts, self-worth, self-realization, and so forth, is depicting latent variables that emerge because of observed indicators. Thus, SEM is ideal for verification of these models.
- Cutting-point models. Cutting-point models, which illustrate boundaries, cutting points, and benchmarks, suggest the use of managed models in Amos. GT researchers can show how hypothesized paths alter between two situations, including any two significant junctures.
- Means–goal models. Means–goal models depict a means to reach an end, and in Amos, this is a simple path diagram between two observed variables. Figure 8 depicts a model is which a person's perceived value in a task is a means to reach a performance goal:

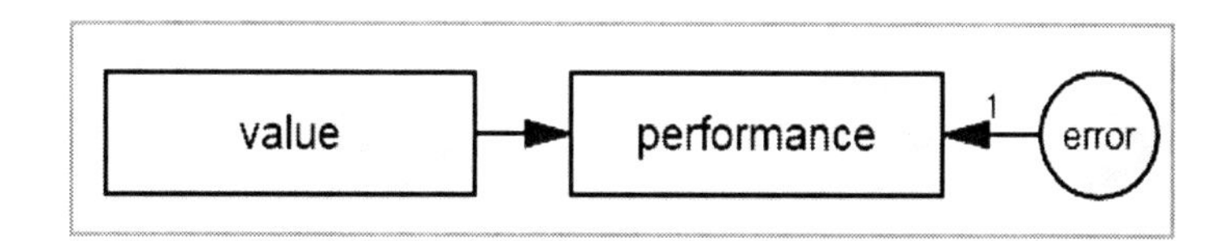

Figure 8: Amos graphical interface of a means–end model

- Consensus models. Consensus models, which depict clusters, uniformities, and conformity and nonconformity between or among two or more groups, can be illustrated in Amos using the "specification search" option. This option permits researchers to analyze relationships that are either the same or different between or among two or more groups.
- Mainline models. Mainline models depict social control (e.g., keeping people in line), social recruitment (e.g., bringing people into the group), social training, social stratification, and social passage. If these latent variables are represented by at least three observed variables, they can be depicted in Amos confirmatory factor analyses models.
- Ordering/elaboration models. Ordering/elaboration models depict a structural order of influence. In Amos, researchers can test both unidirectional hypothesized relationships (i.e., recursive) and bidirectional relationships (i.e., nonrecursive).
- Temporal ordering model. Temporal ordering models illustrate the progression of one thing to another. Figure 9 illustrates this type of model:

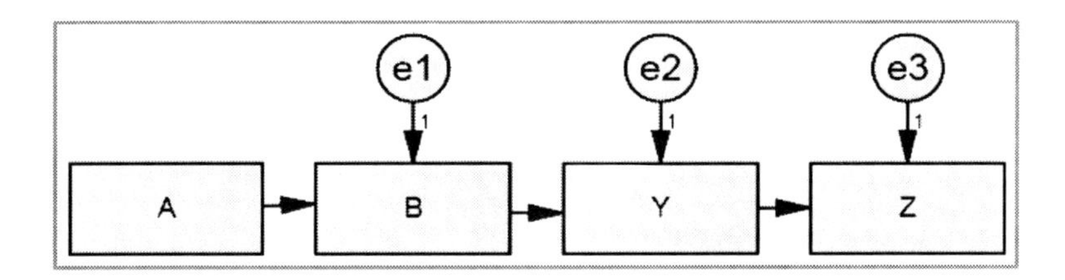

Figure 9: Amos graphical interface of a temporal ordering model

- Conceptual ordering models. Conceptual ordering models reveal how concepts, as opposed to observed variables, relate to each other in a progression. Although Glaser (1978) may consider these models complex, again the Amos graphical interface reveals the simplicity of conceptual progressions. Figure 10 illustrates an example of hypothesized ordered relationships in which family life influences a person's success in marriage and at work, which in turn influences financial success, which in turn helps lead to a comfortable marriage.

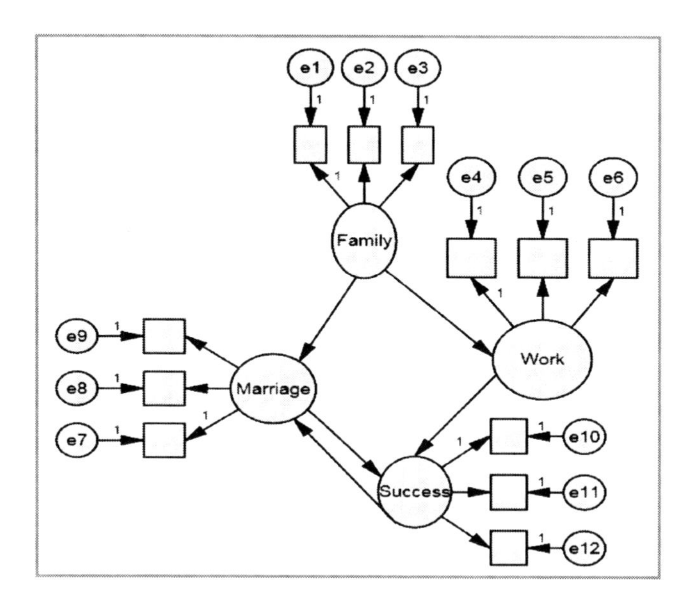

Figure 10: Amos graphical interface of a conceptual ordering

Some Final Thoughts

At this point, the GT researcher should be convinced that theory generation and verification can effortlessly, albeit with SEM education and practice, be aligned in two consecutive studies. Indeed, Glaser and Strauss (1967, p. 18) noted that "in many instances, both forms of data (generation and verification) are needed as "mutual verification." This chapter advances this perspective by offering readers some initial thoughts about following theoretical creation with theoretical verification using Amos SEM software. However, GT researchers who opt to follow a GT study with a second verification study may use other SEM software; including popular software packages such as Lisrel, EQS, SAS, and Mplus.

Although Glaser and Strauss (1967) propose mutual verification, many GT researchers may question the need to complete an additional empirical study merely to verify relationships that emerged from the data. This point remains valid. GT by its very nature is solid because it emerges from extant data. Theoretical generation, per se, does not require simultaneous verification, and doing so represents a timely and costly endeavor. A GT study may appropriately conclude with the offering of hypothetical propositions or a theoretical discussion, which serve as the basis for a future empirical study.

Yet mutual verification remains a bona fide reality and is often a necessity for GT researchers pursuing grants and academic tenure. Mutual verification is a consequence of theoretical creation, not a replacement for creation. Furthermore, rather than leave theoretical verification to future researchers, GT researchers who possess the ability, time, and necessary resources are encouraged to engage in theoretical verification of their own GT studies using SEM. The author's personal stance is that verification buttresses the validity of a proposed GT theoretical discussion by showing that proposed propositions or hypotheses hold true in a new sample. Indeed, GT researchers may ponder whether theoretical verification is another type of theoretical sampling method, albeit, employing a sampling method that chooses to limit variance.

Many GT researchers may find that their careers are influenced by others who possess a positivist mind-set. Indeed, mind-sets that emphasize observable facts and exclude humanistic speculation about conceptual origins or ultimate causes tend to dominate academia. Several dissertation attempts and grant proposals have failed simply because of empiricists, who often dominate academic and research committees. If GT researchers propose a triangulation study, from GT creation through verification, the positivists may question the need for theoretical creative endeavors; yet they would most likely welcome opportunities for researchers to verify a new model that extends and organizes a disparate set of extant articles, model, frameworks, and theories. Perhaps, only the GT researcher will realize that he or she is in reality momentarily verifying a momentary theoretical product.

Another major hurdle to overcome is how to become adept at both GT, when many GT researchers are minus-mentored, and SEM, including Amos software. Although learning SEM skills requires time and effort, many GT models are actually straightforward graphical models that are easy to understand in Amos. Most universities now offer beginning SEM courses, and some courses are now available online. In addition, many universities, including the University of Michigan, University of Connecticut, University of Kansas, and Texas A&M University, offer intensive summer courses on the topic. Finally, websites such as www.amosdevelopment.com, www.statmodel.com, and www.structuralequations.com offer a plethora of training videos, examples, and literature, and books such as Structural Equation Modeling with Amos (Byrne, 2010) and Latent Variable Models (Loehlin, 2004) simplify the SEM learning process for GT researchers who lack statistical training to learn with applications.

GT is an inductive methodology that uses qualitative and/or quantitative data to generate theories that are meaningful, have conceptual grab, explain social phenomena, and extend knowledge regarding substantive areas. Deductive methodologies offer the discipline further verification of a theory or, quite often, the verification of extant theories by offering new combinations of known theories. In essence, deductive methodologies offer researchers assurance that their understanding of extant knowledge remains steadfast, albeit with small incremental changes permitted over time. By moving from theoretical creation to verification in consecutive studies, GT researchers will generate theories that are substantive, important, and interesting, while assuring pundits that these new ideas can be empirically observed. Thus, the futile qualitative–quantitative debate will cease when a GT researcher employs triangulation in a GT study.

References

Arbuckle, J. L. (2007). *Amos 16.0 user's guide.* Chicago: SPSS.

Byrne, B. M. (2010). *Structural equation modeling with AMOS.* New York: Routledge.

Glaser, B. G. (1978). *Theoretical sensitivity.* Mill Valley, CA: Sociology Press.

Glaser, B. G. (2005). *The grounded theory perspective III: Theoretical coding.* Mill Valley, CA: Sociology Press.

Glaser, B. G. (2008). *Doing quantitative grounded theory.* Mill Valley, CA: Sociology Press.

Glaser, B. G., & Strauss, A. L. (1967). *The discovery of grounded theory: Strategies for Qualitative Research.* Mill Valley, CA: Sociology Press.

Lei, P. W., & Wu, Q. (2007). Introduction to structural equation modeling: Issues and practical considerations. *Instructional Topics in Education Measurement*, 26(3), 33-43.

Loehlin, J. C. (2004). *Latent variable models: An introduction to factor, path, and structural equation analysis.* Mahwah, NJ: Lawrence Erlbaum.

Rosenbaum, M. S. (2006). Exploring the social supportive role of third place. *Journal of Service Research*, 9(1), 59-72.

19

THE POWER OF AN ENDURING CONCEPT

Vivian B. Martin
Central Connecticut State University

The *awareness context* is grounded theory's past and future. To some, it might seem that the concept, which formed the core of Glaser and Strauss's 1965 work, *Awareness of Dying*, and is probably grounded theory's most referenced and utilized theory, has been pushed to its limit. But I will argue that the awareness context is also classic grounded theory's future because the work, alongside some early papers Glaser authored independently, contains suggestions that have yet to be mined in ways that would better utilize grounded theory. From the outline for a theory of awareness to an essay calling for greater attention to secondary analysis of data collected for other purposes, Glaser's work from the 1960s continues to ring with ideas that are still vital and instructive for grounded theory and the social sciences. To illustrate the type of work still ahead for grounded theorists I draw on my own theorizing of awareness processes, in particular ongoing work on a formal theory of discounting awareness, to demonstrate both the enduring power of a concept and the possibilities for theory and knowledge-building when researchers follow these concepts as they emerge across disciplines. I advance the earlier-expressed view (Martin, 2006) that, when utilized to full potential, grounded theory continues to be a subversive methodology due to its power to cut through often underdeveloped but extensive extant literature in a field and bring more cohesion to the understanding of substantive issues within disciplines, and also transcend disciplinary boundaries. The latter is especially important in these times in which problems are ever more complex and need the attention of people working across disciplines. I discuss the classic method as providing some tools for cutting through the large amounts of data that need to be connected to other data through conceptualization. As was the case when Glaser and Strauss were first publishing the seminal works in grounded theory in the 1960s, structures within academia today continue to keep analysts cordoned off in disciplinary silos. Classic grounded theorists, who are often ensconced in practical professions interested in a limited range of theory, have not extricated themselves to follow the intellectual paths needed for the grounded theory method to meet its potential. I close the chapter with a short discussion of the need to claim the distinctive features of grounded theory, particularly the techniques that would allow for more cumulative theorizing across disciplines and societal problems.

Awareness

The awareness context emerged from long-term study of processes surrounding dying in hospitals. Conducted at a time when physicians were less transparent about terminal diagnoses, Glaser and Strauss's *Awareness of Dying* provided a framework for understanding how an awareness context shapes the interactions between the medical professionals and patients. Four awareness contexts were discovered: closed, suspicions, mutual pretense, and open. Each type of awareness defines the pattern of interactions toward which the professionals and patients orient their behavior. For example, in contrast to a closed awareness context, where the diagnosis is not shared and medical professionals never discuss the inevitability with the patient, open awareness contexts allow treatment to proceed with an acknowledgment of impending death, which is assumed and referenced during interactions. A patient in suspicion awareness attempts to confirm suspicions that he or she is dying, while patients and medical staff in a mutual pretence context collude to cloak impending death in mutual pretense awareness. This dance of sorts (the authors called it a ritual drama) continues until one side decides to no longer take part in the charade. Importantly, the awareness state is not fixed; as part of their explication of the theory the authors discuss how types of awareness can be foreground and then recede as others emerge.

The theory was a major contribution to medical sociology and nursing research, in particular. Awareness contexts are a staple of health studies, and nursing research in particular, applied to areas from training to the role of awareness in dementia, to name just a few. It is important to note that the concept has not been received uncritically. The work of some researchers has been to address perceived shortcomings and attempt to extend the theory (for ex. Morrissey, 1997, tries to bring in more of what he calls the complexity of what people know and communicate in terminal care), but the concept has proved to have both longevity and continued relevance in the nursing literature (Nathaniel & Andrews, 2010).

The embrace of the awareness context in health research aside, Glaser and Strauss did not intend it as the last word on awareness. In the final chapter of *Awareness of Dying*, the authors laid out what could be viewed as a research program for a broader theory of awareness for anyone looking to take up the challenge. To get there, researchers needed to analyze data, much of it probably already collected for other purposes, and study the concept across disciplines. The work has not been built up this way, however. The awareness context, though a mainstay in the academic literature on dying and almost canonical in some branches of health research such as nursing, has not traveled much outside of medical and health-related studies. Contemporary terms such as "context aware" in artificial intelligence or "situation awareness," which is used in the military, do not originate with the awareness context.

Sprinkled through the literature are efforts to expand the reach of the concept. Ekins(1997) applied the awareness context to "male femaling," an examination of cross-dressing and sex changes among males. The work dovetails with some of the discussion of stigmatized and masked identities in *Awareness of Dying*. The authors had used examples of blacks who pass as white, as well as the heavily publicized work of a white journalist who passed as black, to illustrate dimensions of the awareness context and its role in everyday interpersonal interaction.

Strauss (1994), in a presentation on revisiting the awareness context to build a formal theory, indicated that in addition to data from the initial studies he had been sampling literature on gay lifestyles, con men, and others who would appear to operate in an awareness context where open, closed, suspicious, and mutual pretense would have significant implications. Although Strauss's expressed interest would also seem to have drawn on some of the suggestions from the original work and also be in keeping with his and other symbolic interactionists' ongoing use of awareness contexts in identity research, he asked for people to share data from multiple areas of life. Strauss, who died in 1996, did not complete a formal theory of awareness contexts.

Awareness remains a fundamental social process that people take for granted both in everyday interaction and throughout the social sciences. As a society, we speak about and plan public awareness campaigns to stop drug abuse, domestic violence, drunk driving and, more recently, driving while talking or texting on the cell phone, just to name a few of the behaviors for which people are called to be more aware. Culturally, awareness is treated as both a starting point (people need awareness to act) and a residual phenomenon (limited awareness is assumed to be at work in everything from the latest gadget in need of marketing to childrearing practices). Yet despite its ubiquity as an assumed fact of life, awareness is a highly undertheorized concept. While theories in fields such as cognitive psychology and communications address aspects of awareness and its kin, attention, there is not a fully integrated theory of awareness and the processes that bring it into being or disrupt it. A commitment to the distinctive features of classic grounded theory could assist in this unfinished work.

Discounting Awareness

Discounting awareness is a pattern I discovered as part of a broader theory of news-attending (2004, 2007, 2008). In the theory of purposive attending, there was a loop consisting of awareness, which increased relevance, which in turn increased attending, sometimes recalibrating awareness in such a way that these so-called awareness spirals created higher levels of relevance and attending. But those were the more optimal situations. Awareness is not a force that moves forward unabated. Often I found awareness was blocked because news and information had to travel across social networks and other structures that impeded awareness. If one's social network has little knowledge or interest in an issue, one's awareness may also remain slight; yet some of the lack of awareness was willful. People avoided serious or upsetting news; they had strategies for avoiding it when it came on the television—a few years after the event, a New York woman was still leaving the room to avoid news reports about the 9/11 attack—or they simply took note but did not engage with the information in front of them. Through interviews and observations, I also came to see that people deliberately avoided having conversations about controversial items in the news or political topics sure to cause tension if discussed among friends with dissimilar views.

Normative disattending, a socially learned practice, was evident across the data and the literature, where scholars probed phenomenon such as political disagreement in everyday political conversation or the unacknowledged "elephant in the room" and "learned blindness" Zerubavel (2006) explored. Although my initial theory of purposive attending gave some attention to disattending and what I came to see as discounting awareness, it was clear that awareness structures deserved deeper examination on their own terms. While there is much to be written of the emergence of awareness and how it is related to higher levels of knowledge and consciousness, the area I chose to develop as a theory across a number of fields has been discounting awareness. It is a phenomenon I have since come to see as an important dimension of everyday communication, from the most innocuous decision-making, such as how much credence one should give a weather forecast of rain, to behaviors that marginalize others and poison public discourse.

In *Awareness of Dying*, Glaser and Strauss devoted a chapter to what they describe as discounting awareness, a process where medical professionals assume a lack of awareness on the part of patients. In the most obvious cases, the staff discounted the awareness of the "hopelessly comatose" patient, premature babies, the senile, and certain dying patients for whom staff accord little attention or respect, often due to status markers (p. 108). When awareness was discounted, staff talked in front of patients assuming they could not hear or make out the implications of what was being said. Their routines and interactions around the patient did not attempt to hide impending death as was the case in other closed awareness situations. Discounting awareness in these settings is interactional, as people in face-to-face contact enact behaviors exhibiting the negation of awareness. The behaviors could occur in a variety of scenarios; the authors speak of situational discounting to underscore the ways medical staff sometimes created space away from the patient to discuss aspects of his or her medical circumstances, assuming (not always correctly) that they were out of earshot of the patient. Glaser and Strauss, however, also identified situations in which the medical professionals failed to discount, which was often observed in cases where people had just died and staff acted as if he or she were still present. This type of failure to discount was indicated in the ways staff might prepare the post-mortem body, especially when nurses had been "deeply involved" with patients (p. 113)

My conception of discounting awareness continues to include the interactional patterns discovered by Glaser and Strauss, but it conceives discounting awareness more broadly as a communicative behavior that works on the intrapersonal, interpersonal, and macro communicative level. I use discounting in its verbal form, but it is also a modifier to describe a type of awareness, or mindset, people have brought to a situation. Discounting awareness is a day-to-day form of triaging attention so that awareness of an issue or person might be only *inadvertent awareness* or as involved as an aggressive form of defensive processing in which the *talkdown* attempts to summarily dismiss or berate an object, idea, or person. Some of the dismissive behaviors that occur when people reject warnings against alcohol abuse or other behaviors are indicators as well. Data from academic literature provide insight into how some of the people who would be likely targets for public service announcements warning against smoking, alcohol, drug abuse, or other harmful habits can go into defensive processing due to anger, guilt, and shame that cause them to ignore or discount the warning (see Kunda, 1987; Liberman & Chaiken, 1992, write of defensive processing and health messages).

But discounting awareness is not enacted solely in fits of anger or foulness. The phenomenon is evident in the behavior of people who want to look on the bright side, the eternal optimists whose *oddsmaking* is always in their favor, discounting the possibility of bleaker, more realistic outcomes. On the larger societal landscape, discounting of the swine flu virus as media hype led many to forgo vaccinations, and debates over matters from food safety to climate control attract many who discount the awareness of others without careful consideration. Trust allocations are an important dimension driving discounting. As the theory suggests, and growing cognitive research is showing, there is a predisposition to discount information, situations and people different from one's own views or group. Kahan (2010) summarizes the direction of the data thusly, "People endorse whichever position reinforces their connection to others with whom they share important commitments."

When discounting awareness, people can discount what they "know" or "see" and discount others' capacity for awareness. In a not completely dissimilar manner to the medical staff that Glaser and Strauss observed talking over comatose people they assumed were not aware, the politically partisan are engaging in a discounting process when they complain about people who have differing political orientations. They assume the opponents (Republicans, Democrats, Fox News viewers—fill in the blank) don't know better; certainly, they aren't as aware as the critic. This particular type of discounting intersects with elements of a theory, the so-called Third Person Effect, which is one of several that I have linked as part of my developing theory of discounting awareness and will address shortly.

Discounting awareness can describe the ways in which people erase the unpleasant or threatening from view and act as if it does not exist or is of marginal importance. Across many spheres of life, discounting awareness is an observable communicative behavior. It is self-protective, but the indicators can be quite ugly outbursts, or something more innocuous and inadvertent when people employ some discounting tactics to make it through the day.

As an intra-communication process, we can recognize discounting awareness in the many different ways people disregard advice and warnings. My coeditor, a mother of a teenager, observed that teenagers are quite adept at this form of discounting, and many of us might recognize the inner talkdown we do when our body is sending health warning signals. After becoming aware of how I can almost literally turn the hearing in my inner ear off when there is a news report or public service announcement warning against behaviors that can lead to diseases for which genetics and lifestyle might put me at risk, I did a grounded theory memo about my reactions. I recognized that I always go through some quick oddsmaking that somehow allows me to completely ignore the warning—until the next time around. Discounting awareness, I have come to see, has some connections to Optimism Bias, a theory for which researchers have identified some of the same type of denial patterns evident in discounting awareness. We might describe someone as bringing a discounting awareness to a situation—only partially accepting what is before him or her. That is the mindset described by people who tell me they read certain types of news with a certain amount of skepticism, depending on the subject and the source.

Grounded Theory as Subversive

"Man is a data-gathering animal." Glaser, 1962

This is a good place to acknowledge an underutilized benefit of grounded theory, the ability to dip into the vast amounts of data collected for other purposes. Quantitative researchers are accustomed to re-analyzing data from large national surveys and other publicly available databases, but such repositories are a small pool of what could be available to grounded theorists, who often limit themselves to interviews and perhaps a few manuals related to the area of study. Foraging for data to extend constant comparisons may well be grounded theorists' birthright. Like other features of grounded theory that were evident in Glaser's work some years before the method was named and codified, secondary analysis of "knowledge from research elsewhere," as a Glaser article (1962) named it, is a plan for reusing qualitative research to expand theorizing. Heaton(2004) stated that Glaser's articles (1962, 1963) are among the first to explore the possibility of secondary analysis of qualitative research, something that is still seldom done in research. Glaser, of course, had conducted secondary analysis of survey data for his index on comparative failure and research scientists for his dissertation, which was quickly published (1964). Although grounded theorists like to invoke the dictum of "all is data," few move to the next step to analyze existing data elsewhere. In a knowledge society where the flow of new data is endless, with much of it getting no more than a one-time analysis, it is a duty of scholars to make more use of the mounds of data society creates.

The data need not be academic data. Because grounded theory wants to know what is going on right on the ground, the data that emerge from everyday routines are also there for a second look and coding. I started my expanded study of awareness processes by reviewing memos I had done for my substantive theory on news-attending. I had begun to discover that issues around awareness can be found throughout social science literature, the news, and routines of everyday life even if not explicitly. Glaser and Strauss, in discussing how researchers might go forward to develop a formal theory of awareness noted that "most interactional studies inevitably collect data bearing on awareness; the data need only be analyzed in terms of awareness"(p. 284). My experience theorizing on awareness leads me to think this is true of most of the fundamental social processes one would discover.

One of the first efforts to understand awareness beyond my dissertation work was to code transcripts of jury voir dire. At the time, the high-profile insider trading case against Martha Stewart, the domestic arts guru and media mogul, had just begun and lawyers were questioning prospective jurors about their perceptions of the defendant, starting with their knowledge, "awareness" really, of her. I managed to get transcripts from the court and began coding the first few days of voir dire. The exercise did not provide a lot of new useful codes to pursue for my purposes, but it did help me begin to theorize the type of mental triage people do in everyday life as they take quick note of certain phenomena but are ambivalent or disinterested. Several jurors were aware of Martha Stewart, but she could have been a mannequin in a window for all they knew or cared to know about her. They showed an inadvertent awareness but disattended news reports about her, displaying some of the necessary discounting people need in everyday life. We can not attend everything. My larger point here, however, is the many different types of data grounded theory frees researchers to consider—if only they would let it.

The ability to follow concepts where they lead through theoretical sampling allows grounded theorists to be subversive in ways that can help build knowledge more productively. Gatekeepers in sociology began drawing lines in the sand soon after *The Discovery of Grounded Theory* was published. While some reviewers saw glimpses of the future of research in the book's pages, there were others who warned against the challenge the book was making to the status quo (primarily quantitative, hypothesis-testing sociological research). Sniffed Jan Loubser, a reviewer writing in the *American Journal of Sociology* (May 1968), "We are asked to accept as grounded theory…what we have hitherto regarded as *post-hoc* explanation, proto theory, or at best, hypotheses for further research." The tone and misstatements of a review in such a major sociology publication moved Anselm Strauss, who co-authored *Discovery* with Barney Glaser, to counter in a letter to *ASJ* (Jan.1969): "I would just like to have had him{the reviewer}actually read the book with about four times as much care so as not to entirely misunderstand it."

Grounded theory did not unseat hypotheses-testing theory-building, and that was not the intent. Nevertheless, if used to full potential grounded theory would force academics out of some of the narrowness of scope that stunts the development of knowledge. Working on the theory of discounting awareness made me especially aware of the ways in which several disciplines grapple with the same ideas, churning out partially developed theories that never cohere within the discipline or neighboring fields. When we embrace the view of literature as data, however, and give ourselves freedom to roam, connections between many theories that remain out of communication with one another become more natural. One need not move to the formal theory level to discover this. The problem of underdeveloped theories that are actually speaking on the same topic, though often dressed up as something else, is quite evident on the substantive level. Many grounded theorists work on such "tiny topics," as Glaser calls them, that they may not encounter these families of concepts in need of a master concept to pull them together

In my substantive area, news consumption, there are several traditions, some more focused on political science, others cultural studies, sociology, or social psychology, that operate without reference to the other. Grounded theory's strictures against starting with extant literature helped me fuse some of these traditions for theorizing on news-attending (Martin 2004, 2008), and the method has proved similarly potent when building a theory on discounting awareness, traces of which can be found across my home field of media and communication studies, as well as the rest of the social sciences. Earlier, I referenced the Optimism Bias, a process driven by the discounting of uncomfortable information, and the Third Person Effect, a theory positing that people are apt to assume that others are more influenced by mass media than they, as examples of established theories with traces of discounting awareness in them. Selective Exposure, another series of theories about the avoidance of disconfirming information is another example. Goffman's (1974) seminal discussion of framing, a very active area of media research, made direct links between framing and discounting: to put something in frame means excluding something else, discounting it. What the theories I have cited generally lack is a cohesive series of connected propositions explicating actual processes surrounding these phenomena. That's what grounded theory can offer even when the field of theories seems crowded.

Follow the concept

While grounded theory's ability to bring new perspectives to stagnant literature in a substantive area is generally appreciated by those who work the method, less explored is the ability of grounded theory to cut a swath through disciplines with a strong concept that can illuminate undertheorized processes. This is where grounded theory is at its most subversive and poised to produce new knowledge. The challenge is to follow a concept where it leads. More typically, the conceptualizing and theory-building has stopped short of crossing borders into neighboring disciplines and subject areas. A detour I took while working on the theory of discounting awareness revealed some of the obstacles as well as the way right through or around those challenges.

I coded *The 9/11 Commission Report*, a more than 600-page examination of the September 11, 2001 attacks conducted by a bi-partisan commission of past and present leaders from several spheres of public life. Members of the appointed body had access to documents and people, including President Bush and members of his Cabinet, as they investigated the decision-making prior to, during, and after the tragic events. In the aftermath of the attacks there was much talk about the failure to communicate across government agencies and "connect dots" that, critics believe, might have averted the attacks.

A question that worked its way into my memo-writing was whether discounting awareness might be evident in the testimony the commission collected. I plowed through hoping that I would also learn more about the dynamics underlying discounting awareness. As a data source, the commission report was laden with possibilities and ideas for theory-building in a couple of areas. Discounting awareness, awareness processes in general actually, had a critical presence throughout the report. Sometimes awareness is an in vivo code: Donald Rumsfeld, in his testimony to the commission, spoke of trying to obtain "situational awareness" the morning of the attacks; the military-influenced term is an important dimension in settings where people are responding to surprise events such as disasters. But, on page after page, it was clear that discounting awareness, particularly the marginalizing or dismissing of data because there was not contextual information, played a role in how government agencies framed the information they were receiving in intelligence reports. During the summer of 2001, "the system was blinking red," Director of Central Intelligence George Tenet would later tell commission investigators (p. 277). But territorial partitions made it difficult for agencies to collaborate to bring together the various pieces of the puzzling information each might have been holding. Agency staffers interviewed by the commission revealed how the discounting of people and information contributed to the general lack of awareness in the intelligence ranks.

Of more immediate interest to this discussion on crossing disciplines was coding I did on data showing how government officials responded to the attacks that September morning. From the moment the first plane hit, US officials were caught in the middle of a crisis they could not understand or control. They were always behind events. For a brief period, President George W. Bush could not contact his key advisers, including his Secretary of Defense Donald Rumsfeld; first responders on the ground in New York faced similar disconnects; and, as the report noted, it appears some air force pilots had instructions to shoot down United Flight 93, but couldn't find it.

States the report, "The flight had already crashed by the time they learned it was hijacked"(p. 31). Throughout, a process I came to identify as *escalating contingency*, a spiraling series of reactions in response to sudden crisis, was evident. More important, escalating contingency appears to be the process that unfolded in other high-profile disasters and crises. Tentatively, one can see the move through a succession of flawed responses during the BP explosion and subsequent dumping of oil in the Gulf of Mexico. One hesitates to simply start listing disasters in general without getting deeper into the data and comparing responses to events such as the Gulf oil leaks or Hurricane Katrina, but the concept of extenuating contingency is potentially fruitful..

A study of escalating contingency is not something I am exploring right now. I share it to make the point that grounded theorists, in part because of the reward systems that keep them and others in academia focused on making contributions in the one or two substantive areas in which they are credentialed, do not typically follow the concept, so to speak. There are some logical reasons for their reluctance. Many would see the substantive area and be intimidated by the fact that the sociology of emergency response or disasters is in itself a specialty, a sub discipline inhabited by experts who focus solely on the topic. It is this realization, I believe, that makes grounded theorists hesitant to push forward with formal theory. I am careful not to make too confident an assertion, but I have come to think that following a strong concept and using classic grounded theory protocols would allow theorists to avoid some of the pitfalls they fear. Some familiarity with the related literature in the disciplines they are traveling through will certainly make theorists more sensitized to the possibilities of intersections and the use of the literature as data; but following the concept, say one like awareness context, or, maybe someday, escalating contingency if it patterned out, would reduce the number of fault lines because the theorist's expertise is situated around the concept she is watching unfold, not the particulars of the substantive area. In some ways, grounded theory executed this way comes quite close to being the kind of transdisciplinary knowledge Klein (1991) advocated as a desirable and ultimate outcome of true interdisciplinarity, where scholars work in service to ideas or the problem in question rather than a discipline.

Grounded theory as a distinct method

The awareness context, with its deep implications for everyday life far beyond its original substantive area, remains the best exemplar of how grounded theory can help break the "data overwhelm" that besets many institutions and disciplines. But classic grounded theorists are going to need to elevate their practice in a few ways. First, they must insist on classic grounded theory as a distinct method. A few years ago a reviewer responding to an article I had submitted had a lot of positive things to say about the work, a grounded theory on news-attending. She noted, however, that, while she was willing to let me call my work a grounded theory, she did not think grounded theory was a distinct method. She felt that my process was similar to what others were doing and did not need to be identified as a particular method. The tendency to look at grounded theory in general as a collection or family of methods, is endorsed at varying degrees in influential discussions such as Bryant & Charmaz (2007) and Corbin's (2007) revision of the qualitative research text she originally wrote with Strauss.

Classic grounded theory, however, is a distinct method, with a protocol that needs to be followed to produce the kind of theory first envisioned in Glaser's early writings. Although I get impatient with those grounded theorists who do not recognize the method's limitations for certain areas of study, I would argue that classic grounded theory is the only method in the social sciences that can create the kind of cumulative theory (what GT calls formal theory) across many different areas of inquiry. There is not another established method for building such theory. Smith (2008) argued that there has been more of a focus on novelty than the building of cumulative inquiry in the social sciences. His attempt to provide a method acknowledged Glaser and Strauss's initial work as having promise, but he doesn't pursue that lead, preferring his own formula. However, Gynnild and Martin (2007), noting the discussions within political science and media/communications studies on limitations in cross-national research, where researchers often feel stymied by apple and orange comparisons or prefer to study countries as intact cases instead of integrating the data, used comparisons of newspaper ombudsmen experiences to demonstrate how cross-national data could be broken down, compared and used to build an integrated theory.

Grounded theorists have to get beyond the "tiny topics" that can sometimes make it seem as though people are doing the same study over and over again. Studies on how people manage their diabetes, asthma medicine, and other chronic ailments are often just another study of regimen compliance, but there are no formal theories on these matters. Other clusters of concepts get replayed across several fields where people do grounded theories. At the very least, there are pools of data to share and reanalyze, or projects to mount collectively. The method is there; we just need to use it more boldly.

References

Bryant, A. & Charmaz, K. (2007) Grounded theory research: Methods and practices. In *The Sage handbook of grounded theory*, A. Bryant & K. Charmaz (Eds.), pp. 1-28. London, England and Thousand Oaks, CA: Sage.

Corbin, J. A. & Strauss, A.L (2007). *Basics of qualitative research. Techniques and procedures for developing grounded theory* (3rd ed.). Thousand Oaks, CA: Sage.

Ekins, R. (1997). *Male femaleing: A grounded theory approach to cross-dressing and sex-changing.* New York, NY: Routledge.

Glaser, B.G. (1962). Secondary analysis: A strategy for the use of knowledge from research elsewhere. *Social Problems*, 10(1), 70-74.

Glaser, B.G. (1963). Re-treading research materials. The use of secondary analysis by the independent researcher. *American Behavioral Scientist, 6*(10), 141-14.

Glaser, B.G. (1964). *Organizational scientists: Their professional careers.* New York, NY: Bobbs Merrill

Glaser, B. G. & Strauss, A. L. (1965). *Awareness of dying.* Chicago IL: Aldine Publishing.

Goffman, E. (1974/1986) *Frame analysis: An essay on the organization of experience.* With a new foreword by Bennett Berger. Lebanon, NH: Northeastern University Press.

Gynnild, A. & Martin, V.B. (2007, May). "Bridging Media Industries and the Academy Using Classic Grounded Theory Methodology" Paper delivered at pre-conference for International Communication Association annual conference, San Francisco, CA.

Heaton, J. (2004). *Reworking qualitative data.* London, England: Sage.

Kahan, D. (2010, Jan. 20). Fixing the Communications Failure. *Nature, 463*, 296-297.

Klein, J. T. (1991). *Interdisciplinarity: History, theory, and practice.* Northumberland, England: Bloodaxe Books.

Kunda, Z. (1987). Motivated inference: self-serving generation and evaluation of causal theories. *Journal of Personality and Social Psychology, 53* (4), 636-647.

Liberman, A. & Chaiken, S. (1992). Defensive processing of personally relevant health messages. *Personality and Social Psychology Bulletin, 18*(6), Dec 1992, 669-679.

Martin, V.B. (2004). Getting the news from the news. PhD Dissertation, Union Institute and University. UMI *Dissertation Abstracts International*, 3144535

Martin, V. B. (2006). The relationship between an emerging grounded theory and the literature: Four phases for consideration. *The Grounded Theory Review, 5*(2/3), 47-50.

Martin, V.B. (2007). Purposive attending: How people get the news from the news. In *The Grounded Theory Seminar Reader.* B. G. Glaser & J. A.Holton (Eds.). Mill Valley, CA: Sociology Press.

Martin, V.B. (2008). Attending the news: A grounded theory about a daily regimen. *Journalism: Theory, Practice & Criticism* , *9*, 76-94.

Morrissey, M. (1997). Extending the theory of awareness context by examining the ethical issues faced by nurses in terminal care. *Nursing Ethics, 4*(5), 370-379.

Nathaniel, A. K. & Andrews, T. (2010). The Modifiability of Grounded Theory. *The Grounded Theory Review, 9* (1), 65-77.

National Commission on Terrorist Attacks on the United States. (2004). *The 9/11 Commission Report*: New York, NY: Norton.

Smith, R. (2008) *Cummulative social theory: Transforming novelty into innovation.* New York, NY: Guilford Press.

Strauss, A. L. (1994). Awareness Contexts and Grounded Formal Theory. In *More Grounded Theory Methodology: A Reader.* B.G. Glaser (Ed.), pp. 360-368. Mill Valley, CA: Sociology Press.

Zerubavel, Z. (2006) *The elephant in the room: Silence and denial in everyday life.* New York, NY: Oxford University Press.

Contributors

Tom Andrews received his PhD from the University of Manchester, UK in 2003 and currently lectures in nursing at University College Cork, Ireland. He specialises in the field of Intensive Care nursing. He facilitates seminars on classical Grounded Theory and currently supervises a number of PhD students using this methodology. Tom has taught on under-graduate and post-graduate programmes of study. He is a peer reviewer for a number of international publications and an editorial board member of two journals including the Grounded Theory Review journal. Research interests include clinical decision making and how patients and relatives respond to a worsening progression.

Kathy Charmaz is Professor of Sociology and Director of the Faculty Writing Program at Sonoma State University, a program that supports faculty members' scholarly writing. She has written, co-authored, or co-edited nine books including *Constructing Grounded Theory: A Practical Guide through Qualitative Analysis*, which received a Critics' Choice award from the American Educational Studies Association and has been translated into Chinese, Japanese, Polish, and Portuguese. Her recent multi-authored books are Five Ways of Doing Qualitative Analysis: Phenomenological Psychology, Grounded Theory, Discourse Analysis, Narrative Research, and Intuitive Inquiry and Developing Grounded Theory: The Second Generation. She has just received the Goldstein award for scholarship from Sonoma State University.

Barney G. Glaser received his PhD at Columbia University in New York in 1961. He currently runs the publishing company Sociology Press and the non-profit, web based organization The Grounded Theory Institute. His lebenswerk/life work, the development of grounded theory methodology, began while at Columbia and was further developed during his fruitful research collaboration with Anselm Strauss at the University of California, San Francisco in the 1960's. The collaboration resulted in the much cited books *Awareness of Dying* (1965) and *The Discovery of Grounded Theory: Strategies for Qualitative Research* (1967). Glaser received an honorary doctorate from Stockholm University in 1998. He teaches grounded theory to PhD candidates from all continents.

Evert Gummesson is Emeritus Professor of Marketing and Management at Stockholm University, Sweden. He graduated from the Stockholm School of Economics; received his PhD from Stockholm University; is Honorary Doctor of Hanken School of Economics, Helsinki; and a fellow of the University of Tampere, Finland. He has written or contributed to over 50 books and published numerous articles. His interests embrace service, relationship marketing and networks, which is reflected in his book *Total Relationship Market-*

ing, (3rd, revised ed., 2008). Dr. Gummesson takes a special interest in research methodology and the theory of science. He has received two awards from the *American Marketing Association;* and the *Chartered Institute of Marketing*, UK, lists him as one of the 50 most important contributors to the development of marketing. He has spent 25 years in business and is a frequent speaker around the world.

Wendy Guthrie has a PhD in Marketing (2000), from the University of Strathclyde. Her thesis reveals a theory of client control. She is a fellow of the Grounded Theory Institute and is currently a research associate at the Innovative Manufacturing and Construction Research Centre, Loughborough University. Her recent work has been with futures scenario generation within the context of the built environment, understanding informalizing behavior in design team interactions and opportunizing in construction organizations. At present she is exploring cross disciplinary practice and possibilities within the construction context and beyond.

Astrid Gynnild is a researcher at the Department of Information Science and Media Studies, University of Bergen, Norway. She received her PhD from the same university in 2006, with a grounded theory of creative cycling of news professionals. Astrid specializes in innovation processes in computational and audiovisual journalism online, and in journalism education. Her background is in journalism and newsroom development and also counseling and supervising. She is the author and co-editor of five books, including two on grounded theory. She is a fellow of the Grounded Theory Institute and sits on the editorial board of the Grounded Theory Review.

Cheri Ann Hernandez is a registered nurse, certified diabetes educator and associate professor of nursing at the University of Windsor (Windsor, Ontario), and associate editor of the journal, The Grounded Theory Review. She has PhDs in education (University of Toronto, 1991) and nursing (Case Western University, 1997). Her grounded theory thesis research discovered the theory of integration which guides her program of research in diabetes. Dr. Hernandez has published grounded theory results as well as methodological papers on grounded theory. She wishes to acknowledge Dr. Glaser for his mentorship through his grounded theory workshops and during her 2002-2003 sabbatical.

Judith A. Holton is the Editor-in-Chief of The Grounded Theory Review and a Fellow of the Grounded Theory Institute. A faculty member of Mount Allison University, Canada, she frequently collaborates with Barney Glaser in both publications and seminars.

Andy Lowe previously was a faculty member for Strathclyde University Business School in the UK, where he was the director of studies for the cross faculty PhD research methodology programme. Currently he is resident in Thailand and is about to successfully conclude the supervision of the 8th PhD researcher all of whom used the grounded theory method in their theses. Recently Andy was invited by Glyndwr University in Wales UK to become a visiting professor in research methodology specializing in grounded theory.

Vivian B. Martin directs the Journalism Program at Central Connecticut State University, where she is an associate professor. A former newspaper journalist and magazine writer, Martin continues the grounded theory research on news in everyday life she began as a doctoral student, publishing several well-cited articles related to her theory of purposive attending. She teaches journalism and has taught qualitative methods, including grounded theory, to graduate students. She has organized troubleshooting seminars in New York City for the Grounded Theory Institute and is currently developing a formal theory on *discounting awareness.*

Antoinette McCallin is an Associate Professor in the Faculty of Health and Environmental Sciences at AUT University in Auckland, New Zealand. Antoinette has taught health professionals from wide-ranging disciplines at master and doctoral level for many years. Research interests include grounded theory and interprofessional collaboration in various practice settings. She has written extensively about interdisciplinary teamwork, interprofessional collaboration in professional-client relationships and health professional education, and the practicalities of using the grounded theory method. As a Fellow of the Grounded Theory Institute in San Francisco Antoinette has co-facilitated grounded theory workshops at Oxford in the United Kingdom.

Alvita K. Nathaniel, PhD, is a nurse, educator, and ethicist. She is Associate Professor at West Virginia University School of Nursing where she is Coordinator of the Family Nurse Practitioner program. In addition to her teaching and administrative roles, Alvita also maintained clinical practice. In 1998 Alvita co-authored the nursing textbook, *Ethics & Issues in Contemporary Nursing.* The book is currently in its third edition and continues to be popular in the US and internationally. Writing the ethics textbook led to her grounded theory research on moral reckoning, which she continues to pursue along with additional publications focusing on nursing ethics.

Lisbeth Nilsson, PhD, is a specialist in occupational therapy and an associated researcher of the Division of Occupational Therapy and Gerontology at Lund University, Sweden. She designed and implemented her own research study and received her PhD in the field of Occupational Therapy at the Institution of Health Sciences, Lund University, in 2007. She is the author of several internationally published articles and the facilitator of numerous academic work shops in her field of expertise. She has a special research interest in the use of grounded theory approaches in learning processes, in particular those of people with cognitive disabilities.

Mark S. Rosenbaum is a Fulbright Scholar, Assistant Professor of Marketing at Northern Illinois University, and Research Faculty Fellow, Center for Service Leadership, Arizona State University. His research has focused on services issues such as the impact of friendships in commercial and not-for-profit settings on customer health, ethnic consumption, tourism, and structural equation modeling. His work has been published in leading services and hospitality journals and presented at domestic and international conferences. He is also an editorial board member of *Journal of Services Marketing*. He received his doctorate from Arizona State University in 2003. Dr. Rosenbaum resides in Naperville, IL.

Helen Scott is a web editor and online communicator, an online educator and a researcher. Helen gained her PhD in 2007 in the field of online learner persistence, during which study she developed a particular interest in Grounded Theory. She has attended many, and co-ordinated several, of Barney Glaser's seminars. She is also a Fellow of the Grounded Theory Institute, under the auspices of which she has co-presented two further seminars. In 2009, Helen and 12 of the Fellows set up Grounded Theory Online (http://www.grounedtheoryonline.com), a website designed to support those minus-mentors who are unable to attend face-to-face seminars.

Odis E. Simmons studied under Glaser and Strauss in the early 1970s at the University of California, San Francisco (Ph.D. in Sociology 1974). In the 1970s and 1980s, he originated two classic grounded theory based action methods, "grounded action "and "grounded therapy." He has served on the Sociology and Urban Studies Faculties at the University of Tulsa, as Director of the Self-Care Program in the Yale University School of Medicine, and currently as a faculty member and Director of the Grounded Theory/Grounded Action Concentration in the School of Educational Leadership and Change at Fielding Graduate University.

Massimiliano Tarozzi is professor of Qualitative Research Methods at the University of Trento, Italy, and founding director of the Master degree in "Research Methodology in Education". He teaches courses and seminars on Grounded theory at PhD schools in many universities including Verona, Milano, Roma, Firenze, Urbino, and Seattle. He is currently editor of "Encyclopaideia. Journal of Phenomenology and Education," and member of the editorial board of many international scholarly journals. In addition to several scientific articles, he has written or edited about 10 books, including *Phenomenology and Human Science Today* (Eds. With L. Mortari, 2010); *Che cos'è la Grounded theory* [What is grounded theory], 2008).He was the Italian translator of Glaser & Strauss, "The Discovery of grounded theory" (Roma, 2009) and published a conversation with Barney Glaser "Forty years after Discovery. Grounded Theory worldwide" (*The Grounded Theory Review* 11/2007).

Michael K. Thomas is an assistant professor in the Department of Curriculum and Instruction at the University of Wisconsin-Madison in the Educational Communications and Technology program. A former English teacher, he received his Ph.D. in Instructional Systems Technology and Language Education from Indiana University, Bloomington. His research focuses on the notion of culture in instructional design and problems related to the nuanced implementation of technology-rich innovations in schools. He teaches qualitative research methodologies, methods of integrating technology in instructional contexts, and philosophy of technology.

Hans Thulesius has a PhD in Society Medicine from Lund University, Sweden. He works 50/50 as a family physician and researcher analyzing both quantitative and qualitative data, using statistics as well as classic grounded theory. Hans Thulesius has published several studies based on classic grounded theory analysis. Meeting Dr. Barney Glaser in 2000 was crucial for his scientific career which includes translating *Doing Grounded Theory* into Swedish. Hans Thulesius sits on the editorial board of Grounded Theory Review and two other scientific journals.

CPSIA information can be obtained at www.ICGtesting.com
Printed in the USA
BVOW010956130313

315434BV00009B/224/P

9 781612 335155